**Bioseparation and
Bioprocessing
Volume 1**

*Edited by
Ganapathy Subramanian*

1807–2007 Knowledge for Generations

Each generation has its unique needs and aspirations. When Charles Wiley first opened his small printing shop in lower Manhattan in 1807, it was a generation of boundless potential searching for an identity. And we were there, helping to define a new American literary tradition. Over half a century later, in the midst of the Second Industrial Revolution, it was a generation focused on building the future. Once again, we were there, supplying the critical scientific, technical, and engineering knowledge that helped frame the world. Throughout the 20th Century, and into the new millennium, nations began to reach out beyond their own borders and a new international community was born. Wiley was there, expanding its operations around the world to enable a global exchange of ideas, opinions, and know-how.

For 200 years, Wiley has been an integral part of each generation's journey, enabling the flow of information and understanding necessary to meet their needs and fulfill their aspirations. Today, bold new technologies are changing the way we live and learn. Wiley will be there, providing you the must-have knowledge you need to imagine new worlds, new possibilities, and new opportunities.

Generations come and go, but you can always count to Wiley to provide you the knowledge you need, when and where you need it!

William J. Pesce
President and Chief Executive Officer

Peter Booth Wiley
Chairman of the Board

Bioseparation and Bioprocessing

A Handbook

Volume 1

Edited by
Ganapathy Subramanian

Second, Completely Revised Edition

WILEY-VCH Verlag GmbH & Co. KGaA

The Editor

Dr. Ganapathy Subramanian
Littlebourne
60B Jubilee Road
Canterbury, Kent CT3 1TP
UK

Library of Congress Card No.:
applied for

British Library Cataloguing-in-Publication Data
A catalogue record for this book is available from the British Library.

Bibliographic information published by the Deutsche Nationalbibliothek
Die Deutsche Nationalbibliothek lists this publication in the Deutsche Nationalbibliografie; detailed bibliographic data are available in the Internet at <http://dnb.d-nb.de>.

Typesetting   SNP Best-set Typesetter Ltd., Hong Kong
Printing   betz-druck GmbH, Darmstadt
Binding   Litges & Dopf GmbH, Heppenheim
Wiley Bicentennial Logo   Richard J. Pacifico

Printed in the Federal Republic of Germany
Printed on acid-free paper

ISBN: 978-3-527-31585-7

Contents

Volume 1

Preface *XXIII*
List of Contributors *XXV*

Part I   Strategy and Development

1   Process Development – When to Start, Where to Stop   *3*
         Glenwyn D. Kemp
1.1   Introduction – What is Process Development?   *3*
1.2   The Challenges of Process Development   *4*
1.2.1   Is the Purity High Enough?   *4*
1.2.2   Is the Process Robust?   *5*
1.2.3   Are the Process Specifications Valid?   *6*
1.2.4   Is the Process Scalable?   *7*
1.2.4.1   Considerations for Scale of Operation   *8*
1.2.5   Is the Process Economically Viable?   *9*
1.3   Strategies to Develop a Downstream Process   *10*
1.3.1   The Bigger Test Tube Approach   *10*
1.3.2   The Template Process   *12*
1.3.3   Process Development by Gradual Evolution   *12*
1.3.4   The "Me Too" Process   *13*
1.3.5   The Clean Sheet   *14*
1.4   Process Optimization   *15*
1.4.1   Cell Removal/Clarification   *15*
1.4.2   Sterile Filtration   *17*
1.4.3   Chromatography   *18*
1.4.3.1   Binding Capacity and Column Loading   *19*
1.4.3.2   Throughput as a Chromatography Optimization Parameter   *21*
1.4.4   Ultrafiltration   *21*
1.4.4.1   Optimizing Tangential Flow Ultrafiltration   *22*

Bioseparation and Bioprocessing. Edited by G. Subramanian
Copyright © 2007 WILEY-VCH Verlag GmbH & Co. KGaA, Weinheim
ISBN: 978-3-527-31585-7

VI | *Contents*

1.4.5   Virus Removal   *23*
1.4.6   Specific Considerations for Virus Removal   *24*
1.4.6.1   Protein A-affinity Chromatography   *24*
1.4.6.2   Virus Removal by Filtration   *25*
1.4.7   Lifetime Studies   *26*
1.5   Future Trends in Process Development   *27*
1.5.1   Disposable Process Lines   *27*
1.5.2   Nanoscale Screening   *27*
1.5.3   High-titer Feedstocks   *27*
1.5.4   High-concentration Formulations   *28*
           Further Reading   *28*

2   Strategies in Downstream Processing   *29*
           Yusuf Chisti
2.1   Introduction   *29*
2.2   Overview of Process Considerations   *29*
2.2.1   Possible Recovery Flowsheets   *29*
2.2.2   Designing a Recovery Process   *32*
2.3   Product Quality and Purity Specifications   *35*
2.3.1   Endotoxins   *36*
2.3.2   Residual DNA   *36*
2.3.3   Microorganisms and Viruses   *37*
2.3.4   Other Contaminants   *37*
2.4   Impact of Fermentation on Recovery   *38*
2.4.1   Characteristics of Broth and Microorganism   *38*
2.4.2   Product Concentration   *40*
2.4.3   Combined Fermentation–Recovery Schemes   *41*
2.5   Initial Separations and Concentration   *42*
2.6   Intracellular Products   *45*
2.7   Some Specific Bioseparations   *47*
2.7.1   Precipitation   *47*
2.7.2   Foam Fractionation   *47*
2.7.3   Solvent Extraction   *48*
2.7.4   Aqueous Liquid–Liquid Extraction and its Variants   *48*
2.7.5   Membrane Separations   *49*
2.7.6   Electric and other Field-assisted Bioseparations   *51*
2.7.7   Chromatographic Separations   *51*
2.7.8   Separation of Optical Isomers   *53*
2.8   Recombinant and Other Proteins   *53*
2.8.1   Inclusion Body Proteins   *55*
2.9   Conclusions   *57*
           References   *58*

Part II   Bioprocess and Early DSP

**3   Processes Development and Optimization for Biotechnology Production
            – Monoclonal Antibodies**   *65*
            Jochen Strube, Sven Sommerfeld, and Martin Lohrmann
3.1   Introduction   *65*
3.2   Monoclonal Antibody Production   *67*
3.2.1   Fundamentals   *67*
3.2.2   Market/Potential   *68*
3.2.2.1   Antibodies on the Market   *68*
3.2.2.2   Product Demand   *70*
3.3   Process Development and Optimization   *71*
3.3.1   Regulations and Quality Assurance   *71*
3.3.2   Analytical and Experimental Complexity   *72*
3.3.3   Fermentation – Upstream   *73*
3.3.3.1   Cell Types   *73*
3.3.3.2   Serum   *74*
3.3.3.3   Contaminants   *75*
3.3.4   Downstream   *76*
3.3.4.1   Chromatography   *76*
3.3.4.2   Filtration   *77*
3.3.4.3   Virus Clearance   *78*
3.4   Production Processes   *79*
3.4.1   Examples   *79*
3.4.2   Generic Process   *80*
3.5   Process Analysis: Optimization Potential   *81*
3.5.1   Needs   *81*
3.5.2   Upstream/Downstream   *82*
3.5.2.1   Cost Distribution between Upstream and Downstream   *84*
3.5.2.2   Influence of Production Rate on Specific Downstream Costs   *85*
3.5.2.3   Influence of Process Step Yield on Downstream Costs   *85*
3.5.3   Process Modeling Tool Concept   *87*
3.5.4   Sensitivity Study   *88*
3.5.5   Affinity Chromatography   *88*
3.5.5.1   Resin-binding Capacity   *89*
3.5.5.2   Resin Lifetime   *90*
3.5.6   Comparison with Ion-exchange and Hydrophobic-interaction
            Chromatography   *90*
3.5.7   Comparison between Affinity and Ion-exchange Chromatography as
            Capture Step   *91*
3.5.8   Ultrafiltration   *93*
3.5.8.1   Membrane Area   *94*
3.5.8.2   Filtrate Flux   *95*

3.5.8.3 Single- versus Multi-use Membranes *95*
3.6   Conclusions *97*
         Acknowledgments *97*
         References *97*

4   **Dynamics of Cellular Response to Recombinant Protein Overexpression in *Escherichia coli* *101***
         Balaji Balagurunathan and Guhan Jayaraman
4.1   Introduction *101*
4.2   Global Analysis of the Cellular Response to Recombinant Protein Overexpression *106*
4.3   Metabolic Consequences of Recombinant Protein Overexpression *106*
4.3.1   Inhibition of Growth During Recombinant Protein Overexpression *106*
4.3.2   Alteration of the Energy Generation Systems *107*
4.3.3   Alteration of the Biosynthetic Machinery (Fig. 4.2) *108*
4.4   Strategies to Overcome the Metabolic Consequences of Recombinant Protein Overexpression *110*
4.4.1   Redirection of Metabolic Flux *111*
4.4.2   Antisense Downregulation of Acetate Production *111*
4.5   Consequences of Recombinant Protein Overexpression on the Protein Maintenance Machinery in the Cell *112*
4.5.1   Cellular Protein Folding *112*
4.5.2   Cytoplasmic Folding Modulators *113*
4.5.3   Protein Export *113*
4.5.4   Periplasmic Folding Modulators *115*
4.5.5   Cellular Stress Response *115*
4.5.6   Stringent Response *116*
4.5.7   Unfolded Protein Response *117*
4.5.8   Cytoplasmic Response (Fig. 4.3) *117*
4.5.9   Extracytoplasmic Response *119*
4.6   Strategies to Overcome Cellular Stress Response Induced due to Recombinant Protein Overexpression *119*
4.6.1   Coexpression of Folding Modulators *119*
4.6.2   Cell Conditioning *121*
4.7   Concluding Remarks *121*
         References *122*

Part III Preparative (Chromatographic) Methods

5   **Ion-exchange Chromatography in Biopharmaceutical Manufacturing *125***
         Lothar Jacob and Christian Frech
5.1   Introduction *127*

5.2   Ion-exchange Chromatography in the Downstream Processing of
         Proteins   *128*
5.2.1   Resins: Commonly Used Functional Groups and Recommended
         Buffers   *128*
5.2.2   Principles of Ion-exchange Chromatography   *131*
5.3   Development and Optimization Strategies   *136*
5.4   Scale-up of Ion-exchange Chromatography   *143*
5.5   Application Areas   *144*
         References   *146*

6   **Displacement Chromatography of Biomacromolecules**   *151*
         Ruth Freitag
6.1   Introduction   *151*
6.2   Background and Basic Principle of Displacement
         Chromatography   *153*
6.3   Modeling and Simulation of Displacement Chromatography   *158*
6.3.1   The Ideal Model of Displacement Chromatography   *158*
6.3.2   Kinetic Models for Displacement Chromatography   *159*
6.3.2.1 The Equilibrium-dispersive (ED) or Transport-dispersive (TD) Model in
         Displacement Chromatography   *160*
6.3.2.2 Complex Kinetic Models for Displacement Chromatography   *161*
6.3.3   The Shock Layer Theory   *162*
6.3.4   Isotherm Models   *164*
6.3.4.1 The Langmuir Algorithm   *165*
6.3.5   The SMA Model   *166*
6.4   Technical Aspects and Process Development   *171*
6.4.1   The Stationary Phase/Interaction Mode   *171*
6.4.2   Composition of the Mobile Phase/Flow Rate   *173*
6.4.3   Column Length/Sample Size   *175*
6.5   Displacers for Biopolymer Displacement Chromatography   *175*
6.5.1   Protein Displacers   *176*
6.5.2   Theoretical Considerations in Displacer Behavior and Design   *178*
6.5.3   The Rational Design of Protein Displacers   *180*
6.6   Special Variants of Displacement Chromatography   *181*
6.6.1   Spacer Displacement Chromatography   *182*
6.6.2   Complex Displacement Chromatography   *182*
6.6.3   Selective Displacement Chromatography   *182*
6.6.4   Thin-layer Displacement Chromatography (TLDC)   *183*
6.6.5   Analytical Aspects of Displacement Chromatography   *183*
6.6.6   Miscellaneous   *185*
6.7   Applications of Displacement Chromatography for Separations in
         Biotechnology   *185*
6.7.1   Separation and Isolation of Amino Acids, Peptides and
         Antibiotics   *185*
6.7.2   Protein Separation   *195*

6.7.3   Separation of Isomers   *197*
6.7.4   Miscellaneous   *198*
6.8   Conclusions   *199*
         References   *200*

7   **The Purification of Biomolecules by Countercurrent Chromatography**   *205*
         Ian J.Garrard
7.1   A Description of Countercurrent Chromatography   *205*
7.2   Countercurrent Chromatography Compared to Solid-phase Chromatography   *208*
7.2.1   The Advantages of Countercurrent Chromatography for the Purification of Biomolecules   *209*
7.3   Solvent System Selection Process   *210*
7.4   Countercurrent Chromatography of Polar Biomolecules   *212*
7.4.1   pH Adjustment of the Aqueous Layer   *213*
7.4.2   Addition of Salts   *213*
7.4.3   pH Zone-refining Chromatography   *215*
7.4.4   Affinity Ligands   *216*
7.4.5   Room Temperature Ionic Liquids (RTILs)   *218*
7.5   Aqueous Polymer Systems   *219*
7.6   Conclusion   *222*
         References   *222*

8   **Continuous Chromatography in the Downstream Processing of Products of Biotechnological and Natural Origin**   *225*
         Michael Schulte, Klaus Wekenborg, and Jochen Strube
8.1   Introduction   *225*
8.2   SMB Chromatography   *226*
8.2.1   Basic Principle   *226*
8.2.2   SMB Equipment   *230*
8.2.3   Layout and Optimization   *231*
8.2.4   SMB Modifications   *232*
8.2.4.1 Operation 1: Isocratic Modifications   *232*
8.2.4.2 Operation 2: Gradient Modifications   *233*
8.2.4.3 Design: Number of Columns/Sections   *234*
8.2.5   Application to Biochromatography   *235*
8.2.5.1 Adsorption Chromatography in Normal Phase and Reversed-phase Mode   *240*
8.2.6   Countercurrent Fractional Aqueous Two-phase Extraction   *242*
8.3   Continuous Annular Chromatography   *243*
8.3.1   Basic Principle   *243*
8.3.2   Application to Biochromatography   *245*
8.4   ISEP   *247*

8.4.1 Basic Principle *247*
8.4.2 Application to Biochromatography *249*
8.5 Conclusions and Outlook *250*
References *251*

9 Continuous Annular Chromatography *257*
Frank Hilbrig and Ruth Freitag
9.1 Introduction *257*
9.2 Continuous Annular Chromatography – The Basic Principle *258*
9.3 Development of Instrumentation and Operation *260*
9.4 Modern Continuous Annular Chromatography Systems *263*
9.5 Theory and Modeling of Continuous Annular Chromatography *265*
9.6 Issues of Continuous Annular Chromatography Application in Bioseparation and Bioprocessing *271*
9.6.1 The Annular Column *271*
9.6.2 Method Transfer and Scale-up *273*
9.6.3 Throughput and Productivity *279*
9.6.4 Partial Recycling Schemes *280*
9.7 Applications of Continuous Annular Chromatography Separation in Biotechnology *282*
9.8 Continuous Reactor/Separators *283*
9.9 Conclusions *286*
References *286*

10 Principles of Membrane Separation Processes *289*
Yusuf Chisti
10.1 Introduction *289*
10.2 Membrane Separation Processes *289*
10.3 Membrane Characteristics *290*
10.3.1 Materials of Construction *291*
10.3.2 Hydrophobic and Hydrophilic Membranes *291*
10.3.3 Permeation and Retention *292*
10.3.3.1 Integrity Testing and Characterization of Membrane Pore Size *294*
10.4 Filtration Basics *295*
10.4.1 Driving Forces for Flow *297*
10.4.2 Permeate Flux *298*
10.4.3 Volume Concentration Factor *299*
10.4.4 Solute Rejection *299*
10.4.5 Resistances to Flow Through the Membrane *301*
10.4.6 Flux versus Transmembrane Pressure *302*
10.4.7 Relationship between Concentration Polarization and Mass Transfer of Solute in the Gel Layer *304*
10.4.7.1 Mass Transfer Coefficient *306*
10.4.7.2 Experimental Estimation of Solute Concentration in the Gel Layer *308*
10.5 Membrane Modules *308*

10.5.1 Plate-and-Frame Configuration *309*
10.5.2 Spiral-wound Membranes *309*
10.5.3 Tubular Membranes *310*
10.5.4 Hollow Fiber Bundles *311*
10.6 Operation of Membrane Filtration Systems *312*
10.6.1 Batch Filtration *312*
10.6.2 Continuous Flow Filtration *313*
10.6.3 Continuous Feed and Bleed Recycle *315*
10.6.4 Diafiltration *315*
10.7 Membrane Fouling and Cleaning *316*
10.7.1 Fouling *316*
10.7.2 Cleaning of Membranes *317*
10.8 Economics of Membrane Processes *319*
10.8.1 Membrane Replacement Costs *319*
10.8.2 Other Operational Costs *320*
10.9 Conclusions *320*
References *320*

11 Affinity Precipitation *323*
Frank Hilbrig and Ruth Freitag
11.1 Introduction *323*
11.2 Primary-effect Affinity Precipitation *325*
11.3 Affinity Precipitation by Stimuli-responsive Materials *328*
11.3.1 Stimulus: pH *331*
11.3.2 Stimulus: Salt Addition *334*
11.3.3 Stimulus: Temperature *335*
11.3.4 Stimulus: Photo *339*
11.4 Application of Affinity Precipitation in Bioseparation *339*
11.4.1 Application Protocol *340*
11.4.2 Scale-up and Technical Realization *343*
11.4.3 Protein Purification *344*
11.4.4 Nucleic Acid Purification *346*
11.5 Perspectives *348*
References *349*

Volume 2

Part IV Specific Bioprocesses and Separation Methods

12 Biological Fuel Cells: Processing Substrates to Electricity by the Aid of Biocatalysts *355*
Aarne Halme and Xia-Chang Zhang
12.1 Introduction *355*
12.2 Microbial Fuel Cells *357*

12.2.1   Basic Structure of a Novel Microbial Fuel Cell   *358*
12.2.2   Experimental Properties of the Cell   *360*
12.2.3   Electron Production through Substrate Bioprocessing   *362*
12.2.4   Potential Applications   *363*
12.3   Enzymatic Fuel Cells   *364*
12.3.1   An Enzymatic Fuel Cell for Methanol   *364*
12.3.2   Details of the Cell   *365*
12.3.2.1 Enzyme – Methanol Dehydrogenase   *366*
12.3.2.2 Fuels, Mediator and Additional Stabilizer   *366*
12.3.2.3 Electron Production through Substrate Bioprocessing   *368*
12.3.3   Experimental Properties of the Cell   *372*
12.3.3.1 General Performance of the DMBFC   *373*
12.3.3.2 Dynamic Study of the Enzymatic Fuel Cell   *373*
12.3.4   Electrical Models   *377*
12.3.4.1 Constant Electromotive Force Model   *378*
12.3.4.2 Constant Current Model   *379*
12.3.5   Potential Application of Enzymatic Fuel Cells   *380*
12.4   Summary and Conclusion   *381*
          References   *381*

13   **Removal and Analysis of Contaminants and Impurities**   *383*
          Andreas Richter and Bettina Katterle
13.1   Introduction   *383*
13.2   Potential Risks Associated with Impurities and Contaminants   *384*
13.2.1   Host Cell DNA   *384*
13.2.1.1 Oncogenicty of Residual Host Cell DNA   *385*
13.2.1.2 Activation of Viruses by Host Cell DNA   *385*
13.2.2   Host Cell Protein and Other Protein Impurities   *386*
13.2.3   Product-related Impurities   *386*
13.2.4   Bacterial Endotoxins   *387*
13.2.5   Mycoplasma   *387*
13.2.6   Process-related Impurities such as Leachables   *387*
13.3   Safety Concepts and Acceptance Limits for Impurities and
          Contaminants   *388*
13.3.1   Safety Concepts for Impurities   *388*
13.3.2   Acceptance Limits for Impurities and Contaminants   *388*
13.3.2.1 Limits for Residual DNA   *388*
13.3.2.2 Limits for Residual Host Cell Protein and Other Protein
          Impurities   *389*
13.3.2.3 Limits for Product-related Impurities and Aggregates   *389*
13.3.2.4 Limits for Mycoplasma   *389*
13.3.3   Limits for Bacterial Endotoxins   *389*
13.4   Strategies for the Removal of Impurities and Contaminants   *390*
13.4.1   Strategies for the Removal of Host Cell DNA   *390*
13.4.2   Strategies for the Removal of Host Cell Protein   *391*

13.4.3 Removal of Aggregates *391*

13.5 Assays to Quantify Residual Impurities and Contaminants *391*

13.5.1 Residual DNA *391*

13.5.1.1 Hybridization Method for Residual DNA Analysis *392*

13.5.1.2 Assays Based on DNA-binding Proteins Using the Threshold Instrument *393*

13.5.1.3 Assays Based on PCR Technology *393*

13.5.1.4 Reference Material *394*

13.5.1.5 Sample Pre-treatment *394*

13.5.1.6 Validation Data for Residual DNA Assays *394*

13.5.1.7 Guidelines for the Choice of Residual DNA Analysis Assays *396*

13.5.2 Residual Host Cell Protein *396*

13.5.2.1 Antigen and Antibody Production *396*

13.5.2.2 Development of a Quantitative Host Cell Protein Assay *397*

13.5.2.3 Generic and Specific Host Cell Protein Assays *399*

13.5.3 Assays for Protein Impurities Other than Host Cell Protein *399*

13.5.4 Assays for the Analysis of Product-related Impurities *400*

13.5.4.1 SDS–PAGE *400*

13.5.4.2 Size-exclusion HPLC *401*

13.5.4.3 Capillary Gel Electrophoresis *402*

13.5.5 Endotoxin Measurement in Biotech Processing *405*

13.5.6 Mycoplasma Testing *405*

13.5.7 Examples of Testing for Process-related Impurity *406*

13.6 Conclusion *407*

Acknowledgments *407*

References *408*

14 High-throughput RNA Interference: Emerging Technology for Functional Genomics *411*

Yerramilli V. B. K. Subrahmanyam, and Eric Lader

14.1 Introduction *411*

14.2 RNAi: Current Understanding of the Process *411*

14.3 RNAi as a Technology in Functional Genomics *413*

14.4 siRNA Design *414*

14.5 Homology Analysis *417*

14.6 BLAST, Smith–Waterman and QIAGEN Homology Searches *418*

14.7 High-throughput Synthesis *419*

14.8 Genome-wide Synthetic siRNA Libraries: Content and Design Process *420*

14.9 Small-molecule Screening and Genome-wide RNAi Screens: The Difference *422*

14.10 Genome-wide siRNA Screens in Mammalian Systems *424*

14.11 siRNA Delivery and Monitoring Gene Knockdown *425*

14.12 Concluding Remarks *425*

References *427*

15 Developing an Antibody Purification Process *431*
Alexander Jacobi, Christian Eckermann and Dorothee Ambrosius
15.1 Introduction *431*
15.1.1 Biopharmaceutical Industry Market Size and Total Sales *431*
15.1.2 Current Challenges for Downstream Processing *432*
15.1.3 The Mission of Downstream Processing *435*
15.1.4 Biochemical Challenges in Downstream Processing *435*
15.1.5 Up-scaling and Economy *435*
15.2 Approaches for Downstream Purification *436*
15.2.1 Primary Recovery *436*
15.2.1.1 Harvest *436*
15.2.1.2 Concentration/Conditioning (Ultrafiltration/Diafiltration) *438*
15.2.1.3 Capture Step *438*
15.2.2 Virus Clearance *442*
15.2.2.1 Virus Safety Evaluation *442*
15.2.2.2 Virus Inactivation *443*
15.2.2.3 Virus Removal *443*
15.2.3 Purification and Polishing *444*
15.2.3.1 Hydrophobic-interaction Chromatography or Gel Filtration? *444*
15.2.3.2 Ion-exchange Chromatography *447*
15.2.4 Formulation *449*
15.3 Integrated Development Strategy *449*
15.4 Future Perspectives *451*
15.4.1 Protein A-affinity Chromatography *451*
15.4.2 Alternative Separation Techniques *452*
15.4.3 Novel Affinity Resins *453*
15.4.4 Accelerating the Development Process by Automation and Miniaturization *454*
15.5 Concluding Remarks *454*
References *455*

16 The Many Ways to Purify Plasmid DNA: Choosing the Right Purification Strategy Based on the Downstream Application *459*
Thorsten Singer and Markus Müller
16.1 Introduction *459*
16.2 Plasmid Purification Methods *460*
16.3 Plasmid and Propagation *461*
16.3.1 Lysis *464*
16.3.2 Purification *465*
16.3.3 Column-based Methods *466*
16.3.3.1 Noncolumn-based Methods *473*
16.4 Quality Aspects and Analytical Issues *474*
16.5 Process-scale Plasmid Isolation for Pharmaceutical Applications *478*
16.5.1 Plasmid Construction *479*

16.5.2   Selection of the Bacterial Strain for Production   *480*
16.5.3   Fermentation and Harvest   *481*
16.5.4   Bacterial Lysis and Lysate Clearing   *482*
16.5.5   Downstream Processing   *483*
16.6   Example of a cGMP pDNA Production Process   *483*
16.6.1   The Future: Commercially Feasible Manufacturing of Plasmid-based
          Drugs   *485*
16.7   Summary   *487*
          Acknowledgments   *487*
          References   *487*

**17   Advanced Fractionation Methods and Analysis for Proteomics
          Applications**   *491*
          Scot Weinberger and Egisto Boschetti
17.1   Introduction   *491*
17.2   Proteome Complexity and Current Challenges   *493*
17.3   Sample Prefractionation Methods   *494*
17.3.1   Chromatographic Techniques Applied to Proteomics   *495*
17.3.2   Electrophoresis-based Methods   *499*
17.3.3   Immunoprecipitation   *501*
17.3.4   Depletion of High-abundance Proteins   *502*
17.3.5   Reduction of Dynamic Concentration Range   *503*
17.4   Low-abundance Protein Focus   *508*
17.5   Analytical Methods   *509*
17.5.1   Top-down Methodologies   *511*
17.5.1.1 2-D Gel Electrophoresis-MS   *511*
17.5.1.2 Virtual 2-D Gel Analysis   *512*
17.5.1.3 SELDI   *513*
17.5.1.4 LC-MS   *514*
17.5.1.5 Top-down Fourier Transform Ion Cyclotron Resonance (FT-ICR)-MS
          Analysis   *514*
17.5.2   Bottom-up Methodologies   *516*
17.5.2.1 Shotgun LC-MS and LC-MS/MS Analysis   *516*
17.5.2.2 Multidimensional Protein Identification Technology (MudPIT):
          Towards Higher Shotgun Resolution   *516*
17.5.2.3 Quantitative Challenges in Protein Discovery Shotgun and
          MudPIT   *518*
17.5.2.4 Stable Isotope Labeling for Improved Quantitation   *519*
17.5.2.5 Accurate Mass Tags (AMT)   *519*
17.5.3   Protein Identification and Characterization   *520*
17.5.3.1 Peptide Mass Fingerprinting (PMF)   *521*
17.5.3.2 Sequencing and Protein Identification via MS/MS   *522*
17.6   Future Prospects   *523*
          References   *524*

Part V   Safety, Quality Control, Validation and Regulatory Considerations

18   Biosafety   *533*
           Yusuf Chisti
18.1   Introduction   *533*
18.2   Risk Assessment   *533*
18.2.1   Recombinant Microorganisms   *536*
18.2.2   Animal Cells   *537*
18.3   Containment Levels   *538*
18.4   Risk Management   *546*
18.4.1   Biological Safety Cabinets   *546*
18.4.2   Spill Management   *550*
18.4.3   Buildings and Facilities   *552*
18.4.3.1   Layout   *552*
18.4.3.2   Air Handling   *553*
18.4.3.3   Construction, Finishes and Practices   *554*
18.4.4   Process Equipment   *556*
18.4.4.1   Fermentation Plant   *557*
18.4.4.2   Downstream Processing   *562*
18.4.4.   Other Systems   *565*
18.4.5   Personnel Protective Equipment   *566*
18.4.6   Personnel Training   *567*
18.4.7   Medical Surveillance   *567*
18.4.8   Biowaste   *568*
18.5   Concluding Remarks   *569*
           References   *569*

19   The Manufacture of Gene Therapy Products and Viral Vaccines   *575*
           Andrew Bailey
19.1   Introduction   *575*
19.2   The Regulatory Climate for the Production of Gene Therapy
           Materials   *576*
19.3   Virus Vector Design Issues   *577*
19.3.1   Reducing the Likelihood of Replication-competent Virus   *577*
19.3.2   Deletion of Virus-coding Sequences   *577*
19.3.3   Vector Mobilization   *578*
19.3.4   Broadened Cell Tropism   *578*
19.3.5   Production Systems – Transfection or Established Cell Lines   *579*
19.3.6   Replication-competent Virus Vectors and Attenuated Virus
           Vaccines   *579*
19.4   Cell-based Therapy Design Issues   *580*
19.4.1   Modified Cells   *580*
19.4.2   Endogenous Retrovirus and other Virus Concerns   *581*
19.5   GMPs for Early-phase Manufacturing   *582*
19.5.1   Facility Design   *582*

19.5.2   Materials and Material Control   *584*
19.5.3   Staff Training and Documentation   *584*
19.5.4   Equipment   *585*
19.5.5   Cleaning and Sanitization   *586*
19.5.6   Pest Control   *586*
19.5.7   Quality Assurance/QC Testing   *587*
19.5.8   Vaccine Production and Tumorigenicity Testing   *588*
19.5.9   Product Release and Control   *588*
19.6   Quality Assurance/QC Testing of Cells and Vectors   *588*
19.6.1   Source Tissue, Cell Bank and Virus Bank Testing   *589*
19.6.1.1 Tissue Sourcing   *589*
19.6.1.2 Adventitious Agent Testing of Tissues and Cells   *589*
19.6.2   General Lot-release Testing   *590*
19.6.3   Replication-competent Vector Testing   *591*
19.6.4   DNA-related Testing   *592*
19.6.4.1 Residual DNA Testing   *592*
19.6.4.2 Tumorigenicity and Infectivity Testing   *593*
19.6.5   Biodistribution Studies   *594*
19.7   Vaccine-specific Production Issues   *594*
19.7.1   Live or Attenuated Vaccines   *595*
19.7.2   Killed or Purified Vaccines   *595*
19.8   Future Perspectives for Vaccine and Gene Therapy Products   *596*
       References   *597*

20   **The Virus and Prion Safety of Biopharmaceuticals**   *601*
       Andrew Bailey
20.1   Introduction   *601*
20.2   Acceptability Criterion for Virus Contamination of Start Material   *602*
20.2.1   Retrovirus Contaminants   *604*
20.2.1.1 Tests for Endogenous Retroviruses   *604*
20.2.1.2 Infectivity Tests for Endogenous Retroviruses   *605*
20.2.1.3 Electron Microscopy Tests for Retroviruses   *606*
20.2.1.4 RT Assays   *606*
20.2.2   Zoonotic Viruses   *606*
20.2.3   Nonzoonotic Viruses   *607*
20.3   Testing Strategies for Adventitious Agents   *608*
20.3.1   *In Vivo* Tests for Adventitious Viruses   *608*
20.3.2   *In Vitro* Tests for Adventitious Viruses   *608*
20.3.3   PCR Tests   *610*
20.4   Validation of Virus Inactivation and Removal   *611*
20.4.1   How Many Virus Inactivation/Removal Steps Do I Need?   *612*
20.4.2   General Process Design Considerations   *613*
20.4.3   General Considerations in the Design of Virus Clearance
       Studies   *614*
20.4.3.1 Cytotoxicity and Interference Studies   *614*

20.4.3.2 Influence of the Spike on Performance of the Process   *614*
20.4.3.3 How Many and Which Viruses to Select?   *614*
20.4.3.4 Mass Balance   *615*
20.4.3.5 Robustness   *615*
20.4.3.6 Specific Considerations for Established Virus Clearance Steps   *616*
20.4.3.7 Infectivity-versus Molecular-based Virus Tests   *618*
20.5   Transmissible Spongiform Encephalopathies and
          Biopharmaceuticals   *619*
20.5.1   BSE Risk Management – The Current State of the Art   *620*
20.5.2   TSE Clearance Studies: The Nature of the Agent   *620*
20.5.3   Agents and Forms Available for Spiking   *621*
20.5.4   Methods of Detection   *621*
20.5.5   BSE and vCJD – The Future   *622*
20.6   Future Perspectives: More or Less Testing?   *623*
          References   *625*

21   Protein Glycosylation: Analysis and Characterization   *631*
          Susan T. Sharfstein and Jong Hyun Nam
21.1   Introduction   *631*
21.2   Overview of Protein Glycosylation   *632*
21.2.1   *N*-linked Glycosylation   *632*
21.2.2   *O*-linked Glycosylation   *638*
21.2.3   GPI Anchors   *639*
21.2.4   Other Less-common Forms of Glycosylation   *639*
21.3   Effects of Glycosylation on Proteins   *640*
21.3.1   Structure and Stability   *640*
21.3.2   Activity   *641*
21.3.3   Immunogenicity and Biological Clearance   *641*
21.4   Techniques for Separating Glycoproteins and Glycopeptides   *642*
21.4.1   Lectin-affinity Capture   *643*
21.4.2   LC-MS of Glycopeptides   *643*
21.4.3   Identification of Glycosylation Sites   *644*
21.4.4   Techniques for Deglycosylation   *645*
21.4.4.1 Chemical Methods   *645*
21.4.4.2 Enzymatic Methods   *646*
21.5   Strategies for Analysis and Characterization of Glycans   *647*
21.5.1   MS   *647*
21.5.2   Chromatographic Profiling   *649*
21.5.2.1 Normal-phase LC   *649*
21.5.2.2 Anion-exchange Chromatography   *649*
21.5.2.3 Reversed-phase Chromatography   *650*
21.5.2.4 Multidimensional LC   *651*
21.5.2.5 Frontal Affinity Chromatography   *652*
21.5.3   Capillary Electrophoresis   *652*

21.5.4   Detection Methods for Chromatography and Capillary
          Electrophoresis   *654*
21.5.4.1 Derivatization and Fluorescent/UV Detection   *654*
21.5.4.2 Direct UV Detection Following Capillary Electrophoresis   *654*
21.5.4.3 Indirect UV Detection Following Capillary Electrophoresis   *655*
21.5.4.4 MS Detection   *655*
21.5.5   Electrophoresis   *655*
21.5.6   Sequential Enzymatic Digestion   *656*
21.5.7   Sialic Acid and Monosaccharide Composition   *656*
21.5.7.1 Sialic Acid Analysis   *656*
21.5.7.2 Monosaccharide Composition Analysis   *658*
21.6   Conclusions   *658*
          Acknowledgments   *658*
          References   *658*

**22   High-throughput Glycoanalysis for Use in Biopharmaceuticals
          Development and Manufacturing   *663***
          *Rakefet Rosenfeld, Revital Rosenberg, Roberto Olender, Inbar Plaschkes,
          David Dabush, Chanan Himmelfarb, Sabine Boehme, Kurt Forrer,
          and Ruth Ben-Yakar Maya*
22.1   Introduction   *663*
22.2   Glycosylation and the Biopharmaceutical Industry   *664*
22.2.1   Characteristics of Glycosylation   *664*
22.2.2   Importance of Glycosylation in the Biopharmaceutical Industry   *664*
22.2.3   Glycosylation of Recombinant Monoclonal Antibodies   *665*
22.3   Lectin Array Platform   *665*
22.4   Glycoanalysis of Recombinant Monoclonal Antibodies   *667*
22.4.1   Distinct Glycosylation of Recombinant Monoclonal Antibodies
          Produced by Different Cell Lines   *667*
22.4.2   Changes in Glycosylation during Fermentation   *667*
22.5   Glycoanalysis Directly on Nonpurified Supernatant Samples   *668*
22.5.1   Differences in Glycosylation between Bioreactors   *669*
22.5.2   Trends of Glycosylation Change during Fermentation   *670*
22.6   Discussion   *671*
          References   *672*

23   Quality Control of Implantable Drug Delivery Systems   *675*
          Steven S. Kuwahara
23.1   Introduction   *675*
23.2   Device-like Qualities   *677*
23.3   Color and Physical Appearance   *679*
23.4   Dissolution and Disintegration   *680*
23.5   Microbiology   *685*
23.6   Matrix Problems   *687*
23.7   Stability   *689*
23.8   Conclusion   *689*

Part VI   Analytical Methods and Technologies

**24   Polymerase Chain Reaction (PCR): An Analytical Tool in
            Bioprocessing**   *693*
            Dirk Loeffert
24.1   PCR: History and Principle   *693*
24.2   Good Laboratory Practice   *694*
24.2.1   Equipment Required to Conduct PCR   *694*
24.2.2   Preventing PCR Contamination   *694*
24.2.3   Preventing Carryover with Uracil-*N*-glycosylase (UNG)   *695*
24.3   Essential Components of the PCR   *696*
24.3.1   Thermostable DNA Polymerases   *696*
24.3.2   Hot-start Technologies   *696*
24.3.3   Reaction Buffer Chemistry and dNTPs   *698*
24.3.4   Primers   *699*
24.3.5   Template Nucleic Acid   *701*
24.4   Qualitative PCR   *702*
24.4.1   RT-PCR   *703*
24.5   Quantitative PCR   *704*
24.5.1   Different Approaches to Quantify Nucleic Acids by the PCR   *705*
24.5.2   Principle of Real-time PCR   *705*
24.5.2.1 Detection of PCR Products in Real-time PCR   *706*
24.5.2.2 Increased Quantification Accuracy by Using Multiplex, Real-time
            PCR   *709*
24.5.3   Real-time PCR Quantification Strategies   *710*
24.5.3.1 Absolute Quantification   *711*
24.5.3.2 Relative Quantification   *713*
24.6   Use of Real-time PCR for Quality Control in Bioprocessing and
            Therapeutics Development   *713*
24.6.1   Overview   *713*
24.6.2   Essential Prerequisites for Assay Validation   *714*
24.6.2.1 Real-time PCR Components   *715*
24.6.2.2 Controls   *715*
            References   *715*

25   Electrokinetic Separations   *719*
            Zaki Megeed, Kaushal Rege, Arul Jayaraman, and Martin L. Yarmush
25.1   Introduction   *719*
25.2   Gel Electrophoresis   *719*
25.2.1   PAGE   *720*
25.2.2   IEF   *722*
25.2.3   Two-dimensional Gel Electrophoresis   *723*
25.2.4   Agarose Gel Electrophoresis (AGE)   *724*
25.2.5   Analysis of Gels after Electrophoresis   *724*
25.2.6   Pulsed-field Gel Electrophoresis (PFGE)   *725*
25.3   Capillary Electrophoresis (CE)   *725*

25.3.1   Modes of CE   *726*
25.3.2   Modifications of CE   *728*
25.3.3   Emerging Applications of CE Techniques   *728*
25.3.3.1 Genomics and Proteomics: Biomolecules   *728*
25.3.3.2 Biological Particles: Viruses, Bacteria and Mammalian Cells   *729*
25.3.3.3 Multidimensional Techniques   *729*
25.4   Microscale Electrophoresis   *730*
25.5   Conclusion   *731*
          References   *732*

**26   Capillary Isoelectric Focusing Methods for Charge-based Analysis of
          Biotech Pharmaceutical Samples**   *735*
          Jiaqi Wu and Tiemin Huang
26.1   Introduction   *735*
26.1.1   Capillary Isoelectric Focusing (cIEF) Performed on General-purpose
          Capillary Electrophoresis (CE) Instruments – Conventional cIEF   *735*
26.1.2   Difficulty of Conventional cIEF   *736*
26.1.3   Whole-column Detection cIEF   *737*
26.1.4   Purpose of the Chapter   *738*
26.2   Initial Conditions in cIEF Method Development   *738*
26.2.1   Selection of Carrier Ampholytes   *739*
26.2.1.1 Manufacturers (Brand Names)   *739*
26.2.1.2 Use of Narrow pH Range Carrier Ampholytes   *740*
26.2.1.3 Carrier Ampholyte Concentration   *744*
26.2.2   Selection of Electropholytes   *745*
26.2.3   Additives   *745*
26.2.4   Sample Concentration   *745*
26.2.5   Focusing Voltage   *746*
26.2.6   Focusing Time   *746*
26.2.6.1 Focusing Time and Focusing Current   *746*
26.2.6.2 Factors Affecting the Focusing Time   *747*
26.2.6.3 Optimal Focusing Time   *747*
26.3   Some Issues in Method Development   *750*
26.3.1   pI Determination and Peak Identification in cIEF using pI
          Markers   *751*
26.3.1.1 Conditions for pI Determination   *751*
26.3.1.2 pI Markers Used in cIEF   *752*
26.3.1.3 Why pI Markers are Needed   *752*
26.3.1.4 pH Gradients of Commonly Used Carrier Ampholytes   *752*
26.3.1.5 Peak Identification Using Two pI Markers   *754*
26.4   Examples   *755*
26.4.1   Example 1: A Monoclonal Antibody (Mab1)   *755*
26.4.2   Example 2: A Glycosylated Protein (Gly1)   *756*
          Acknowledgments   *757*
          References   *758*

          Index   *759*

Preface

Over the past decade we have witnessed significant advances in the field of biotechnology, mainly due to the development of novel applications of advanced technology that have arisen to meet the various challenges faced by the research community. This book aims to present recent dynamic advances in technology and R & D in the fields of bioprocessing and bioseparation.

I am indebted to the many authors and co-authors of the 26 chapters from both the academic and industrial sectors from all over the world who have agreed to share their experience and knowledge. Chapters 1 and 2 in Part I deal with the when to start and where to stop process development, and the various strategies in downstream processing. Chapters 3 and 4 in Part II cover process development and optimization for the production of antibodies, and the dynamics of cellular responses to recombinant protein overexpression in *Escherichia coli*. Part III covers preparative methods. Ion-exchange chromatography in biopharmaceutical manufacturing is dealt with Chapter 5. Chapters 6 and 7 deal with displacement chromatography of biomacromolecules and the purification of biomolecules by countercurrent chromatography. Continuous chromatography in downstream processing of compounds of biotechnological and natural origin, continuous annular chromatography, the principles of membrane separation processes, and affinity precipitation are explained in Chapters 8, 9, 10 and 11, respectively. Part IV covers specific bioprocesses and separation methods. Biological fuel cells are dealt with in Chapter 12, removal and analysis of contaminants and impurities in Chapter 13, high-throughput RNA interference as an emerging technology for functional genomics in Chapter 14, developing an antibody purification process in Chapter 15, the many ways to purify plasmid DNA in Chapter 16, and advanced fractionation and analysis for proteomics applications in Chapter 17. Chapters 18–23 make up Part V on safety, quality control, validation and regulatory considerations. Part VI on analytical methods and technologies contains Chapter 24 on the polymerase chain reaction, Chapter 25 on electrokinetic separations and Chapter 26 on capillary isoelectric focusing methods for charge-based analysis of biopharmaceuticals.

For comprehensive and up-to-date information the reader should refer to specialized review articles.

Bioseparation and Bioprocessing. Edited by G. Subramanian
Copyright © 2007 WILEY-VCH Verlag GmbH & Co. KGaA, Weinheim
ISBN: 978-3-527-31585-7

This book will be an invaluable companion for biotechnologists, pharmaceutical engineers and others who are involved in the manufacture of biologically active biotechnological products.

I wish to express my sincere thanks to Dr. Frank Weinreich for inviting me to edit this volume, and to Rainer Münz and his colleagues at Wiley-VCH for their enthusiasm and support throughout the production of this book.

Canterbury, Kent, UK   *G. Subramanian*

List of Contributors

Dorothee Ambrosius
Boehringer Ingelheim Pharma GmbH
& Co. KG
Dept. Process Science, Downstream
Development
Birhendorfer Str. 65
88397 Biberach
Germany

Andrew Bailey
ViruSure GmbH
Veterinärplatz
1210 Vienna
Austria

Balaji Balagurunathan
Department of Biotechnology
Bhupat and Jyoti Mehta School of
Biosciences Building
Indian Institute of Technology Madras
Chennai 600036
India

Ruth Ben-Yakar Maya
Procognia (Israel) Ltd
3 Habosem Street
77610 Ashdod
Israel

Sabine Boehme
Biotechnology Development
4002 Basel
Switzerland

Egisto Boschetti
Bio-Rad Laboratories
c/o CEA-Saclay
DSV
91191 Gif-sur-Yvette
France

Yusuf Chisti
Institute of Technology and
Engineering
Massey University
Private Bag 11 222
Palmerston North 4442
New Zealand

David Dabush
Procognia (Israel) Ltd
3 Habosem Street
77610 Ashdod
Israel

Bioseparation and Bioprocessing. Edited by G. Subramanian
Copyright © 2007 WILEY-VCH Verlag GmbH & Co. KGaA, Weinheim
ISBN: 978-3-527-31585-7

Christian Eckermann
Boehringer Ingelheim Pharma
GmbH & Co. KG
Dept. Process Science, Downstream
Development
Birkendorfer Str. 65
88397 Biberach
Germany

Kurt Forrer
Novartis Pharma AG
Biotechnology Development
4002 Basel
Switzerland

Christian Frech
University of Applied Sciences
Department of Biotechnology
Windeckstr. 110
68163 Mannheim
Germany

Ruth Freitag
University of Bayreuth
Process Biotechnology
95440 Bayreuth
Germany

Ian J. Garrard
Brunel University
Institute for Bioengineering
The Brunel Uxbridge
West London
UB8 3PH
UK

Aarne Halme
Helsinki University of Technology
Department of Automation and
Systems Engineering
Otaniementie 17, PO Box 5500
02015 TKK, Espoo
Finland

Frank Hilbrig
University of Bayreuth
Process Biotechnology
95440 Bayreuth
Germany

Chanan Himmelfarb
Procognia (Israel) Ltd
3 Habosem Street
77610 Ashdod
Israel

Tiemin Huang
Convergent Bioscience Ltd.
27 Coronet Road
Toronto
Ontario M8Z 2L8
Canada

Lothar Jacob
Merck KGaA
Performance & Life Science Chemicals
Frankfurter Str. 250
64293 Darmstadt
Germany

Alexander Jacobi
Boehringer Ingelheim Pharma
GmbH & Co. KG
Dept. Process Science, Downstream
Development
Birkendorfer Str. 65
88397 Biberach
Germany

Arul Jayaraman
Texas A&M University
Department of Chemical Engineering
TX
USA

Guhan Jayaraman
Department of Biotechnology
Bhupat and Jyoti Mehta
School of Biosciences Building
Indian Institute of Technology Madras
Chennai 600036
India

Bettina Katterle
NewLab BioQuality AG
Max-Planck-Str. 15a
40699 Erkrath
Germany

Glenwyn D. Kemp
Biotechnology Consulting Services
82 Village Heights
Gateshead
Tyne and Wear
NE 8 1 PW
UK

Steven S. Kuwahara
GXP BioTechnology, LLC
PMB 506, 1669-2 Hollenbeck Avenue
Sunnyvale, CA 94087-5402
USA

Eric Lader
QIAGEN Inc.
19300 Germantown Road
Germantown, MD 20874
USA

Dirk Loeffert
QIAGEN GmbH
Research & Development
QIAGEN-Str.
40724 Hilden
Germany

Martin Lohrmann
Bayer Technology Services GmbH
BTS-PT-PT-CEM, Build. B310
51368 Leverkusen
Germany

Zaki Megeed
Center for Engineering in Medicine
Massachusetts General Hospital
Shriners Burns Hospital
Harvard Medical School
Boston, MA 02114
USA

Markus Müller
QIAGEN GmbH
40724 Hilden
Germany

Jong Hyun Nam
Rensselaer Polytechnic Institute
Department of Chemical and Biological
Engineering
Troy, NY 12180
USA

Roberto Olender
Procognia (Israel) Ltd
3 Habosem Street
77610 Ashdod
Israel

Inbar Plaschkes
Procognia (Israel) Ltd
3 Habosem Street
77610 Ashdod
Israel

Kaushal Rege
Center for Engineering in Medicine
Massachusetts General Hospital
Shriners Burns Hospital
Harvard Medical School
Boston, MA 02114
USA

Andreas Richter
NewLab BioQuality AG
Max-Planck-Str. 15a
40699 Erkrath
Germany

Revital Rosenberg
Procognia (Israel) Ltd
3 Habosem Street
77610 Ashdod
Israel

Rakefet Rosenfeld
Procognia (Israel) Ltd
3 Habosem Street
77610 Ashdod
Israel

Michael Schulte
Merck KGaA
PLS R&D LSS
Frankfurter Str. 250
64293 Darmstadt
Germany

Susan T. Sharfstein
Rensselaer Polytechnic Institute
Department of Chemical and
Biological Engineering
Biotech 2nd floor
110 8th Street
Troy, NY 12180
USA

Thorsten Singer
QIAGEN GmbH
40724 Hilden
Germany

Sven Sommerfeld
Bayer Technology Services GmbH
BTS-PT-PT-CEM, Build. B 310
51368 Leverkusen
Germany

Jochen Strube
Clausthal University of Technology
Institute for Separation and Process
Technology
Leibnizstr. 15
38678 Clausthal-Zellerfeld
Germany

Yerramilli V.B.K. Subrahmanyam
QIAGEN Inc.
19300 Germantown Road
Germantown, MD 20874
USA

Scot Weinberger
GenNext Technologies Inc.
657 George Street
PO Box 370645
Montara, CA 94037
USA

Klaus Wekenborg
Merck KGaA
ZVE-I
Frankfurter Str. 250
64293 Darmstadt
Germany

Jiaqi Wu
Convergent Bioscience Ltd.
27 Coronet Road, 7
Toronto
Ontario M8Z 2L8
Canada

Martin L. Yarmush
Center for Engineering in Medicine
Massachusetts General Hospital
Shriners Burns Hospital
Harvard Medical School
Boston, MA 02114
USA

Xia-Chang Zhang
Enfucell Ltd.
Tekniikantie 12
02150 Espoo
Finland

Part I Strategy and Development

Bioseparation and Bioprocessing. Edited by G. Subramanian
Copyright © 2007 WILEY-VCH Verlag GmbH & Co. KGaA, Weinheim
ISBN: 978-3-527-31585-7

1
Process Development – When to Start, Where to Stop

Glenwyn D. Kemp

1.1
Introduction – What is Process Development?

Process development provides the vital link between R & D and manufacturing. It takes a drug discovered in the laboratory and shown to be clinically promising, and allows it to be produced in sufficient quantities to be used on the patient population. However, this is not the total extent of process development. The purification method developed also has to meet the critical quality parameters of purity and activity within a very proscriptive regulatory framework. Moreover, it has to do this reliably and reproducibly for batch after batch, and at such a cost that the drug is commercially viable.

Hence, the process development scientist has to balance the often conflicting demands of low cost and high quality within a highly constrained timeframe. This aim of this chapter is to discuss ways of balancing these conflicting demands, to look at areas where emphasis should be placed and to discuss areas where investment of time may be less productive than hoped.

A fundamental challenge for process development is that it inevitably has to occur within a constrained timeframe. This is an important paradigm and is in contrast to classical scientific training in which the desired outcome is to determine a definitive answer to a problem, however much time and effort is required. In the case of process development, the timeframe is dictated by the needs of clinical trial phases and economic management. In this context, the desired outcome is not necessarily a definitive answer, but rather a workable solution. This approach can be both counterintuitive and uncomfortable to scientists new to the field of process development. Fortunately, experience eventually accustoms scientist to that environment where best compromises are often the only achievable outcome.

Bioseparation and Bioprocessing. Edited by G. Subramanian
Copyright © 2007 WILEY-VCH Verlag GmbH & Co. KGaA, Weinheim
ISBN: 978-3-527-31585-7

1.2
The Challenges of Process Development

The single fundamental question with which to challenge any process destined for the manufacturing floor is not whether it is the best process possible, but rather to ask "Is it good enough?", i.e. is it fit for purpose. As with all good science a reliable answer can only come from fully understanding the original question. Therefore, it is worth spending a little time to consider in more detail what makes a process "fit for purpose". In order to qualify for this accolade the process should be pass the following criteria:

1.2.1
Is the Purity High Enough?

The key term here is "enough". There is no economic benefit in producing a higher purity if this is not required. However, the typical purity would be expected to be routinely somewhat higher than the minimum specification, to allow for process robustness as we shall see below. In general, a higher purity is obtained by adding more purification or product-handling steps. However, the greater the number of these steps, the lower the recovery of the product will be. It is essential to consider the overall recovery of product for all the process steps rather than to look at individual process steps: 90% recovery may seem acceptable, if this is applied to three chromatography purification steps the overall recovery becomes a reasonable 73%. Consider, however, that each centrifugation, filtration, chromatography or even hold step may potentially cause a loss and a more realistic number of operations for the process may be eight steps, giving an overall recovery of only 43%. If the average recovery drops to 85%, the overall recovery after eight steps will only be 27%. A few extra percentage lost in the odd step or two may seem trivial, but for a billion dollar blockbuster drug every 1% loss of drug will represent a potential $10 million in lost revenue! The addition of extra unwarranted purification steps will have serious economic implications. There will also be implications for the process robustness by having effectively included unnecessary steps.

The question of what is a "high enough" purity is itself fraught with difficulties. With an experimental drug it instinctively feels safer to aim for the highest purity possibly, with absolute purity being the ideal result. As discussed above, this is usually the most expensive option. Drugs produced in research laboratories are often the result of an intensive labor of love, hand crafted by highly skilled scientists with little care for cost, time or difficulty. The preparation of the drug is in itself the scientific challenge. Such a handmade bespoke drug is ideal for initial evaluation, but not necessarily an appropriate model for a mass production. There is a danger that, in setting purity criteria using the initial data from the development laboratory, this will be used as the lower baseline for purity. This may result in an impossibly high standard to be set for the final drug. It should also be borne in mind that it is relatively easy to convince the regulatory authorities to accept a revision of purity specification for a drug if the new specification is tighter, e.g. higher purity, but it is almost impossible to argue with the regulatory authorities for an easing of specifications, especially without recourse to effectively repeating a phase III study.

The outcome of these considerations is that the drug should have a minimum purity against which it is validated, and that this purity should be based on pragmatic consideration of the potency and efficacy of the drug, and the nature and level of likely contaminants. It may also be necessary to identify and separate impurities, if only to demonstrate that they are clinically neutral. An example of this would be isomeric forms of a monoclonal antibody. In some cases these may be clinically inactive or neutral, in which case the isoforms can be copurified with the drug itself. If, however, the isoforms are shown to have an adverse clinical activity, perhaps by competitive inhibition of the active drug, then these isoforms will need to be removed. This can be a difficult separation to perform and will inevitably add to the cost, either as a result of requiring more time, additional steps or lost recovery of the active drug.

It may even be worth considering returning to the original drug screen to find an alternative antibody without the isoforms if the problem is picked up early enough in the development cycle.

1.2.2
Is the Process Robust?

As we noted above, the purification method used in the research laboratory may typically be the work of a single highly skilled and capable individual. It may require precise peak cutting, high-efficiency columns, temperamental buffer systems and be performed in a cold room. All of these factors may well be detrimental to the robustness of the process. Put simply, the robustness of a process is how far from the ideal set of conditions the process can operate and still yield product of the required purity.

It should be the aim of the process development scientist to make a process as "bomb proof" as possible. Certainly routine variations in the process parameters should not present a challenge to the process.

Process robustness can be thought of as path along a cliff top. If the path is set too far back from the cliff, the benefit of the breathtaking views will be lost; if, on the other hand, the path is set to close to the cliff, one careless footstep may lead to a catastrophic loss of enjoyment! The ideal is to place the track close enough to the cliff edge to reap the benefits of the view, but far enough away from the edge to be safe. Hence, the ideal process will be operating near to the limits of the process for maximum efficiency, e.g. sample load mass, flow rate or holding time, without undue risk of loss due perhaps to low recovery, purity or activity. The control of process robustness is by a combination of step selection, i.e. the actual methods and media used for each specific step and the parameters defined in the operation of those steps – the process specifications.

1.2.3
Are the Process Specifications Valid?

Unfortunately, it is often the case that the process development scientists are closer to the research laboratory then the production floor in their outlook. The result is usually overly tight specification which cannot be reliable repeated at full scale. For example, if the resolution required for an ion-exchange chromatography purification is very precise, the step will be inherently more sensitive to variations in, for example, column bed height, buffer flow rate, buffer pH, salt gradient mixing. While 1–2 L of buffer in a flask can easily be checked and adjusted for conductivity, pH and even temperature, this is a more challenging task for 10 000 L of buffer. If a buffer falls outside the specified combination of pH and conductivity it will have to be reworked or discarded – both expensive options.

Thus, it is imperative to know, for example, how precisely a buffer can be prepared; what are the acceptable ranges of pH and conductivity. This can then be used as a basis for setting a range value to the specification. Indeed if a buffer cannot be used within the typical range generated on the production floor it would be better to investigate alternative buffers at the process development stage than to try to reengineer the manufacturing equipment or procedures.

Another common problem with process specifications is that of the "rangeless parameter". When a process is being defined all conditions must be given not to a specific value, but to a value within a range, e.g. pH 7.4 ± 0.2. The act of deciding what this range is will also produce support data for the robustness of the process. Not only should ranges be provided for all critical parameters specified, but it is helpful to have data to indicate a range for parameters not specified as critical, to support the assumption that they are not critical. It may also be borne in mind that ranges do not have to always be plus or minus a median value, they can also be given as "no more than" or "no less than" values. In this case it is still a good idea to have an upper or lower limit, respectively, if only for practical purposes.

In general, it is easier to reduce the range of the specification after multiple full-scale cycles than to increase the range. Therefore, it is good practice to specify ranges as large as possible from the outset. However, operation at the extremes of these ranges must still result in a final product of acceptable quality. Ideally range values would be derived from a series of experiments in which the parameters are increased or decreased successively until the process fails. This should then be repeated for combinations of parameters at the determined limits. In reality there is rarely enough time or material to allow such extensive studies. A degree of pragmatism has to be applied. The critical parameters are usually selected and become the ones to be investigated in depth. The critical parameters are usually selected from past experience or published data. The danger here is that there may be "hidden" parameters which are specific to the product or process. For example, temperature may affect the binding capacity and resolution of a hydrophobic-interaction chromatography column, but would not be expected to change either for a Protein A-affinity step. Hence, it is always a worthwhile exercise to take a step back from each process, consider all of the potential variable parameters present, and actively decide which are likely to be critical to the process and which are not. This exercise also allows a third group of parameters to be identified – the "maybe critical" variables. In this case a few experiments at extreme values will usually indicate if the variable is critical or not.

In determining the process specifications the interaction of individual parameters must also be considered. Often the interactions of parameters are obvious and can be predicted with reasonable certainty, e.g. the effect of bed height and flow rate on resolution of a gel-filtration column. Sometimes, however, it is not obvious which parameters even interact, let alone the nature of the interaction. The areas of statistical design of experimentation (DOE) approaches are often cited for identifying interactions between different parameters. When well designed and executed these studies can indeed drastically reduce the amount of experimentation required to identify parameter interactions and to cooptimize these parameters within the process. However, DOE should be approached with an element of caution – it is essential that the assumptions made at the outset of the experimental design are understood and are sound in order to interpret the data generated correctly. Nonlinear relationships, in particular, are liable to distort the data and lead to false conclusions. Training should be undertaken before embarking on either the design or interpretation of "DOE" studies as there is great potential to make unfounded "leap of faith" conclusions.

1.2.4
Is the Process Scalable?

A fundamental question to ask at the outset of any process development exercise is "what is the anticipated final scale for this process?". This is usually a much easier question to ask than to answer. The earlier the development phase, the less accurate the answer will be; however, it is still a valid question even at the research phase. The answer will depend on a number of factors.

(i) *The potential patient population.* A drug to treat a rare congenital condition will clearly have a much lower target population of perhaps a few thousand. This can be contrasted with a drug for, say, male pattern baldness or chronic obesity, which would have potential target population of tens or even hundreds of millions. A prophylactic treatment such as a vaccine may potentially have a patient population of billions. The size of the target population will be mostly guesswork, but it should be reasonable to estimate an order of magnitude at least.

(ii) *The required dosage – single dose or chronic treatment.* The required dosage can usually only be estimated based on the required dosage for already established or tested drugs. Within this framework it is clear that an antibody will typically require up to several grams per patient per treatment, whereas a regulatory molecule such as a hormone or interferon will probably require several orders of magnitude lower dose. Similarly, a single dosage form such as a vaccine will require a much lower amount of product per patient compared with a drug required for a lifelong chronic illness such rheumatoid arthritis or multiple sclerosis. Hence, the size the dose, the frequency of dosage and the duration of treatment will all be critical in determining the final requirements for the drug.

(iii) *Other potential applications.* In the past, drugs have been registered initially to treat unmet needs in small patient populations or for rare conditions in order to qualify for orphan drug status. Receiving orphan drug status gives the drug developer the benefits of protected rights to sell the drug for 7 years, tax breaks, subsidies and expedited US Food and Drug Administration (FDA) review, allowing the drug to reach the market faster and via smaller-scale, and thus lower-cost, clinical trials. Once the drug has been approved a number of much larger scale applications can then be licensed for the same drug. In extreme cases this can result in a drug that was ostensibly only going to be produced at a small or moderate scale becoming a major blockbuster. Examples include Amgen's Epogen® (erythropoietin), which was initially licensed for anemia due to kidney dialysis, and Genentech's human growth hormone. In both these only small doses are required per patient; for some other orphan drugs such as Genentech's Rituxan® this is not the case, with a current production scale at the tonne level. If the patient population expands greatly beyond the originally anticipated clinical need the assumptions made for the final process scale may be seriously compromised.

1.2.4.1 Considerations for Scale of Operation

Once a rough estimate of the final scale has been decided, the final scale of operation can be deduced, based on previous or published experience of expression levels and purification yields. Clearly, this will have to match the anticipated final need. In the initial period after the launch of Amgen's Enbrel® the greatest challenge was to produce enough drug to meet the high market demand. The frustrations of getting a drug licensed are nothing compared to the frustration of not being able to produce enough of a successful drug to meet demand! The final scale of production can help decide what purification steps would be feasible, e.g. centrifugation is a common method for the clarification of cell cultures in the research laboratory. In this case the centrifugation is in discreet batches. This is achievable because the volumes being handled are relatively low. Once the process is scaled up beyond a few tens of liters, batch centrifuges become less efficient. Unfortunately this is also still a range where disk-stack centrifuges are either not available or represent an overly complex and inefficient solution to the problem. At this scale, filtration is an excellent alternative. Multilayer filters are now available starting with coarse filter pads to remove whole cells followed by subsequently finer layers designed to remove cell wall fragments and colloids by progressively finer filters. These combined membranes are even available in large-scale self-contained units for ease and handling and operator safety. At pilot scale and up to a few hundred liters these represent a cost-effective solution for initial clarification. However, for very large-scale processes handling 1000 L or more filtration pads now become unwieldy and expensive, and centrifugation once again becomes the method of choice, albeit using disk-stack centrifuges. Although filtration membranes are still required for postcentrifugation polishing, the combined centrifuge and filter train will still outperform a filtration train alone. It can be seen from this example that the optimum solution may change depending on scale and that extensive development work performed at the intermediate scale may be wasted or, worse still, lock the process into an inefficient solution. By careful forethought and consideration of the final potential scale, the process can be tailored, from the very beginning of process development, to provide the most efficient manufacturing process.

1.2.5
Is the Process Economically Viable?

It is clear that the process scientist has a challenge on their hands with each process they are given. Namely, to select the correct set of operations and the operating conditions to allow the purification of the target molecule to a sufficiently high purity, and to do so within the most proscriptive regulatory framework of perhaps any industry. So it is perhaps not too surprising that the additional challenge of cost analysis and cost reduction is not usually a high priority. It goes without saying that the yield should be as high as possible as this will make the process more economic. Fewer steps will also usually reduce costs too. However, this is often as far as a process scientist will get in cost modeling to evaluate the economic viability of a process.

There has been a tradition within the biotech industry that the products are so specialized and so exquisitely effective that they can justify the high price tag that usually accompanies a "biotech" drug. This, in turn, allows for a high cost of manufacture – after all, the market will pay the premium. This comfortable approach to cost of goods is changing. These changes are being driven both internally and externally. Internally there is by a desire to maximize profits. This becomes especially prominent as the scale of manufacture increases. Until relatively few years ago most biotech companies would have found it difficult to give an accurate breakdown for their manufacturing costs. Although the "big ticket" items would be known, such as the cost of affinity media in a production-scale column, the cost of a liter of buffer or the hourly overhead rate would be a mystery. Now many biotech companies are operating as manufacturing companies as much as research companies. This has forced them into a new mindset in which their products have become commodities. It has also resulted in senior mangers recruited from the classical pharmaceutical industry and more used to incremental cost savings through process improvements than the "eureka" moments which previously dominated biotech.

Externally, cost reductions are being driven by a public and political pressure to reduce drug costs, and by the imminent acceptance of generic biotech drugs or "biosimilars". These pressures are leading to the "commoditization" of biotech drugs. An example of the effects of this can be seen in "old biotech" in the production of antibiotics. Many antibiotics or their precursors are produced by fermentation, and there is fierce competition within the industry to reduce costs and increase yields, driven by market forces. The barriers to entry for antibiotic production are relatively low. The molecules are well defined, and there is a large amount of generic data on production, safety and efficacy. As a result these biotech companies have to differentiate themselves, and gain a competitive edge by marginal increases in yield through strain selection/development, fermentation optimization and purification process improvements. It is not unreasonable to expect similar market forces to prevail in the future for some biotech drugs

It is within this new management paradigm that process development scientist now find themselves working. Considerations of manufacturing costs and next-generation process evolution are rapidly becoming as integral to process development as the selection of purification conditions.

Specific areas where cost reductions can be gained have been discussed already. To recap – reduce the number of purification steps, avoid overly high purity and activity specifications, and select operational steps and conditions suitable for use at the anticipated manufacturing scale. All these approaches will all have a positive impact in final cost of goods.

1.3
Strategies to Develop a Downstream Process

As discussed above, the requirement for any successful process is that it will produce enough of the target drug, at the required purity and activity in a cost-effective way that will be accepted by the regulatory authorities. In the following sections some strategies for achieving this requirement will be discussed. First, the actual methodology of process development will be considered. There are several ways to develop a process – each has it merits and its drawbacks.

1.3.1
The Bigger Test Tube Approach

This basically entails taking what has been developed already in the research laboratory and simply reproducing it at progressively larger scales. The advantage of this method is that it is very fast, the process is effectively already complete and all that needs to be sourced are larger vessels/equipment of the same nature. Also, the data previously generated will be applicable for validation of the process. Although this may seem too naive an approach, there are cases where it can have merit. If the aim is to very rapidly generate samples of drug for early clinical or preclinical trials this is probably the most cost-effective way of completing such proof-of-principle studies. It can allow several potential drugs to be produced in parallel for final selection. If the final amount of drug required is likely to be small, e.g. cytokines, interferons or hormones, or if the process is likely to be relatively simple, as seen with some vaccines, then the larger test tube approach may suffice even as far as the final manufacturing scale. For the first years of production of Epogen Amgen continued to use the roller bottle reactors used in the original process. Scale-up was simply by adding another rack and more roller bottles. This is another advantage of this approach – scale-up is often simply a matter of replicating the production process.

The bigger test tube approach does have some significant drawbacks, however. These are more obvious the larger the disparity grows between the research and the final manufacturing scales. As noted above, some methods, such as batch centrifugation, simply do not scale-up without becoming unwieldy and inefficient. Some techniques such as preparative electrophoreses have never successfully made the transition from laboratory to pilot scale, let alone manufacturing. As equipment is scaled up, a range of challenges will occur in terms of mass transfer, homogeneity and thermal transfer. For some technologies such as fermenter design and operation, these parameters are well studied and guidelines published on how to scale-up effectively. For some technologies, especially newer technologies, such as disposable "fermenter in a bag" designs, scale-up is less clearly defined. Another caveat from this approach is in the use of esoteric or nonscalable chromatography media. An example would be an affinity media using a specific monoclonal antibody as the affinity ligand. This technique can be applied in the research laboratory with great success. However, for a production-scale process the affinity antibody would have to be produced in bulk and to almost the same degree of purity as the therapeutic target molecule itself. The cost of such a chromatography media would almost certainly outweigh the potential benefits.

Another problem can occur if the chromatography media used in the research laboratory is either not available in bulk or is simply not scaleable. Many prepacked chromatography columns sold for laboratory use contain media which is different to that provided in bulk for packing large-scale columns. The most common differences are that the analytical-scale media are less cross-linked or may have smaller particle size. The former occurs with compressible media such as agarose and results in media which are softer. These can be used successfully in small columns, where the column wall support helps protect the media from compression. On transfer to a large size column where the wall support is no longer significant, typically around 100 mm diameter or larger, the media may no longer be mechanically strong enough to support the required flow rate. This results in bed compression which, in turn, results in a higher backpressure. This then causes further media compression and a further increase in backpressure, thus creating a positive feedback loop which can quickly raise the backpressure and damage either the media or the column. The only solutions are to reduce the flow rate, which will significantly increase the process time and therefore reduce productivity, or change to a more cross-linked version of the media. This can cause problems in later scale-up stages due to the subtle changes in the background interactions between the media and feedstock components. For example, when agarose media are more highly cross-linked they also become slightly more hydrophobic in nature, which may lead to differences in the impurity profiles of the eluted proteins.

The use of increasingly smaller particle size media is common with rigid media such as silicas where the smaller particles sizes afford greater resolution. However, bulk packing of small particle size media is more difficult in preparative-scale columns and the general solution is to use larger particle sizes. This also has advantages in both lower cost and lower operational backpressure. However, it is critical that the separation be developed on media with the same particle size to ensure that the resolution is still sufficient to achieve the desired separation. Problems can arise in obtaining suitable analytical-scale columns – usually these will only be available as custom columns from the manufacturer of the bulk media.

1.3.2
The Template Process

Most established process development laboratories use this approach as a starting point. The advantages of using a standardized template process are a reduction in process development time, familiarity with the techniques being optimized and confidence that the final process will be validatable. However, the template approach can potentially contain hidden traps. It should be stated that there is, in reality, no such thing as a true template process. That it is a process which can be simply repeated for any and all target molecules of a certain type. This was the original aim and desire of companies working with monoclonal antibodies. It was felt that if a process could be developed once, that process would then be applicable to all other monoclonal antibodies, thus generating huge time and cost savings in bringing monoclonal antibodies to the market. Unfortunately, the reality has proven more intractable. By its very nature each monoclonal antibody clone is unique and different to other monoclonal antibody clones. Even an apparently minor change in the antibody such as single amino acid variations or changed glycosylation patterns can generate an antibody that will behave differently during the purification process. These changes can be as subtle as a small shoulder on an elution peak or as dramatic as precipitation during elution. Due to this variation the use of a template process must be seen as a short cut to part way down the process development line, rather than a means of bypassing it altogether.

1.3.3
Process Development by Gradual Evolution

The shortcomings of the pure template approach usually lead to a more pragmatic approach in which a process is developed as a process of gradual evolution from a standard starting point, i.e. the template, but with the acceptance that the final process will have some specific character of its own. These changes can be driven both by the characteristics of the specific target molecule and also by changes in the regulatory environment which may require further or more stringent processing than was previously required. An example of the latter is the increased emphasis on virus inactivation and removal since the problems experience by the blood product industry with human immunodeficiency virus (HIV) contamination. It should also be noted that the FDA and other regulatory bodies are now more open to scale-up and postapproval changes (SUPAC). This is a distinct change from earlier approaches validating biotechnology products.

Despite that fact that each process will have its individuals twists and quirks, it is reasonable to say that the majority of therapeutic monoclonal antibodies currently being produced use an almost identical process. Starting with Protein A-affinity chromatography as a capture step, followed by cation exchange. The main variation comes in the final step. Mostly this is an anion-exchange step using either chromatography columns or charged membranes, but processes using hydrophobic-interaction or hydroxyapatite chromatography are also in place.

This observation leads on to the next method of process development.

1.3.4
The "Me Too" Process

This should be differentiated from the pure template process approach in that a "me too" process would be developed using a template from outside the current process development teams experience. Although apparently a simple approach, it is usually very difficult to obtain sufficient details to allow a complete process to be reproduced.

It is one of the greatest challenges for the process development scientist to keep up to date with the latest developments in the field. This is especially problematical for smaller or start-up companies. The key reason for this is that little or no information tends to be published on the development of current processes. From the perspective of the biotechnology companies, this may be a sound commercial decision. Certainly in the more commoditized pharmaceutical industry incremental increases in yield or reductions in overhead costs, through more efficient processes, will directly impact the economic viability of a process and therefore be regarded as a commercially sensitive. As a result, most information published on processes tend to be for "previous-generation" obsolete processes or processes for drugs which have failed in clinical trials. This information is also usually only available in presentations at conferences, rather than in scientific journals. A crossover area can be found in the commercial trade journals, where solicited articles discuss current issues. However, these are also interspersed with articles which are designed to promote new technologies or products which may not yet have widespread use or acceptance. For these reasons a "me too" process can be very difficult to achieve unless through acquiring expertise through the natural movement of staff between companies.

One advantage of reproducing a current process is that it should already be familiar to the regulatory authorities. However, this should not be seen as an opportunity to shortcut process validation. It is still the case that each process must be validated for its own performance. As noted above, individual variations in the nature of different clones of monoclonal antibodies will still produce enough variation in response to process steps to necessitate further optimization even for a "me too" process. In this respect it would be similar to the problems of trying to implement a template process.

1.3.5
The Clean Sheet

Perhaps surprisingly, the least-encountered method for process development is the clean sheet approach, where the characteristics of the target molecule are considered and a rational theoretical purification train is postulated. This approach is the most scientifically "pure" and perhaps closest to the approach seen in a research laboratory. The key difference for a process development laboratory is that this approach allows the use new or "esoteric" large-scale techniques which may not be valid at the small scale. The downsides to this approach are clear. Novelty is not usually encouraged in process development. Being the first company to present a new purification technique to the regulatory bodies is fraught with risk. Potentially, the approval could be delayed or even not granted if there is doubt over the reliability and reproducibility of the purification process. The only justification for such a risk would be if all standard methods of purification have failed.

Adopting a conservative and risk averse approach, while making business sense, does have serious implications for research into new and novel methods or purifying proteins. It discourages process research within the biotechnology industry. Few companies are in a position to provide resources for research on novel processes which in all probability will not be implemented. On the other hand, the companies which provide the purification tools, such as chromatography media and filters, will be disinclined to develop truly novel technologies that will either not be adopted or, in the case of lower-cost methods, will harm their current sales. For entrepreneurs and inventors the situation is even less promising, as the potential adoption of a new technique is likely to require significantly longer than their funding will allow. Probably the last significant "new" technique in large-scale protein purification over the last 20 years was expanded-bed chromatography. Although widely used in the chemical industry, this method was only developed for use in protein purification in the late 1980s and was first introduced commercially around 1990s. It was almost 10 years later in 1999 when the first FDA approved process using expanded-bed chromatography was reported by GE Amersham. This was for a relatively small-scale application, requiring only around 20 L of media, for the production of a lipoprotein component of a human vaccine by SmithKline Beecham. Another supplier of expanded bed media, Upfront Chromatography, also report, on their website, the use of the technique at multitonne scale for the purification of lactoferrin and IgG from cheese whey. Notwithstanding this application, the technique remains rare in large-scale protein purification processes. It can be seen from this that the potential payback period for a successful new process is around 15 years. As a result of these commercial pressures there is significant lack of research into new and novel methods of large-scale protein purification. This is especially the case for low-cost methods or adaptation of current technologies from other fields where the opportunities for commercial exploitation are low. Although there are some very good academic institutions working on new purification methods, the costs involved in producing suitable feedstock and operating at large scale present a real problem. Without closer cooperation and significant funding from the biotechnology industry, the academic study of large-scale protein purification is unlikely to yield any new techniques in the foreseeable future.

1.4
Process Optimization

Having discussed general approaches to process development it is appropriate, at this point, to consider some specific examples of how process steps can be optimized. These techniques and approaches are relevant to all the previously discussed approaches to process development.

1.4.1
Cell Removal/Clarification

Over recent years there has been a progressive trend towards higher titers from fermenters, and this has resulted in higher cell densities and lower cell viabilities than previously experienced. This trend is often seen in process development. In the early discovery and preclinical stages of a developing a biological drug the fermentation will not usually have been optimized. The aim at this stage is simply to produce enough of the potential drug to allow evaluation of its effects with *in vitro* and animal models. The problems tend to arise when the drug is further progressed, especially as it heads towards phase II clinical development. Typically, the upstream cell culture groups and downstream purification groups are in different laboratories, different departments, different buildings and, in some cases, different countries! The target of the upstream group is to produce the highest possible titer of product in the shortest possible time. The aim of the downstream group is to take the feedstock from the upstream group and develop a suitable purification process. The problem arises because these two groups are working in parallel simultaneously. If the upstream group are doing their job, the feedstock is likely to progressively have higher cell densities and lower cell viabilities. This will also be accompanied by changes in the impurity profiles. For example, lower cell viabilities can result in increased levels of DNA and proteases in the feedstock as the nonviable cells degrade. Physically, the feedstock may also change as a result, becoming more viscous. If a process is developed, using the original feedstock, which is not sufficiently robust it may run into serious problems when challenged with the "improved" feedstock which may manifest as increased backpressures, slower unit operation times, reduced ability to clear impurities or increased product degradation.

New feedstock components can also be a source of problems. These may have different binding characteristics to those for which the process was originally developed resulting in a failure to reach desired purity and also can interact with target protein in different ways, e.g. promoting aggregation or even copurifying.

The challenge is clearly to develop a process capable of handling a feedstock which is constantly changing. The answer lies in two areas. (i) The direction of potential changes should be anticipated and the process developed with extra robustness built in to hopefully accommodate this. (ii) There must be good communication between the upstream and downstream groups. It is this latter which provides perhaps the greatest challenge. Having a suitable management structure and means of cross-communication could be one of the best investments made in to assist process development.

Clarification itself cannot commonly be achieved in a single step. Thus, multi-train operations will be required. These can be subdivided into primary, secondary and sterile steps. The primary step can be via centrifugation or filtration. As noted already, centrifugation is best suited to the small scale and very large scale, with filtration being preferred at the intermediate stage. Scaling down centrifugation steps is fraught with difficulty. This is because the discreet container format of small-scale centrifugation is so different to the disk-stack centrifuge. The results of a disk-stack centrifuge can be modeled by selectively reintroducing some of the pellet from a batch centrifuge back into the process stream to mimic the higher level of solids; however, this will not truly model the redistribution of small and large solids and colloids observed. Colloids, especially, can have a severely detrimental effect on subsequent filter membranes.

It is possible to replace the centrifuge with a filter. Either in tangential flow mode or as a normal flow depth filter. For a tangential flow filter membrane, an open screen type must be selected to cope with whole cells and large cell debris. Depth filters are available in a bewildering variety of sizes and formats. Although it may be tempting to think of all depth filters as equivalent, there can be differences in performance that when scaled up can have a major impact on the process economics.

Therefore, it is best to screen several clarification filters. This is especially the case for multilayer membranes. The use of multilayer membranes gives a great improvement in ease of use, especially where fully enclosed "cartridges" are available at process scale. However, the relative porosity and capacity of each filter layer can massively impact performance. Screening is essential for these types of membranes. One approach to screening such clarification membranes is the P_{max} test. For this simple procedure, the filter being tested has a pressure gauge placed inline at the inlet and the feedstock is pumped though the membrane at a constant flow rate. As the depth filter blocks the pressure will increase until the maximum pressure rating of the filter or some other nominal limit pressure is reached. Since the flow rate is constant, either the time or the volume passed through the filter can be plotted against the pressure increase. This can be done for a number of membranes and the membrane with the highest capacity can be determined. As a further check, samples of the output from the membrane should be taken at intervals and checked for the presence of impurities, typically done by simply monitoring turbidity. This can also be plotted against the time or volume. If a membrane is too open it may allow a greater volume of feedstock through, but this feedstock may have too high an impurity content for the subsequent filter step. Hence, this complementary check should be performed on at least the two most promising membranes.

The clarified solution produced by the above test should be retained for use, selecting the sterilizing filter in the final part of the clarification process. As with depth filters, sterilizing filters are now available in a bewildering variety of type and sizes. Sterilizing filters should be coselected with the clarification filters. The relative area of sterilization filter required is a function of the efficacy of the clarification filter layer(s). Traditionally, sterilizing filters have a higher cost than depth filters. Where this is the case, the clarification membranes should be oversized to protect the more expensive sterile filters. Multilayer depth filters and cartridge-based systems have greater ease of use, save time, ensure system integrity and have much lower hold-up volumes. These advantages make them a good choice for selection; however, they have the penalty of higher costs. In such cases it may be economically better to optimize the clarification for the multilayer membranes at the expense of a greater area of sterilizing membrane.

1.4.2
Sterile Filtration

Sterile filtration has widespread use throughout a typical process – not just for filtration of product, but also for sterilization of buffer and growth media. Typically, a sterilizing filter will have a nominal pore size of 0.22 μm. Smaller pore sizes of 0.1 μm are also available; however, the reduced pore size also reduces the process flux, reducing throughput, and will foul more quickly, necessitating a higher membrane area for a given application. Hence, these membranes tend to only be used where it is felt Mycoplasma contamination may be a potential problem. Recent advances in membrane technology have produced high flow rate sterilizing filters based on polyethersulfone chemistries. These membranes have high flux, high retention and are robust enough to be sterilized, by γ-irradiation, prior to use. Because of these advantages they are becoming the default starting point for any membrane screening study. However, older membrane products made from regenerated cellulose are still the most common sterilizing membranes in current use in full-scale processes and do have the advantage of many years of validation data behind them. Again, it should be noted that membranes from different manufacturers, made with the same base chemistry and pore sizes, might have different behaviors in terms of flux and fouling for the same feedstock. Indeed, it is the author's experience that even using the same membrane may produce different results for different target process streams. Again, it is wise to screen a few alternative membranes to see if such anomalies are present rather than simply using the same membrane used previously. If there is only marginal difference in performance, there is probably more to be gained in manufacturing efficiency and cost reduction by consolidation of processes around a preferred membrane or supplier, although validation of a sterilizing membrane from a second supplier, if time permits, will allow for a fall back position if required in the future.

Sterile filters are usually screened and sizing data generated using the V_{max} method, based on the theoretical maximum volume which can be passed through a unit area of membrane before flow is reduced to an unacceptably low level as a result of membrane fouling. In this test the process fluid is passed through a sample of the membrane under test at a fixed pressure, usually achieved using a pressure vessel attached to an air supply via a regulator. The changing flow rate is monitored by measuring the cumulative mass of liquid passing through the membrane over a period of time. This can then be plotted as a graph of (time/ volume) versus time.

In contrast to the P_{max} method described above, the V_{max} method does require slightly more specialized equipment and, moreover, the analysis of the resulting data requires a mathematical model to be used which makes assumptions about the mode of fouling present. Fortunately, most membrane manufacturers can provide help in using the V_{max} method for membrane screening and scale-up studies. The V_{max} value gives an indication for future scale-up requirements. However, this will still need to interpreted within the context of the process requirements. The V_{max} value for a given membrane can vary from as little as a $200\,L\,m^{-2}$, for a serum containing culture media which has not been prefiltered, up to well over $10\,000\,L\,m^{-2}$ for a clean buffer. In the case of the former, a larger area of membrane will be required simply to complete the filtration. In the latter case, extra membrane may well be used simply to shorten the buffer preparation time, despite the filter membrane itself being underutilized.

Often when a process is developed at the bench sterile filtration is used generously to ensure that the process fluid is not contaminated during the purification process. Ideally, in an aseptic environment, this should no longer be an issue. Despite this, it is not unusual to see many of the original "bench-top" sterilization steps incorporated into the final process. This adds unnecessary time, cost and risk to the process. Before attempting to optimize any sterile filtration step in a process, it is always worthwhile to question whether the step is still required.

1.4.3
Chromatography

Chromatography lies at the core of all biotechnology purification processes. Despite this, there remains a lack of suitable texts on process-scale chromatography. Almost all the published literature on chromatography has been written for analytical applications. It is important to appreciate that analytical chromatography has a completely different set of aims compared to preparative chromatography. Usually when the literature speaks of optimizing a separation, it means achieving the maximum number of resolved peaks. In preparative chromatography there is only one peak or real value, i.e. that of the target molecule. If all other molecules were eluted in only two peaks, one before the product and one after, this would represent an optimized manufacturing process. The emphasis in process chromatography is merely to ensure that the target peak is sufficiently separated from the closest peaks on either side. This should be the central aim of any changes made in elution conditions. One potential opportunity from this approach is the use of ion-exchange media in isocratic mode. If the conductivity and pH are carefully selected it is possible to load an ion-exchange media with feedstock and have the target protein retarded, while nonbinding proteins pass through the column and more tightly binding proteins are retained. Although requiring more development effort, this approach removes the need for gradient formation and can allow much higher loads of target protein to be separated in each purification cycle.

1.4.3.1 Binding Capacity and Column Loading

Column loading is an area which often causes problems in developing a chromatography separation. Again, in analytical chromatography, the problem is very simply solved by massively underloading the column. In large-scale operation this is not an economically desirable solution. Published data on binding capacities may give a rough indication of the load which can be reasonably expected, but these figures should not be used as the basis for a process. The usable dynamic binding capacity of a media must be determined for the specific target protein. For example, different monoclonal antibodies can exhibit large variance in dynamic binding capacity on the same Protein A media, even if the antibodies share 90% or more of a common amino acid sequence. Care should also be taken to differentiate between dynamic and static binding capacity. The static binding capacity represents the total amount of a specified protein that can be bound by a unit volume of the media. This figure is often quoted for ion-exchange media. Achieving saturation is not practical in the real-world; instead, a comparison of dynamic binding capacities under similar conditions using the same protein is the only reliable way to compare different media. Unfortunately, there is no standardized test used by all media suppliers. Thus, comparison of media solely based on data published manufacturers should be done with great caution.

For affinity chromatography media, such as Protein A capture columns, the question of how much to load can be answered reasonable simply. The column can be loaded to breakthrough point to determine the dynamic binding capacity and then loaded to around 80–90% below this value. In determining the breakthrough capacity, there a few parameters which need consideration. First, the dynamic binding capacity will be dependent upon the loading flow rate. The faster the loading flow rate, the less time the target molecule will have in the column to diffuse into the pores of the media. There will be a critical flow rate above which the dynamic binding capacity will rapidly decrease. The actual value of this flow rate will vary depending upon the mass transfer properties of the media being used. Media with very efficient mass transfer properties, such as controlled pore glass-based media or some highly cross-linked "open pore" agarose media, have very efficient mass transfer properties. These media can be used at high flow rates.

As a general rule highly cross-linked agarose media can be used at linear flow velocities of up to 500 cm h^{-1} in 20-cm beds. Controlled pore glass, on the other hand, being rigid and mechanically strong, can be operated at much higher flow velocities and in much longer bed heights, up 1000 cm h^{-1} in 40-cm high beds have been reported by the manufacturer. It is not necessary to use such extremes of bed height and flow rate, but the option of increasing either of these parameters beyond the fairly modest limits of agarose media does provide the process developer with much greater flexibility and ensures that the final-scale process can be operated well within the envelope of operation of the media, thus increasing process robustness.

Another consideration in determining the dynamic binding capacity is the selection of a breakthrough endpoint. Dynamic binding capacities are typically quoted at 10% breakthrough. If the shape of the breakthrough curve is steep, this will be only slightly higher than the actual point at which breakthrough occurred. However, for most media the breakthrough is not immediate, but gradual; the more gradual the breakthrough, the more flattering the 10% breakthrough figure will be compared to the actual capacity before target protein is lost. One method reported to increase the available capacity is to initially load a column at a high flow rate and then, as the binding capacity of the column is approached, to slow down the flow velocity by a factor of 2 or more and complete loading at the lower flow rate. This will allow greater time for diffusion of the target molecules in and out of the media in the search for the relatively few remaining binding sites. Although an effective approach to increase the binding capacity, the author is not aware of this method currently being used in a validated process.

For an ion-exchange column the amount to be loaded is a more complex issue. Overloading column will cause changes in the elution profile. Therefore, scale-up should be based on data from smaller columns that have had equivalent loading. Due to the relatively low cost of ion-exchange media, compared say to Protein A-affinity media, the traditional solution has been to oversize ion-exchange columns and only load to around 30–40% of the theoretical capacity. As processes increase in scale, however, this results in ever larger and more unwieldy columns. In the largest commercially available columns, with a diameter of 2 m, the amount of media required now represents a significant cost (630 L for a 20-cm bed). As a result of this one area of growing interest is the use of membrane-based ion-exchange devices.

This is especially true of flowthrough applications for ion-exchange media. An example can typically be seen with anion-exchange columns used in the final polishing step of many monoclonal antibody processes. The p*I* of most monoclonal antibodies allows them to pass directly through the anion-exchange media without binding, whereas several key impurities, such as DNA, endotoxin and most host cell proteins, bind to the anion-exchange media. In this case the relative amount of impurity is very low, typically less than 2%, compared to the amount of monoclonal antibody. Due to this the required binding capacity is also very low and this application is thus well suited to the use of an anion-exchange membrane system.

1.4.3.2   **Throughput as a Chromatography Optimization Parameter**

Determination of the dynamic binding capacity should be regarded as the first step in optimization a chromatography column. Unfortunately is it often also regarded as the final step. As noted above, the dynamic binding capacity will increase as the loading flow rate is decreased. Hence, the highest binding capacity will occur at the lowest flow rates. If time is not critical, the highest capacity can be achieved by loading over a period of hours. Consequently, it is not unusual to see laboratory-scale purification processes reported in the literature with the sample being loaded overnight, usually performed in a cold room to prevent sample degradation. However, this is not the most efficient mode of operation. In a manufacturing process the cycle time should be short enough to allow flexibility in scheduling, preferably allowing the operation to occur within one working shift with time for set-up and cleaning. The question is how to balance the compromise of a high dynamic binding capacity and a short process time. The answer lies in considering the throughput. Throughput is simply the mass of material purified per unit time. It can be seen that a high-capacity process with a very long cycle time will have a poor throughout. Similarly, very rapid process will also have a low throughput if speed is gained at the expense of too great a loss in capacity per cycle. In general, the loss in capacity is marginal compared to the reduction in cycle time until a critical flow velocity is approached. Because of this, the most efficient mode of operation is to have a relatively short residence time in the column, in the order of a few minutes for most media. In order select the optimum residence time, giving the highest throughput, the dynamic binding capacity should be determined over a range of flow rates. For each of these flow rates the cycle time should also be noted or calculated. The throughput can then be calculated as the mass of protein purified per unit time and the optimum value selected to give the most efficient process cycle. Throughput can also be further normalized to yield a values for the mass or protein produced per unit time per unit column volume. This is then a value which represents the productivity of a specific media or column. This value is useful when screening different column geometries (i.e. bed heights versus diameters).

1.4.4
Ultrafiltration

Along with chromatography, ultrafiltration is another ubiquitous process step encountered in biotechnology. It is commonly used for buffer exchange and product concentration, both between purification steps and for final formulation. Where possible, consecutive steps should be selected to minimize buffer changes of the need for concentration to ensure superfluous ultrafiltration steps are not present. Although there is some degree of separation, ultrafiltration has a low resolution, only being able to reliable separate molecules with an order of magnitude size difference, because of this ultrafiltration is not usually selected purely as a purification step.

It is not uncommon to encounter an ultrafiltration step which uses the "typical operating conditions" given in the manufacturer's literature. However, consideration of the desired outcome and some optimization will usually yield benefits in process efficiency either by reducing the cycle time, increasing product recovery or both.

1.4.4.1 Optimizing Tangential Flow Ultrafiltration

The first choice for a tangential flow step is to select the appropriate membrane. The pore size (or molecular weight cut-off size) quoted by a filter manufacturer is a nominal value, i.e. it represents an average pore size for the membrane. However, the distribution of pore sizes will affect the performance of the membrane both in terms of speed and product loss. For example, a process using a nominal 50-kDa molecular weight cut-off membrane from one manufacture may have poor product retention but be much faster than a 30-kDa membrane from the same manufacturer. However, a 30-kDa membrane, having a less homog eneous structure and thus wider distribution of pores sizes, perhaps sourced from another manufacturer, may give a better balance of retention and speed. Unfortunately, most manufacturers do not make such data widely available. It is therefore worth screening membranes from different manufactures even with the same nominal pore size, if time permits.

In tangential flow filtration systems, the early selection of the appropriate recirculation flow rate will have far reaching impacts as the process is scaled up to manufacturing levels. The permeate flux, i.e. the rate at which liquid passes through the membrane, is proportionally related to feed flow rate, i.e. the rate feedstock passes over the membrane.

The appropriate feed flow rate will maximize permeate flux while minimizing pumping requirements and maintaining a gentle environment for the product. It is critical to consider this relationship and its impact on product quality during process development, especially for later process scale-up.

Typically, a feed flow rate near the highest recommended by the filter manufacturer is selected. The main benefit of high feed flow rate is a higher permeate flux at any given transmembrane pressure. This is because the higher flow rate produces a greater sweeping action across the membrane surface, reducing the stagnant gel layer at the membrane and thus increasing mass transfer. The benefit of a higher permeate flux can be either a more rapid process operation or a reduction in the required membrane area for a given process time. Typically, the selected operating conditions will allow the process to be completed within a desired amount of time to fit in with scheduling requirements of the process. Usually the process is developed to minimize the surface area of membrane required. In the case of very long tangential flow filtration processes it may be worth increasing the surface area to reduce the overhead costs and allow more flexible process scheduling. It is worth noting that increasing the surface area will make the process step faster, but this may be at the expense of hold up volumes or reduced recoveries. Another mechanism by which the gel layer depth can be controlled is in the use of screens within the filter device to promote turbulence. With more turbulence resulting in a smaller the gel layer and thus higher permeate flux. However, these screens are also sensitive to the nature of the feedstock, with the most efficient "turbulence promoters" creating a higher backpressure and being more prone to blockage. Therefore, care should to be taken to match the screen type with the feedstock, especially in the case of high concentration feed or feed prone to precipitation.

The disadvantages of using high feed flow rates are that this will increase the number of pump passes during the process cycle, assuming a smaller membrane area has been selected in preference to process time reduction. In practice, few protein are so shear sensitive that they will suffer denaturation from the additional pump passages; however, it is still worth considering, especially for nonglobular, multimeric or particularly large proteins. If it is proposed to use a current system for scale-up or manufacturing care should be taken that the flow rate selected at small scale will still be achievable at all future scales without significant investment in new hardware.

It should also be noted that, as concentration increases, the feed side pressure drop in the membrane device will also increase; this may become excessive in processes where the concentration factor or final product concentration is high.

1.4.5
Virus Removal

In the 1980s, HIV contamination of blood products brought the issue of virus inactivation and removal to the forefront of the process development. Since then the regulatory authorities have progressively tightened the requirements for virus removal. The current position is that phase I clinical material must have a validated virus removal step. In practice this means that there must be data to show that at least one model virus will be safely removed by the process used to manufacture the phase I drug. By the time a drug reaches phase II there must be a more complete set of data to demonstrate reliable removal of a panel of model viruses. The model viruses are selected to present the process with a set of realistic worst-case challenges. They will usually contain at least a model retrovirus, such as murine leukemia virus (MLV), and a small and robust parvovirus, such as mouse minute virus (MMV). Blood plasma products would also have a virus such as Sindbis as a specific model for hepatitis C virus. In order to validate virus removal or inactivation, a three-pronged approach is taken.

(i) *Raw material compliance.* All raw materials are evaluated for potential virus contamination. In the case of some raw material such as mammalian cell lines it is assumed that there will be an intrinsic viral load due to endogenous viruses. Other feedstock may have to undergo specific physical or chemical treatments to give some assurance of viral inactivation. This consideration has also been the driver to remove fetal calf serum and other animal-derived products from growth media recipes.

(ii) *Individual process step validation.* Each process step which is deemed to have a robust virus-reducing effect is validated to demonstrate this effect. Virus reduction can be by virus elimination, physical removal of the virus or virus inactivation. In the latter, the original virus particles remains, but are no longer capable of infection. These approaches can be measured by two different methods. Absolute removal can be measured by polymerase chain reaction (PCR)-based assays in which the amount of viral genome present is measured. Virus inactivation is measured by viral viability assays, in which the titer of active virus is determined by incubation with host cell. In both cases the amount of virus removed is quoted as a log reduction value (LRV). This is $\log_{10}$ of the ratio of the total virus load before clearance and the total virus load after clearance, e.g. a LRV of 4 would indicate that the step reduces the virus load in a test spike by a factor 10 000.

(iii) *Cumulative process clearance.* The virus clearance capabilities of individual steps can be added together to give a cumulative virus clearance capability for a process. To make this more robust the steps being added together should be orthogonal, i.e. to say they should have different modes of action (e.g. a chemical inactivation and physical filtration step). To obtain the overall clearance, the LRVs for each step are simply added together.

1.4.6
Specific Considerations for Virus Removal

1.4.6.1 Protein A-affinity Chromatography

Chromatography can be used as a method of virus removal. The mode of removal can be through passive partition (e.g. by size exclusion), by active partition (e.g. on an ion-exchange column) or by a chemical inactivation because of the chromatography conditions. Protein A-affinity chromatography provides a good example of the use of a chromatography step to remove virus. There are two mechanisms in action: physical partition through the column (monoclonal antibodies will bind to the Protein A ligand, while the virus will pass through the column unhindered) and chemical inactivation, due to the low pH of the elution buffer.

There some points which need to be considered with this step. There is the possibility of the media becoming fouled over prolonged periods of use. This may change the surface of the media, especially near the top of the column, and result in a media which will interact with and retain the virus particle. It would not be unreasonable to perform follow-up virus removal studies on used media to demonstrate no loss in the ability of the column to partition virus. Also, the chemical inactivation step is a result of holding the virus at a low pH, typically around pH

4.0. The hold step must be validated, and then specified at a specific pH or lower and at a given protein concentration or lower. Higher pH and/or higher protein concentration in the elution pool will affect the degree of chemical inactivation which occurs. The minimum hold time must also be specified. All of these parameters should be considered when designing the validation study so they will reflect a suitable safety margin for the final process

The two methods of virus assay mentioned above are particularly useful for the Protein A chromatography step. Assay by PCR will enumerate both active and inactive virus remaining in the elution pool. Comparison of this figure with the amount of virus in the spiking feedstock will show the amount of virus removed by partition alone. Analysis of viable virus in the elution pool will indicate the amount of virus reduction from both the partition and the chemical inactivation. Subtracting the latter from the former will give a log reduction value for the chemical inactivation alone. This method then provides log reduction values for the two separate orthogonal methods of virus removal.

1.4.6.2 Virus Removal by Filtration

A common physical method used to remove virus particles is absolute filtration. The filters used to remove virus obviously have a very small pore size, typically around 20 nm. This gives rise to the two most common problems with this technique – slow speed and tendency to block. Both of these issues are addressed by placing the virus filtration step late in the purification train. At this stage the concentration of product is high, thus reducing the volume to be handled, and also the product is pure, reducing blockage due to nondrug moieties. Although this is a generally valid approach, the volumes being handled may still be appreciable at manufacturing scale, resulting in a significant cycle time. Much work has been performed by the filter manufacturers to try to improve the flux of virus removal membrane, helping to reduce the process time issue. However, the blockage issue remains a problem. This is compounded by the virus spike method used to validate the filters themselves. In order to validate virus removal a high titer spike of virus is passed through the membrane. Such high-titer virus preparations can be highly variable in quality. At worst they can be so heavily laden with particulate cell fragments that they cause filter failure long before the filter itself would fail if presented with the "normal" feedstock. The result will be a significant overspecification of membrane area required for virus removal. Given the high cost of virus removal membranes, this presents a significant threat to the process economics. The answer lies in the use of suitably purified virus spike solutions.

Another consideration for the methodology used in validation of the membrane is the time at which the virus spike is applied. Traditionally, the spike is applied first and then the membrane is used to filter feedstock afterwards. A dirty spike sample can block the membrane and artificially reduce the volume of feedstock the membrane is able to handle. A more representative method is to run the required amount of feedstock through the membrane, to represent a used membrane, and then spike with virus. This has the double benefits of allowing a more representative feedstock to be used to size the area of membrane required and to produce virus removal validation data on "used", i.e. worst-case, membranes.

1.4.7
Lifetime Studies

The real test of a process is not how well it works not on the first run, but how well it performs on the final runs. Repeated exposure to feedstock can have a progressively deleterious effect on chromatography media. The problem is somewhat bypassed for filtration since most filtration media are designed to be used only once. However, ultrafiltration membranes can, and indeed often are, used for repeated cycles too. It is important to validate the potential lifetime of any step of a purification process that will be exposed to multiple cycles of use. The effect of a shorter than expected lifetime can be catastrophic to the overall economic viability of a process. Conversely, if the media can be used for more times than expected this can give an added bonus to the profitability of the process. Either way, it essential to be aware of the expected operation lifetime.

An argument could well be made that the best time to generate validation data is at the end of a series of production cycle. Unfortunately, this is both too late and too expensive an exercise at full scale. During a series of processing runs data can and should be collected to assist in retrospective validation of a process. However, this is at best only support data. Instead, a lifetime study is an essential part of process development. It is widely acknowledged that lifetime studies are labor intensive and, from a scientific point of view rather boring. This is especially the case for a successful process – with a long lifetime and no problems. Lifetime studies are is also not a great way to advance a scientific career. Months of work may effectively generate only one graph showing, hopefully, that not much is happening! Despite this, lifetime studies are absolutely critical to the validation of a process, and to also give and idea of the eventual economic viability of the process.

Lifetime studies are made easier by automated fast protein liquid chromatography systems; however, there still remain some caveats to be aware of. The feedstock should ideally be from multiple sources rather than one single source, although this may be difficult. Using one feedstock throughout a study can skew the results if source represents either a better or worse than average case. Feedstock should be stored and used in aliquots that model the final process. A single lot of feedstock standing for a period of weeks or even months is unlikely to still be representative by the end of the study. Ideally, aliquots to represent single campaign batches should be stored separately and frozen if necessary. It is likely that the real feedstock will not be frozen in the full-scale process. If this is the case the validity of freeze–thawed aliquots of feedstock as representative must be validated too. It is the experience of the author that some feedstock, left on the bench for only a few hours, will show significant signs of aggregate formation which, while not being visible as cloudiness, will affect the elution profile of a chromatographic separation.

1.5
Future Trends in Process Development

There are some clear trends in process development, some of which are closer to general implementation than others.

1.5.1
Disposable Process Lines

The use of disposable bags for buffer preparation has been common for a long time in pilot plants. There has also been strong growth in the use of disposable bag fermenters. Manufacturers are now increasingly able to supply disposable assemblies for even more of the purification train, including disposable pipework and tank liners, and filtration assemblies. It may even be envisaged that disposable chromatography columns will one day be available to allow a fully disposable purification train removing concerns over cross-contamination or post-use cleaning. Although the use of fully disposable manufacturing systems may still be some way off, the ease of use and fast turnaround for different product lines makes this approach highly desirable for pilot process development laboratories.

1.5.2
Nanoscale Screening

High-throughput screening techniques have been developed and are now well established for drug screening. There is a clear move towards adapting these techniques for the subsequent process development stages too. Small-scale screening can be used to determine which media could potentially be used to purify the target molecule and also to get a good indication of the optimum conditions for purification. For example, in a single afternoon the experimenter can determine not only which media to use, but also the best buffer, pH and conductivity for loading the media, and similarly the best buffer, pH and conductivity for elution. This massively reduces the total time typically required to generate a near-optimal purification method. A further advantage of this approach, using multiwell plates or a similar format, is that only very small amounts of sample are required allowing the work to be done even earlier in the development cycle.

1.5.3
High-titer Feedstocks

As noted, there has been a progressive trend towards higher product titers as cell lines and culture conditions are better understood and optimized. In the late 1990s, a monoclonal antibody titer of around $0.1 \, g \, L^{-1}$ in the clarified cell culture supernatant was not untypical. Currently, titers around $1-2 \, g \, L^{-1}$ are the norm and there are reports in most cell culture conferences of titers of $5 \, g \, L^{-1}$ or higher at the development stage.

As the titer approaches such high levels the relative amount of product, antibody, to contaminant, cell culture protein, changes. It is possible at these very high titers that the most common first step used in monoclonal antibody production, i.e. Protein A-affinity chromatography, will become obsolete as the starting product will already be sufficiently concentrated and pure. The role of the first step of the purification process then shifts from capture directly to purification.

1.5.4
High-concentration Formulations

Another area becoming increasingly prevalent is the use higher-titer formulations. Some biological drugs such as antibodies require significant mass to be introduced into the patient in each dose. Concentrating the drugs ever higher makes administration easier. It is already not unusual to see antibody concentrations in excess of $100\,\mathrm{g\,L^{-1}}$. Although the antibody may well be stable at this concentration, this approach does generate some distinct process problems for the final concentration and formulation step. On concentration the titer and the viscosity increases. Very high concentration factors can cause problems with ultrafiltration membranes. A membrane which is optimal for lower viscosity may have such a low permeate flux at high viscosity that it becomes virtually unusable. Also, the dead volume in a system, i.e. the cumulative volume of material in the pipework and inside the membrane housing, may be such that, at high concentration factors, a significant part of the product cannot be recovered. Additional steps may have to be introduced to increase recovery, such as using clean air to flush out the membrane holders and pipework.

From these considerations, it can be seen that the role of the process development scientist is not only a vital bridge between R & D and manufacturing, but also between past conservatism and exciting new trends. Fundamentally, it will be the process development scientists who will have to figure out how to apply ever more cost-effective methods to produce ever-larger amounts of the next generation of biotechnology products.

Further Reading

For further introduction to process development, especially with regard to the purification of monoclonal antibodies, the reader is directed towards the following excellent sources.

Sofer, G., Hagel, L., *Handbook of Process Chromatography – A Guide to Optimization, Scale-up and Validation*. Academic Press, London, **1997**.

Subramanian, G. (Ed). *Antibodies. Volume 1: Production and Purification*. Kluwer, New York, **2004**.

Subramanian, G. (Ed.). *Bioseparation and Bioprocessing (Volumes I and II)*. Wiley-VCH, Weinheim, **1998**.

US Food and Drug Administration Center for Biologics Evaluation and Research. Points to consider in the manufacture and testing of monoclonal antibody products for human use. *J Immunother* **1997**, *20*, 214–243.

2
Strategies in Downstream Processing

Yusuf Chisti

2.1
Introduction

Biological products come from many sources: human and animal tissue (e.g. blood, pancreas, pituitary) and body fluids (e.g. milk of transgenics [1, 2]), plant material (e.g. Taxol® from the bark of *Taxus* species, oils), microbial fermentations, cultures of higher eukaryotes, and raw broths from enzyme bioreactors. Irrespective of the source, crude extracts, fluids and broths invariably undergo separation and purification to recover the product in the desired form, concentration and purity. Processing beyond the bioreaction step is termed downstream processing. Here, the "bioreaction step" includes producing plants and animals.

A recovery process consists of physicochemical operations such as those listed in Tab. 2.1. The steps of a properly engineered downstream process are integrated with each other and with the bioreaction stage to yield an optimal recovery scheme. This discussion is limited to factors that must be considered in developing any economically viable product purification and concentration scheme based on a small selection of the many available processing operations. Individual operations are detailed in other sources [3–6, 29–33] as well as some other chapters in this book.

2.2
Overview of Process Considerations

2.2.1
Possible Recovery Flowsheets

One example of a recovery flowsheet is shown in Fig. 2.1. The process shown is for the recovery of the recombinant protein human α_1-antitrypsin from the yeast *Saccharomyces cerevisiae* [34]. The required yeast biomass is produced in the fermenter. The product is contained within the yeast cells in a biologically active

Bioseparation and Bioprocessing. Edited by G. Subramanian
Copyright © 2007 WILEY-VCH Verlag GmbH & Co. KGaA, Weinheim
ISBN: 978-3-527-31585-7

Tab. 2.1 Bioseparation operations

Solid-liquid separations [1–6]	centrifugation, depth or media filtration, flocculation, ultrasonic flocculation, flotation, sedimentation [7, 8]
Membrane separations [3–5, 9]	diafiltration and dialysis, electrodialysis, ultrafiltration, microfiltration, pervaporation [10–12], reverse osmosis
Extractions	aqueous liquid-liquid extraction [3, 13, 14], extraction and leaching of solids [15], reversed miceller extraction, liquid membrane extraction [16, 17], solvent extraction [3, 18], reactive extraction, supercritical fluid extraction [19, 20]
Chromatographic methods [3–5, 21, 22]	Affinity, gel permeation, hydrophobic interaction, ion exchange, membrane chromatography, radial flow, fluid bed, fast flow, supercritical fluid, preparative HPLC, simulated moving bed chromatography
Thermal operations	distillation, reactive distillation, drying (drum, cabinet, fluid bed and spray drying) [3], evaporation, freeze drying or lyophilization [23]
Miscellaneous	adsorption, cell disruption [24–26], crystallization, electrophoresis and other electrokinetic methods, precipitation (salt, alcohol and organic solvent, polymer fractionations, isoelectric precipitation) [3], comminution or size reduction, separations facilitated by magnetic and ultrasound fields [27, 28]

form. For the case shown, the yeast biomass is first concentrated using a dynamic microfilter. The same filter is then used in a diafiltration mode to wash the cells of the residual fermentation medium. The washed and concentrated slurry of cells is then circulated through a high-speed bead mill to disrupt [24, 26] the cells and release the intracellular α_1-antitrypsin along with the other cellular constituents. The cell homogenate undergoes a flocculation step [35], as flocculated cell debris is easier to remove by centrifugation than are the fine particles produced during cell disruption [34]. The flocculated homogenate of the disrupted cells is centrifuged to remove the suspended matter. The resulting clarified solution of proteins contains the product that is precipitated by salt (ammonium sulfate) fractionation [3, 36]. The crude precipitate is recovered in a centrifuge, redissolved in buffer and subjected to at least two chromatographic polishing steps to recover the product at more than 95% purity. Like the example shown in Fig. 2.1, purification and recovery processing of any pharmaceutical requires a train of several process steps or unit operations.

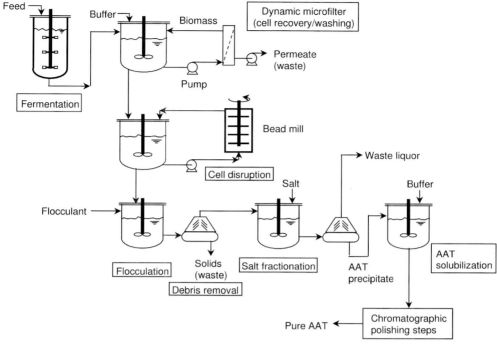

Fig. 2.1 Process flowsheet for producing α_1-antitrypsin (AAT) [34].

Anyone faced with designing a bioseparation scheme can take comfort in the variety of available separation processes (Tab. 2.1); however, the same variety can be a source of much distress. The number of possible recovery flowsheets N_f that can be theoretically devised to completely separate a mixture of C components by using S number of separation operations is given by:

$$N_f = \frac{[2(C-1)]!}{C!(C-1)!} + S^{(C-1)}. \tag{1}$$

Thus, complete separation of a five-component mixture using two separation methods would generate 30 possible flowsheets! Some of these flowsheets can be easily eliminated as being unrealistic, but a large number of potentially workable flowsheets still remains.

As an example, if three components (A, B and C) need separating (i.e. $C = 3$) and one operation (e.g. distillation) is available to separate them (i.e. $S = 1$), Eq. (1) shows that three different possible flowsheets can be devised for this separation. These are shown in Fig. 2.2. Flowsheet 1 (Fig. 2.2) requires a single separation stage that recovers the desired product A. Flowsheets 2 and 3 also recover the product, but require two stages of processing. Although usually only one product (i.e. the pharmaceutical of interest) is wanted in a recovery process, recovering this product typically necessitates separation of several components. For example, in

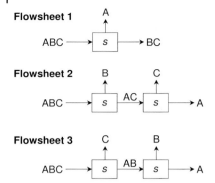

Fig. 2.2 Three possible flowsheets for recovering A from a mixture of A, B and C.

the recovery flowsheet in Fig. 2.1, the product is α_1-antitrypsin; however, to recover this product, the microbial cells must first be separated from the suspending broth (i.e. two components). After the cells are disrupted to release the intracellular α_1-antitrypsin and other proteins, the protein mixture needs to be separated from the cell debris (i.e. another two components). Often, a process must include additional nonseparating steps such as heating and mixing.

Not all possible flowsheets can be exhaustively evaluated; instead, experience and thorough knowledge of individual bioseparations and relevant fermentation must be relied on to narrow the choices to a few practicable options for detailed evaluations and experimental testing.

2.2.2
Designing a Recovery Process

Factors that must be considered in designing a downstream processing scheme include the nature, concentration and stability of the product, the desired purity, and the end use. Due to contamination and supply considerations, there is a distinct trend to move away from direct extraction of human and animal sources to recombinant cells. Thus, for example, microbially produced recombinant human insulin and growth hormone are now available. For vaccines, too, attempts are underway to engineer safer organisms to produce the antigenic material that would otherwise be obtained form pathogens.

How a product is used and its final form both influence its specifications and, therefore, have important implications for the design of the recovery process. The end use of the product may be in research, as an *in vitro* diagnostic, food and animal feeds, soil inoculants and pesticides, medicinals, medical devices, cosmetics, etc. The specific form of the product may include live human cells for medical purposes, live microorganisms, viruses (e.g. for vaccines), spores (e.g. for biotransformations, insecticides and solid-state culture) and higher organisms (e.g. nematodes), bioactive polymers, proteins and enzymes, inactive polymers [e.g. food protein, xanthan, poly-β-hydroxybutyrate (PHB)], smaller organics (e.g. streptomycin, amino acids, citric acid, ethanol, Taxol®), polypeptides (e.g. cyclosporin), and cellular organelles (e.g. nuclei, mitochondria, chloroplasts).

Location of the product, whether extracellular, intracellular or periplasmic, affects how it is recovered. Physical and chemical properties of the product and contaminants need addressing, and biosafety issues must be given attention. A further consideration is price relative to existing sources and other competing products. When no competing products or alternative sources can be identified, the estimated production costs would need to be compared with what the market can be reasonably expected to pay.

As far as possible, the requisite purification and concentration should be achieved with the fewest processing steps; generally, no more than six or seven steps are used – a situation quite different from that in chemistry and biochemistry laboratories, where the number of individual steps is often not a major consideration and the purity of the product is usually more important than overall yield or costs [37]. The overall yield of an n-step process with a step yield of x percent is $(x/100)^n$. Therefore, n must be minimized for a high overall yield. For example, a train of only five steps, each with 90% step yield, would reduce the overall recovery to less than 60% [38]. To minimize reduction of the overall yield, high-resolution separations such as chromatography should be used as early as possible in the purification scheme in keeping with the processing constraints that these steps require (e.g. clean process streams free of debris, particulates, lipids, etc.).

Separation schemes incorporating unit operations that utilize different physical–chemical interactions as the basis of separation are likely to achieve the greatest performance for a given number of steps. Combining two separation stages based on the same separation principle may not be an effective approach. As an example, when two chromatographic steps in series are selected, gel-filtration chromatography, which separates based on molecular size, and ion-exchange chromatography, which separates based on difference in charge on the molecules, may be a suitable combination.

Speed of processing is another factor that significantly affects the design of a recovery scheme. The size of the bioreaction step and the frequency of harvest usually determine the turnaround time for the downstream process train. Sometimes during processing, exposure of material to relatively severe environmental conditions is unavoidable. Very many factors affect stability, including temperature, pH, proteases and other degrading enzymes, mechanical forces, microbial contamination, and oxidants and other denaturing chemicals. In severe environments, the duration of exposure must be minimized and special precautions (e.g. low temperature, addition of chemicals to reduce oxidation, etc.) are necessary to reduce the impact of exposure. The need for speedy processing constrains equipment choice and capacity. For example, the low pH necessary during extraction of penicillins affects stability, hence rapid extraction is essential and thus mixer–settler-type extraction is contraindicated.

Typically, a separation process must operate within the physiological ranges of pH and temperature (pH ~7.0; temperature ≤37°C), but differences from these norms are not unusual. For example, enzymes such as lysozyme, ribonuclease and acid proteases are quite stable at low pH values [37]. Some biologically active molecules, particularly proteins, may be sensitive to excessive agitation; however, enzymes, with the exception of multienzyme complexes and membrane-associated enzymes, are not damaged by shear in the absence of gas–liquid interfaces [3, 39].

Except for the final few finishing operations, downstream processing is usually conducted under nonsterile, but bioburden-controlled conditions; however, prevention of unwanted contamination and cleaning/sanitization considerations require that the processing machinery be designed to the same high standards as have been described for sterile bioreactors [40, 41]. Containment and hygienic processing requirements may severely affect equipment choice.

A commercial recovery scheme must be reliable and consistent. Process robustness is essential to economic production, process validation and product quality. Automation assures consistency and rapid turnaround of the process equipment. Operations such as in-place cleaning are often automated [42].

Additional considerations include biosafety and containment. Bioproducts may be potentially allergenic and they may produce activity-associated reactions in process personnel [15]. In addition, process material may be pathogenic, cytotoxic, oncogenic or otherwise hazardous. Processing of such material requires attention to containment and biosafety both during design and operation of the bioseparation scheme [15, 43]. Certain processing operations are difficult to contain and may pose peculiar operational problems. For example, gasket failures during high-pressure homogenization could create high-pressure sprays [24] and, unless designed with containment features, operations such as centrifugation may generate aerosols.

Small quantities of multiple products are sometimes produced in the same plant – a series of runs or campaigns of one product is followed by another. The risk of cross-contamination is high and adequate safeguards are essential. Experience suggests that cross-contamination with penicillins and penicillin-containing substances cannot be reasonably prevented in a multiproduct facility. As penicillins may produce adverse reactions in some patients, Good Manufacturing Practice (GMP) regulations demand dedicated penicillin-processing facilities that are segregated from nonpenicillin products. Separate air-handling systems are necessary if a building processes penicillins as well as nonpenicillin products.

GMP regulations, including the validation requirements [44], affect all aspects of downstream processing. Requirements depend on the kind of product (e.g. food, bulk pharmaceutical, final dosage form, etc.) and the jurisdiction. Consult Willig [45] for specific guidance.

The final few downstream processing steps include formulation, which is highly product specific. How a product is formulated may critically affect its stability, efficacy and bioavailability. Formulation may involve addition of fillers (e.g. starch, cellulose, sugar, flour), diluants, preservatives, sunlight protectants (e.g. carbon black, dyes, titanium oxide), dispersal aids, emulsifiers, buffers, moisture retainers, adjuvants (e.g. mineral oils and aluminum hydroxide added to improve antigenicity of certain vaccines), flavors, colors and fragrances. Additional finishing operations may include sterile filtration, vialing, granulation, agglomeration, size reduction, coating, encapsulation, tableting, labeling and packaging.

2.3
Product Quality and Purity Specifications

The specifications on product purity and concentration should be carefully considered in developing a purification protocol. Concentration or purification to levels beyond those dictated by needs is wasteful. The acceptable level of contamination in a particular bioproduct depends on the dosage, frequency of use and method of application (e.g. food, drug, oral, parenteral), as well as on the nature and toxicity (or perceived risk) associated with the contaminants [37]. Products such as vaccines, which are used only a few times in a lifetime, may be acceptable with relatively high levels of other than the desired biomolecule. In some cases, contaminating protein levels of about 100 ppm may be acceptable. *In vitro* diagnostic proteins (enzymes, monoclonal antibodies) may tolerate greater levels of contaminants so long as the contaminants do not interfere with the analytical performance of the product. With certain diagnostic proteins, such as blood-typing monoclonal antibodies, cross-contamination causing misdiagnosis is an extreme concern because of possibly fatal consequences of mistyping. Such concerns influence the design and operation of the downstream process, particularly for multi-product plants.

Parenteral therapeutics usually must be purer than 99.99%. A variety of approaches are used to assure quality. Methods typically used with protein therapeutics are summarized in Tab. 2.2. Further details are provided by Anicetti and

Tab. 2.2 Methods for quality assurance of protein therapeutics [46]

Impurity or contaminant	Analytical technique
Protein contaminants (e.g. host cell proteins)	SDS–polyacrylamide gel electrophoresis (PAGE), HPLC, immunoassays (enzyme-linked immunosorbent assay, etc.)
Endotoxin	rabbit pyrogen test, LAL
DNA	DNA dot-blot hybridization
Proteolytic degradation products	isoelectric focusing, SDS–PAGE, HPLC, N- and C-terminal analysis
Presence of mutants and other residues	tryptic mapping, amino acid analysis
Deamidated forms	isoelectric focusing
Microbial contamination	sterility testing
Virus	viral susceptibility tests
Mycoplasma	21 Code of Federal Regulations method
General safety	As per 21 CFR 610.11

coworkers [47] and others [48–50]. Requirements relating to some specific contaminants are discussed below.

2.3.1
Endotoxins

Products derived from bacteria such as *Escherichia coli* will invariably be contaminated with bacterial cell wall endotoxins [51] that can cause adverse reactions (headaches, vomiting, diarrhea, fevers, etc.) in patients unless reduced to very low levels (e.g. less than $5 \times 10^{-13}\,kg\,kg^{-1}$ body weight). Endotoxins are extremely heat-stable lipopolysaccharides that are not easily removed from solutions of macromolecules. Ultrafiltration and reverse osmosis are effective for depyrogenation of water and small solutes. Other pyrogen removal methods are adsorption on activated carbon and barium sulfate, hydrophobic-interaction chromatography, and affinity chromatography. Endotoxins bind to polymixin B-affinity columns, but this method must be combined with detergent treatment for effectively removing protein-bound endotoxins [46]. Chromatography using a *Limulus* amoebocyte lysate (LAL)-affinity matrix also removes endotoxins.

As a guiding principle, processing must aim to minimize endotoxin contamination by controls on process water and other additives. In addition, aseptic and bioburden controlled operation, and frequent cleaning of equipment help to reduce contamination. The equipment cleaning protocol must include procedures proven for depyrogenation. Standard alkali-based cleaning procedures [42] are quite effective in depyrogenation of stainless steel equipment, but other methods are necessary for cleaning chromatographic columns and membrane filters. The depyrogenation step employed during cleaning of membrane filters usually involves a 30 min, 30–50°C treatment with sodium hydroxide (0.1 M), hydrochloric acid (0.1 M), phosphoric acid or hypochlorite (300 ppm free chlorine). Thorough rinsing with pyrogen-free water follows. Similar procedures are used for chromatographic columns.

An endotoxin-free product should be validated using the LAL test. This test is based on endotoxin induced coagulation of amoebocyte lysate of horseshoe crab (*Limulus polyphemus*) at 37°C, pH 7.0. Less than $0.3\,ng\,mL^{-1}$ endotoxin levels are easily detected. Scrupulously clean glassware and water are necessary to prevent false positives. Some known interfering agents are ethylenediaminetetraacetic acid (EDTA), sodium dodecylsulfate (SDS), urea, heparin and benzyl penicillin.

2.3.2
Residual DNA

Residual DNA from producing cells can potentially contaminate the product. DNA fragments from established animal cells were once believed to be potentially oncogenic, which prompted the US Food and Drug Administration to recommended a contamination level of no more than 10 pg DNA per dose [52]. Less-restrictive limits are now accepted because no oncogenic events were observed following injections of large doses of DNA into animals. Nonetheless, DNA is a contaminant and demonstration of its satisfactory clearance is essential to quality assurance of the product [52]. Residual DNA is removed usually by adsorption on strong anion-exchange resins at pH ≥4. Hydrophobic-interaction chromatography is also effective and so is affinity chromatography under conditions that bind the desired protein but not the DNA.

Sometimes DNA needs to be recovered as the target product, e.g. for use in gene therapy and DNA vaccines. Downstream processing considerations for recovering plasmid DNA have been discussed elsewhere [53, 54].

2.3.3
Microorganisms and Viruses

Parenteral products, other than certain vaccines, must be free of microorganisms and viruses [37]. Products derived from potentially contaminated sources, such as human donors, animals and some cell lines, can be especially problematic. For such products, the purification scheme must demonstrate viral inactivation or removal unless the product is terminally sterilized by validated means. Usually, in-series processing with at least two steps, each capable of 6 log virus removal or deactivation, would be necessary. Viruses can be removed by ultrafiltration [55] or deactivated by methods such as heating, treatment with chemicals (e.g. β-propiolactone), solvents and detergents, and ultraviolet or γ-irradiation [56]. In one study with plasma-derived human serum albumin, heat treatment at 60°C for 10h in the final container produced more than 5 log reduction of vaccinia, polio 1, vesicular stomatitis, Sindbis and human immunodeficiency virus (HIV)-1 within 10 min [57]. In another case, freeze-dried coagulation factors were treated at 80°C for 72h in the final vial. For Factor VIII, inactivation of HIV-1 occurred within 24h, without significant deterioration of the product [57]. For a Factor IX preparation, treatment with solvent/detergent combination of tri-(n-butyl) phosphate and Tween-80 for 5h inactivated a range of typical enveloped viruses within 1h [57]. Up to 6 log reduction of some typical enveloped viruses such as herpes simplex-1 and Sindbis could be achieved in spiked samples using Protein G column chromatography with acid elution; however, only 3 log reduction was observed for acid tolerant nonenveloped polio virus [57]. Validation of virus clearance has been discussed by Darling [58].

2.3.4
Other Contaminants

For many biological products, particularly pharmaceuticals, seemingly minor alterations in downstream processing can have important implications on the performance of the product. For example, penicillins may be recovered by liquid–liquid extraction of either the whole fermentation broth or solids-free broth. The latter scheme requires an additional solid–liquid separation step than the whole-broth process. However, the whole-broth extracted product has been known to cause more frequent cases of allergenic reactions in comparison with the other processing alternative. In fact, some pharmaceutical companies now demand of contract suppliers that, in addition to meeting product specifications in terms of measurable contamination, the product they supply must conform to a certain production method, in this case extraction after removal of fungal solids. When raw penicillin is for bulk conversion to semisynthetic penicillins, whole-broth extraction may be acceptable in view of the security afforded by the additional steps involved in making and purifying 6-aminopenicillanic acid from raw penicillin [37].

2.4
Impact of Fermentation on Recovery

Downstream processing should not be considered in isolation from the bioreaction step. Development of a biocatalyst by natural selection, mutation, directed evolution [59, 60] and recombinant DNA technology is a powerful means of influencing downstream processing [61]. Similarly, modification of fermentation feeding strategies, culture media and conditions profoundly affect the downstream process [61].

2.4.1
Characteristics of Broth and Microorganism

Composition of the fermentation medium affects downstream recovery. Relatively poorly defined complex media components are often acceptable for producing commodity chemicals and bulk antibiotics, but usually not for parenteral proteins. Low-serum and protein-free media are commonly employed in animal cell culture to greatly simplify recovery of sparing amounts of proteins produced. Similarly, the type of antifoam and its concentration must accommodate the recovery constraints.

For some processes, the use of alternative microorganisms may be a viable option. Preference should be given to faster-growing, easy to process organisms. Selection of a producer must consider the overall productivity of the process, not just that of the fermentation step. Production of recombinant proteins in *S. cerevisiae* may have important advantages relative to production in genetically modified bacteria such as *E. coli* [62]. *S. cerevisiae* is generally recognized as safe for food and pharmaceutical use. In addition, unlike bacteria, the yeast does not produce endotoxins, and its broths are much easier to process than those of mycelial fungi and filamentous bacteria [62]. Unlike the DNA-laden homogenates of bacteria such as *E. coli*, yeast lysates are not excessively viscous. In yet other cases, it may be possible to naturally select autoflocculating strains, as has been done with certain brewing yeasts and bacteria. Cells may also be genetically modified to flocculating ones.

Genetic engineering of producing organisms and products provide new opportunities for influencing downstream bioseparations. For example, recombinant fusion proteins with added polypeptide "affinity tags" have been produced to facilitate purification [63–65]. Affinity tags have been developed for ion-exchange, hydrophobic-interaction, affinity, immunoaffinity and immobilized metal ion chromatography. Specific cleavage sites between the tag and the protein allow removal of the tag after purification [63]. Some of the available affinity tags and the chromatographic methods applied with those tags are listed in Tab. 2.3. Reagents and enzymes that have been used to cleave the tags, and the specific cleavage sites, are noted in Tab. 2.4.

Small differences in the hydrophobicity of proteins are effective in facilitating separation by hydrophobic-interaction chromatography. The outer surface of most proteins is rich in hydrophilic residues. In contrast, hydrophobic residues are typically deeply buried in the protein structure. Consequently, selective separation of a target protein can be facilitated by engineering it to posses surface-exposed hydrophobic regions [65]. Protein hydrophobicity is easily modified by minor genetic modifications [65].

Another strategy for simplifying downstream recovery is genetic manipulation to enable extracellular secretion of the recombinant protein. Failing outright secretion, it may be possible to achieve secretion into the periplasm of microorganisms such as *E. coli*. Relatively mild disruption or extraction conditions can then be used

Tab. 2.3 Affinity tags and corresponding chromatographic separations [63]

Affinity tag	Chromatography scheme
Polyarginine	ion exchange
Polyphenylalanine	hydrophobic interaction
β-Galactosidase	affinity
Protein A	affinity
Antigenic peptides	immunoaffinity
Polyhistidine	metal ion chelate

Tab. 2.4 Chemicals and enzymes for specific cleavage of fusion proteins [63]

Cleavage reagent	Cleavage site
Cyanogen bromide	Met ↓
Formic acid	Asp ↓ Pro
Hydroxylamine	Asn ↓ Gly
Collagenase	Pro–Val ↓ Gly–Pro
Factor Xa	Ile–Glu–Gly–Arg ↓
Enterokinase	Asp–Asp–Asp–Lys ↓
Rennin	His–Pro–Phe–His–Leu–Leu ↓
Carboxypeptidase A	C-terminal aromatic amino acids
Carboxypeptidase B	C-terminal basic amino acids

for recovery in comparison with products produced in the cytoplasm. Periplasmic secretion has additional advantages: periplasm of *E. coli* contains only seven of the 25 cellular proteases [66]; hence, the likelihood of proteolysis is reduced. Moreover, periplasm contains only 100–200 proteins [66]; therefore, selective extraction of periplasm yields a less complex, easier to purify mixture. In addition, the oxidative environment of periplasm is more favorable to formation of disulfide bonds than is the environment of cytoplasm. Disulfide linkages determine the correct folding of the polypeptide chain and, therefore, its biological activity. Chemicals such as chloroform, Triton X-100 and combinations of lysozyme/EDTA [66] facilitate release of periplasmic proteins. Extraction chemicals should be tested for possible effects on protein stability. In one study, Garrido and coworkers [62] observed loss of β-galactosidase activity even at 4°C when the enzyme was extracted with a mixture of chloroform and sparing amounts of SDS. In larger quantities, SDS is a well-known protein denaturant [67].

Secretion or extracellular leakage of an otherwise intracellular product is sometimes achieved simply by modifying the fermentation conditions. For example, addition of penicillin during growth in certain amino acid fermentations produces cells that leak the amino acid that is recovered by isoelectric precipitation from the extracellular fluid.

2.4.2
Product Concentration

The source material from which the product is recovered needs to be selected carefully. In some cases, the choice of raw material is limited to a single source; however, often the required compound may be available in several biological materials. In deciding on a suitable raw material, the issues needing consideration are the quantities available, consistency of supply, concentration of the active ingredient in the raw material, any potential toxic compounds and other contaminants present, and cost of the raw material. In the absence of other limitations, a raw material that contains the highest concentration of the desired pharmaceutical will generally be the preferred source. The difficulty of recovering a product usually increases as the concentration declines in the source material. This factor can greatly influence the cost of recovery.

For products that are produced by fermentation processes, concentration in the broth is usually quite low. Some values typically seen in culture broths are noted in Tab. 2.5. In addition to the product, the broth contains many contaminants, some of which may be quite similar to the desired product. Contaminants may include toxic or otherwise hazardous substances such as endotoxins and mycotoxins.

For fermentation products, downstream processing typically represents 60–80% of the cost of production of the product. This may misleadingly suggest that the process improvement effort should focus primarily on the recovery operations. However, even small improvements in the yield or purity of the product in the bioreaction or fermentation step can have a significant effect on downstream

Tab. 2.5 Typical concentration of various products in raw fermentation broth

Product	Final concentration (kg m^{-3})
Vitamin B$_{12}$	0.06
Monoclonal antibodies	0.1–0.5
Riboflavin	0.1–7.0
Antibiotics	0.2–35.0
Gibberelic acid	1–2
Amino acids	2–100
Yeast	30–60

recovery costs. As a rough guide, the selling price P ($\$_{1984}$ kg^{-1}) of a product (i.e. a reflection of cost of production) depends on its concentration C_i in the broth or the starting material. This dependence can be described by the equation:

$$P = 528 \times C_i^{-1}, \tag{2}$$

which is based on data compiled by Dwyer [68]. The potential for yield improvement at the bioreaction stage is usually high. Major yield enhancements have been fairly commonly achieved by strain selection, medium development, optimization of feeding strategies and environmental controls.

2.4.3
Combined Fermentation–Recovery Schemes

In keeping with a global approach to process improvement or intensification, schemes that combine the bioreaction stage and parts of downstream processing are potentially attractive [61]. Such schemes include extractive fermentations, fermentation–distillation, fermentation–pervaporation [11], perfusion culture using membranes, inclined settlers or "spinfilters" to retain the cells in the bioreactor, fermentation–adsorption using chromatographic media, as well as other methods. Combining fermentation and recovery not only reduces the number of individual processing steps, but the productivity of the fermentation may also be substantially enhanced by eliminating or reducing the inhibitory effects of certain products.

A novel scheme for retaining particles, particularly animal cells, in perfusion bioreactors relies on standing sound waves applied perpendicular to a vertically aligned harvest flow channel [69–71]. The sound waves concentrate the suspended cells in bands aligned with the flow [72, 73]. Gravity sediments such aggregated particles against the flow once the sound is switched off; hence, a clarified liquor leaves the flow channel, whereas the solids are concentrated in the feed vessel. This type of separation in ultrasonic flow fields provides an effective means of retaining cells in continuous-flow bioreactors. This technique allows easy maintenance of sterility as no mechanical items penetrate the sedimentation chamber. Moreover, there is nothing to clog, foul or breakdown. Process-scale implementation of this method is being developed. In addition to its use in separating particles, ultrasound can be used to substantially improve the productivity of certain fermentation processes [28].

2.5
Initial Separations and Concentration

The first few processing operations in a purification train are aimed at volume reduction to minimize processing costs by reducing the size of the required downstream machinery. Removal of suspended material and substances which might interfere with further downstream operations are additional requirements of some of the early separation steps. Further, because viscous broths are difficult to handle, viscosity reduction should be achieved as early as possible to simplify pumping, mixing, filtration, sedimentation, etc. Removal of suspended solids, digestion of carbohydrates or removal of nucleic acids are some of the operations that may be needed to improve broth handling.

Typically, solid–liquid separation would be among the first processing steps for extracellular as well as intracellular products. For the latter, solid–liquid separations are usually a means of concentrating the biomass, or removing the excess culture fluid prior to cell disruption or other downstream treatment. Cell or other solid product washing operations often employ solid–liquid separation steps. The commonly used methods of solid–liquid separation are filtration and centrifugation. Centrifuges are used also to separate difficult to break emulsions and other liquid–liquid systems. Some examples are recovery of cream from milk, recovery of oil drops, recovery of fats (e.g. in rendering and meat processing plants) and waxes, and liquid–liquid extraction.

Solid–liquid separations can be implemented in a variety of ways that are best suited to particular applications. Thus, as detailed in Tab. 2.6, many different designs of centrifuges are available [3, 74]. Similarly, filtration may be performed in conventional filter presses, horizontal and vertical leaf-type pressure filters, rotary drum pressure or vacuum filters with or without filter aid (or body feed or admix), and using different means of solids discharge. Production-scale rotary drum filters tend to be quite large: 0.9–4.3 m drum diameter and up to 6 m drum width. Sterile operation is usually not feasible and containment is difficult. Alternatively, solids may be recovered by membrane filtration either in dead-end (e.g. in many filter sterilizations) or cross-flow modes; the latter may be implemented in flat-plate, hollow-fiber or spiral-wound static membrane cartridges, as well as in dynamic modes [3]. While the variety of available options helps to ensure that specific need are met, careful consideration of the problem at hand is required for selection of the optimal processing method. Alternatives should be considered whenever possible. For example, rotary drum filters with string discharge usually perform well in separating mycelial solids from penicillin broths, but this discharge mechanism, without filter aids, causes problems with broths of Streptomycetes and other bacteria [37]. Precoat drum filtration may be used with bacterial

Tab. 2.6 Types and applications of centrifuges [3, 37]

Tubular bowl	Tubular bowl machines are capable of high g forces (16 000–20 000 g in industrial devices). Solids accumulate in the bowl and must be removed manually at the end of operation. Bowl capacity limits solids-holding capability. To ensure sufficient interval between bowl cleaning, the solids concentration in the feed should usually be ≤1% volume/volume; higher concentrations can be processed with smaller batches, e.g. in the production of certain vaccines. Good dewatering of solids is obtained.
Multichamber bowl	Similar to tubular bowl machines. Division of bowl into multiple chambers increases solids-holding capacity. Solids must be discharged manually; hence, economic operation is feasible only with feeds with low concentration of solids. Good for polishing of otherwise clarified liquors. Capable of high g forces. Gradation of g forces from inner to outer chamber. Smallest particles sediment in the outermost chamber. Good dewatering of solids.
Disc stack	Lower g forces (around 8000 g) than tubular bowl machines. Solids may be retained, or discharged intermittently or continuously by various mechanisms (e.g. periodic ejection of solids by hydraulic separation of upper and lower parts of the bowl, nozzle discharge under pressure, valves, etc.). Not all discharge methods are suitable for all solids. Solids must flow. Poor dewatering. Not suited for mycelial solids; good for slurries of yeasts and certain bacteria. Depending on the mechanism of solids discharge, may handle feeds with up to 30% (v/v) solids. Provides a large sedimentation surface and short settling depth.
Scroll discharge	Scroll discharge or decanter bowl centrifuges are suitable for slurries with high concentration of relatively large, dense solids. Feed solids concentrations of 5–80% (v/v) can be handled. Solids are discharged continuously. The g forces are low at around 3000 g. Suitable for fungal broths and dewatering of sewage sludge.
Perforated bowl or basket centrifuges	Also known as filtering centrifuges. Useful for low-g recovery of relatively large, mostly crystalline solids. The perforated bowl is lined with filter cloth to retain solids, whereas the liquid passes through. Sedimented cake may be washed and recovered as fairly dry material. Not effective for particles below 5 μm, and loadings of less than 5% (v/v) [74].

broths when biomass is not the desired product. A knife blade (or doctor blade) discharge mechanism is used to continuously remove the deposited solids along with a thin layer of the precoat. Knife discharge without precoat or filter aids is suitable for recovering yeast from the filter cloth on drum filters; however, knife blades are not suited to cleanly cutting away a layer of deposited mycelial fungi because of the stringy nature of solids. Similarly, because of the concentration and the morphology of the solids, the disk-stack centrifuge is not suitable for fungal fermentation broths, but properly selected scroll discharge decanter machines are effective. Leaf filters are generally batch devices that are inexpensive to install, but labor intensive to operate. Leaf filters are suitable for broths with few suspended solids, e.g. in polishing of beer [37]. Gravity sedimentation may be employed as a volume reduction step prior to removal of solids by other means, but sedimentation by itself is not common for biomass removal in processing of high-value products. Gravity sedimentation in thickeners and clarifiers [7, 8] is encountered widely in sludge recovery in biological wastewater treatment. Certain solids may be recovered using hydrocyclones, but this method is rarely used in bioprocessing. Animal cells can be recovered by microfiltration. Disk-stack centrifuges have been successfully used to recover intact animal cells such as hybridomas from broths [75].

When more than one processing option is technically feasible, evaluation of the economics of use in terms of capital expenditure on equipment and its operating costs (processing time, yields, labor, cleaning, maintenance, analytical support) is necessary for optimal process selection. Economic evaluations should be performed over the expected lifetime of the equipment [37]. For example, for separation of solids from fermentation broth, centrifugation and microfiltration may be two competing alternatives [3]. In still other applications, e.g. when very fragile cells are to be separated from suspending liquid, centrifugation may not be an option at all.

Some other concentration steps, applicable to products in solution, are precipitation [3, 76], adsorption, chromatography [21], evaporation, pervaporation [10–12] and ultrafiltration [3]. Some of these operations are equally capable purification steps (e.g. chromatographic separations). Certain steps (e.g. some chromatographic separations and membrane separations) may require a relatively clean process stream, free of debris, lipids and micelles that may cause fouling of the equipment. Such steps are often used downstream of steps that can handle cruder material [37].

Sometimes the characteristics of the fermentation broth or process liquor may be modified by pretreatment to enable processing by a certain method [30]. Major changes in processing characteristics may be achieved by pH and/or temperature shift, use of additives such as polyelectrolytes, other flocculants and enzymes, and changes in ionic strength [37]. Flocculants (e.g. alum, calcium and iron salts, tannic acid, quaternary ammonium salts, polyacrylamide) can enhance sedimentation rates by thousands of fold relative to unflocculated suspension [35]. Aging of protein precipitates and crystals can substantially improve filtration and sedimentation. Addition of salts is sometimes helpful in dewatering difficult to dewater solids such as protein precipitates. Water is drawn out of the pores of the solid into the salt-containing liquid film on the outside. Osmosis or chemical potential difference drives the flow. Among other factors, time of harvest can beneficially alter processing behavior of the broth as well as the stability of the labile product. Culture conditions and methodology influence microbial morphology, product formation and downstream recovery. For example, cells grown in defined media are generally easier to disrupt than those cultured in complex media [24]. Also, high specific growth rates produce less robust cells.

2.6
Intracellular Products

In general, a biological product is either secreted into the extracellular environment or it is retained intracellularly. In comparison with the total amount of biochemicals produced by the cell, very little material is usually secreted to the outside; however, this selective secretion is itself a purification step that simplifies the task of the biochemical engineer. Extracellular products, being in a less-complex mixture, are relatively easy to recover. On the other hand, because a greater quantity and variety of biochemicals are retained within cells, intracellular substances are bound to eventually become a major source of bioproducts [24]. Among some of the newer intracellular products are recombinant proteins produced as dense inclusion bodies in bacteria and yeasts. Recovery of intracellular products is more expensive as it requires such additional processing as cell disruption [24–26], lysis [24], permeabilization [77] or extraction. Intracellular polymers such as PHB may be recovered either by cell disruption [25, 67] or solvent extraction. In principle, selective release of the desired intracellular products is possible, but in practice it is neither easily achieved nor sufficiently selective. Hence, the desired product must be purified from a relatively complex mixture, complicating processing and adding to the cost [37]. Nevertheless, an increasing number of intracellular products are in production. Economics of production may be improved by recovering several products (intracellular and extracellular) from the same fermentation batch [39].

As for other separations, many options exist for disruption of cells (Tab. 2.7). Of these, high-pressure homogenization is apparently the most suitable for bacterial broths, whereas bead mills are more widely used for fungal cultures [3, 24]. For dissolved products, cell disruption conditions (e.g. pressure, number of passes) must be selected to prevent excessive micronization of debris because micronization complicates solid–liquid separation further downstream [24]. However, when the product is an intracellular solid that is undamaged by homogenization, micronization of debris actually favors product recovery. This strategy is useful with protein inclusion bodies, certain cellular organelles and sometimes with granules of bioplastics such as polyhydroxyalkanoates. Nonetheless, overzealous disruption conditions should be avoided in view of the published evidence that suggests loss of intracellular solids by micronization [67].

Disruption of bacterial cells releases large amounts of nucleic acids which increase the viscosity of the broth, often producing viscoelastic behavior. To ease further purification, the nucleic acids are usually removed by precipitation (e.g. with manganous sulfate, streptomycin or polyethyleneimine) [3]; alternatively, viscosity may be reduced by enzymatic digestion of nucleic acids or high-shear processing in high-pressure homogenizers [37]. Another alternative for eliminating nucleic acid polymers is heat shock treatment prior to disrupting the cells. Heat shock treatment would typically require rapid heating to at least 64°C and a holding time of 20–30 min. This treatment should digest almost all DNA/RNA. Lower holding times may be satisfactory if complete degradation is not necessary

Tab. 2.7 Cell disruption options [24–26]

High-pressure homogenization	Frequently used for large-scale disruption of yeasts and nonfilamentous bacteria. Generally not suitable for mycelial broths. Broth must be free of large suspended solids, tight cell clumps and flocs. Maximum acceptable particle size is about 20 μm, but a lower size is preferred. Slurry viscosity should not normally exceed 1 Pa s [3, 24]. Optimal viscosity and solids concentration ranges are narrower than for bead mills.
Bead milling	Bead mills come in vertical and horizontal configurations with different mechanisms for retention of grinding media and different types of agitators. Agitators that reduce backmixing are preferred. Vertical mills are susceptible to fluidization of beads and accompanying loss in performance. Typically three to six passes should achieve complete disruption. Useful for yeasts, mycelial fungi, algae; less efficient with bacteria. Grinding bead size affects disruption. Smaller the microbial cell, smaller the optimal bead size [24, 62].
Autolysis	Under suitable conditions certain cultures would autolyse in the stationary phase upon completion of fermentation. Baker's yeast can autolyse.
Osmotic shock	Useful for animal cells and in specific cases for bacteria. Large dilutions may be necessary.
Thermolysis	Sufficiently heat-stable products may be released by heat shocking the cells. Microbial susceptibility to heat shock treatment varies widely. Monvalent metal ions such as Na^+ and K^+ may aid thermolysis. Suited to specific cases.
Enzymes and chemicals	Detergents, EDTA, solvents (e.g. toluene), antibiotics and lytic enzymes may be used. Sometimes enzymes and chemical additives are used in combination with homogenization or bead milling to reduce the severity of mechanical treatment required. Treatment with acids and alkalis may be useful in specific cases. Especially useful for extraction from periplasm.
Others	Ultrasonication, desiccation, freeze–thaw, extrusion of frozen paste. Applicable only to laboratory scale.

for processability. Rapid temperature rise preferentially destroys proteases relative to RNA-hydrolyzing enzymes. Thermal treatment may be feasible for heat-stable products [67] as well as for those produced as denatured inclusion bodies.

Processing considerations relevant to some specific bioseparations are discussed in the following section.

2.7
Some Specific Bioseparations

2.7.1
Precipitation

Proteins are easily concentrated by precipitation with organic solvents (e.g. ethanol, acetone), polymers [e.g. poly(ethylene glycol) (PEG), poly(propylene glycol) (PPG), dextran] and salts. Fractional precipitation allows for a degree of separation [3]. Fractionation with ammonium sulfate is commonly used. Organic solvents produce a denaturing environment, making low-temperature processing neces-sary [3]. Alcohol precipitation is frequently used in recovering biologically inactive dissolved polymers such as polysaccharides. Examples include precipitation of xanthan and gellan with isopropanol. Precipitation methods can handle large amounts of crude material, are easily scaled up and can be implemented in con-tinuous-processing modes [3, 36]. However, precipitation is generally not useful for recovery from very dilute animal cell culture fluids. Ammonium sulfate pre-cipitation for recovery of recombinant β-galactosidase from *S. cerevisiae* has been detailed by Zhang and coworkers [36].

2.7.2
Foam Fractionation

Foam fractionation, microflotation or froth flotation is potentially useful for con-centrating particles (cells, organelles, other small solids such as granules of PHB) and proteins [78] into a foam phase for further recovery. The technique involves gentle bubbling of air (or other inert gas) at the base of a column of broth or solu-tion. Hydrophobic solids and surface active molecules accumulate at the gas–liquid interface and rise with the bubbles. Collector surfactants and other promoters are often added to improve attachment. Additives such as frothing agents and stabiliz-ers may be necessary. Enrichment in the foam depends on physical collection efficiency of bubbles (that is on bubble size, hydrodynamics, bubbling rate, con-centration of particles) and adsorption chemistry. Empirical investigation is essen-tial for selecting suitable additives, concentrations and hydrodynamic regimes, and for assessing performance, including recovery from the foam phase. Culture con-ditions may be used to influence adsorption behavior. Froth flotation is encoun-tered only occasionally in bioprocessing. Potentially, fermenters used in batch cultivation could subsequently be employed for froth floatation. Airlift bioreactors with gas–liquid separators [79] and added means of skimming the gas-floated biomass are used in activated sludge treatment of wastewater. Part of the harvested sludge is returned to the reactor as inoculum.

2.7.3
Solvent Extraction

Rapid solvent extraction can be carried out in centrifugal extractors such as the Podbielniak and the Alfa Laval machines that are commonly used in antibiotics processing [3, 18]. These devices were originally designed to handle solids-free liquids, but have been adapted to media containing limited amounts of small particles. Other more conventional extractors are banks of mixer–settlers, the York–Scheibel column (suitable for solids-free liquids) and the reciprocating plate Karr column (suitable for whole broths). Supercritical extraction of solids and liquids with carbon dioxide or other solvents (e.g. pentane) may be useful for small organic solutes [19, 20]. In these cases, a concentrated solute is obtained easily by boiling off the solvent. Serum albumin has been extracted into aqueous reverse micelles formed in carbon dioxide using a perfluoropolyether surfactant [80]. This opens up new opportunities for purification of proteins and other large molecules.

2.7.4
Aqueous Liquid–Liquid Extraction and its Variants

Conventional liquid–liquid extraction based on partitioning between an aqueous phase and a water-immiscible organic solvent is not suitable for proteins and protein-based cellular organelles because of low protein stability in organic solvents. A suitable alternative is partitioning between two immiscible aqueous phases [3, 13, 14]. Such phases are obtained by adding two incompatible polymers, e.g. PEG and dextran, to water, or by mixing a relatively hydrophobic polymer solution with salts. Examples of such systems are aqueous mixtures of PEG–poly(vinyl alcohol), PPG–dextran, PPG–potassium phosphate, PEG–ammonium sulfate, etc. Partitioning of solutes is brought about by differences in net charge and hydrophobicity. Higher-polarity molecules solubilize preferentially in the salt-rich phase, whereas the relatively hydrophobic molecules concentrate in the polymer-rich phase. Polymers with attached affinity ligands – hydrophobic and ionizable functional groups – can improve partitioning behavior. Partitioning is strongly affected by pH, composition and type of phases (e.g. molecular weight of polymer, ionic strength, salt, polymer). In addition, the volume ratio of the phase mixture to that of the protein solution should be such that neither phase approaches saturation with protein. Aqueous two-phase systems have been successfully employed for enrichment of proteins, cells, organelles and small molecules. Proteins that extract into the polymer phase are back-extracted into the salt phase for recovery. Phase separation can be slow because of high viscosity and small density differences. Gravity separation is generally satisfactory for PEG–salt systems, but centrifugal separation may be necessary for PEG–dextran. Aqueous two-phase extraction is commercially employed, but it is relatively uncommon.

Among relatively new developments in liquid–liquid extraction is reversed micellar extraction [17], also known as liquid membrane emulsion extraction. Reversed micelles are surfactant-stabilized microdroplets of an aqueous phase suspended in a water-immiscible solvent. Contacting the reversed micelle-laden organic phase with an aqueous mixture of proteins or other solutes results in preferential transfer of one or more species from the aqueous phase to the organic phase and from there to the aqueous core of the reversed micelles. The intervening organic phase constitutes a liquid "membrane". Extraction is influenced by pH and ionic strength of the bulk aqueous phase and the nature of the reversed micel-ler core. Usually, a protein solubilizes in the reverse miceller phase at pH values below its isoelectric pH when the ionic strength is low. Once a component has been extracted, reversed micelles can be back-extracted with buffers to yield a solution rich in the desired substance. Back-extraction is favored by altering the pH and ionic strength. Factors such as hydrophobicity of the protein also contribute to partitioning behavior.

A variation of liquid membrane emulsion extraction is supported liquid membrane extraction [16]. No stabilizing surfactant is necessary in this case; instead, the liquid membrane-forming organic phase is supported in the pores of a porous solid that separates the two aqueous phases. Additives may be employed to enhance mass transfer through the organic phase [16]. Reversed micelles and liquid membranes are not widely used at present.

2.7.5
Membrane Separations

Cross-flow membrane filtration flux typically ranges over $10-120\,L\,m^{-2}\,h^{-1}$ (the exact value depends on the membrane pore size and the viscosity of the suspending fluid). Microfiltration of animal cells and microbial homogenates is done best at transmembrane pressures less than $1.38 \times 10^4\,Pa$. Higher pressures, typically $6.9-34.5 \times 10^4\,Pa$, are used in recovering microbial cells. Due to the small pore size, ultrafiltration membranes invariably require high transmembrane pressures ($13.8-27.6 \times 10^4\,Pa$) for reasonable flux.

Polymer membranes predominate in bioprocessing, but ceramic and sintered metal membranes are used occasionally. Hydrophilic membranes are preferred for liquids. Hydrophobic polymer membranes are easily fouled by silicone anti-foams which may cause as much as 50% decline in flux. Low-molecular-weight PPG- or PEG-based antifoams are usually better. Mechanical foam control [42, 81] during fermentation is sometimes helpful in eliminating or reducing antifoam consumption.

Even without antifoams, membrane performance deteriorates over time, making periodic replacement necessary. Prior experience or experimentation are the only reliable predictors of membrane life [9]. Membranes are not easily cleaned and detectable residues of bioactive material may remain after any reasonable cleaning. Such situations require product-dedicated filters to prevent cross-contamination. Furthermore, polymer-based membrane filters cannot usually be heat sterilized – chemical sanitization and atmospheric steaming are the only options. Chemical cleaning, sanitization and steaming lower membrane life; hence, the choice of chemicals and cleaning conditions need to be carefully assessed.

The major costs associated with ultrafiltration and microfiltration are the initial capital expense and the cost of membrane replacement; energy is not a major expense. The frequency of membrane replacement determines the feasibility of membrane separations. In contrast, in reverse osmosis where high transmembrane pressure is unavoidable, pumping expense and membrane replacement costs are major contributors to operating costs. As with centrifuges, membrane filter selection requires experimental evaluations [3, 9].

Even in cross-flow operation, membrane filters experience performance loss due to concentration polarization or accumulation of a solute layer at the surface of the membrane. Small amounts of relatively large, dense inert solids such as cellulose fibers or polymer beads added to the feed are known to reduce concentration polarization by disturbing the fluid boundary layer on the membrane surface. Cross-flow channels are sometimes also inserted with static turbulence enhancers such as wire screens, but such filter modules are not suitable for mycelial or filamentous biomass, especially at high concentration of solids. Mechanical methods of increasing turbulence are employed in dynamic filters, but few such devices have gained any commercial acceptance. One dynamic configuration utilized two porous concentric cylinders with microfiltration membranes supported on the surfaces of the annulus. The inner cylinder rotated at high speed; differences in angular velocities of the fluid elements along the width of the annular gap produced Taylor vortices that substantially enhanced filtrate flux relative to static cross-flow operation [82]. Nonetheless, limited scale-up potential prevented further development. A variation on the concentric cylinder theme has been introduced by Pall Filters. This design consists of a stack of supported circular microfilter membranes with mechanically agitated circular steel disks mounted in between. Rotation of disks dramatically enhances filtrate flux [83]. The stack supports a membrane surface up to $1.5\,m^2$, but this may be substantially increased in future designs simply by increasing the overall height of the stack. The device is suited to recovering yeasts and nonfilamentous bacteria from relatively less viscous broths.

Membrane filters are used also in the diafiltration mode for buffer exchange, washing of solids, desalting and removing other small molecules from solution of macromolecules.

Pervaporation is another membrane separation that is particularly useful for low-energy recovery of relatively volatile liquids (e.g. ethanol) from fermentation broths [10–12]. Permselective membranes separating the broth from a vapor phase allow only selective permeation of the desired solvent to the other side, where hot air or heat supplied to the membrane continuously evaporates the solvent, maintaining a mass transfer driving force. Membrane chemistry determines permselectivity. Membranes with improved permselectivity for organic solvents are becoming available for use in solvent recovery from aqueous solutions by pervaporation.

A relatively recent development is nanofiltration. Nanofiltration membranes with a molecular weight cut-off in the range of 0.2–1.0 kDa allow retention that is in between the capabilities of conventional reverse osmosis and ultrafiltration membranes [84]. Nanofiltration membranes typically carry a positive or negative charge and are, therefore, ion permselective [84]. Metal-nanotube membranes with electrochemically switchable ion permselectivity have been shown to work [84].

2.7.6
Electric and other Field-assisted Bioseparations

In addition to the already discussed use of acoustic fields for enhancing bioseparations (Section 2.4.3), electric and magnetic fields have been used to enhance some conventional separation processes and devise entirely new processes [27]. Various field-assisted fractionations for analytical use in biotechnology have been discussed by Reschiglian and coworkers [85]. High-gradient magnetic separations can be used to recover entities that are intrinsically magnetic (e.g. erythrocytes and magnetotactic bacteria) and that can be made magnetic through attachment of magnetically responsive objects [27]. Magnetically enhanced separations are attracting renewed interest for several reasons – availability of new methods of producing high-strength magnetic field gradients, reducing cost of superconducting magnets and emerging methods of magnetically labeling of target particles [27]. Magnetically enhanced bioseparations have been reviewed by Karumanchi and coworkers [27].

Electric fields may be used to enhance bioseparations [27, 85–88], but commercial use is limited at present because of the damaging effects of ohmic heating that accompanies current flow. Electrolysis can be another problem. Nevertheless, electrokinetic forces on charged particles have been demonstrated to reduce concentration polarization and membrane fouling during microfiltration and ultrafiltration, thereby enhancing filtration rates [86]. Up to 7-fold enhancement of transmembrane flow has been recorded during microfiltration with direct current electric field strengths of $100–120\,V\,cm^{-1}$ [86]. Some of the problems associated with electric fields may be reduced by replacing the steady direct current fields with pulsed direct current fields [86]. Electric discharges have been used also to instantaneously break foams during processing.

The separation potential of electric field is best illustrated by electrophoresis, which is a well-established, extremely high-resolution method for separation of proteins. Differences in molecular charge and weight are the bases of separation. However, despite attempts to scale-up [88], electrophoresis remains confined mostly to laboratory use. Except for small volumes, rapid removal of the heat generated has proven difficult without using convective mixing that would destroy any separation.

2.7.7
Chromatographic Separations

Enhancing speed has been a major preoccupation with chromatographic processing. Except for bed height-dependent gel permeation, the speed of most chromatographic processes can be enhanced by replacing the usual high-resistance packed vertical columns with radial flow devices [88]. Adsorption media used in conventional columns can still be utilized, but the medium is packed in the annulus between two porous concentric cylinders. Radial flow columns attain 10- to 50-fold greater flow rates than conventional columns [88]. Industrial-scale simulated moving bed chromatographic systems are now available [89].

Among other improvements, better, more rigid yet porous chromatographic media that are less susceptible to bed compression have been developed [12]. Other novel media have enabled extremely high-speed or perfusion chromatography. Unlike conventional media, perfusion media contain through-pores for bulk flow of fluid through the particle. Diffusional pores as in conventional media are also present. Through-pores allow high flow rates – up to 100-fold greater than in diffusive media [88]. Resulting convection within the particles reduces diffusive transport limitations.

Another high-rate chromatographic system is expanded- or fluidized-bed chromatography. The medium bed is expanded or fluidized during loading by upflow of unclarified fermentation fluid or cell homogenate [88]. There is little pressure drop through the expanded bed. Plug flow of fluid is desired and easily attained. After adsorption, the microbial solids are washed away by upflow of water or buffer. The adsorbed product is recovered as in conventional chromatography by downward elution of the settled, packed bed. As this method handles unclarified fluids, some solid–liquid separation steps are eliminated. Fluid-bed chromatography has been demonstrated with numerous fluids, including broths of *E. coli*, yeast, mammalian cells [88], autolysed yeast and blood plasma.

A further rapid chromatographic method that may potentially handle solid-laden fluids is membrane chromatography. This technique employs ion-exchange groups or other high-specificity adsorption ligands attached to the inner surfaces of pores of conventional microfiltration membranes. Rapid flow through pores reduces diffusion limitations, hence speeding adsorption and, later, desorption. Hollow fiber membrane modules that allow compact packing of large membrane areas have been used for membrane chromatography [90].

Some particularly high-resolution chromatographic separations include high-performance liquid chromatography (HPLC) and bioaffinity-based methods. Process-scale HPLC continues to be useful for small batches [68]; however, this method is expensive, slow, and the high-pressure columns appear to have reached an upper limit of about 0.3 m diameter and 2.4–3.0 m height. Bioaffinity chromatography with affinity ligands – receptors, antibodies, enzymes and other active proteins – immobilized onto the support media has been used for quite some time, but it remains expensive. Other problems are often poor stability of the affinity matrix and ligand leakage into the product. With few exceptions [e.g. Protein A-affinity columns can be cleaned with the strong denaturant guanidine hydrochloride (6 M), which solubilizes adsorbed proteins without affecting the ligand], ligand stability limits the column cleaning regimen. Due to these factors, a trend toward replacing labile bioaffinity ligands with inexpensive and robust alternatives (e.g. dyes, metal ions) is apparent.

Note that some of the speed-enhancing techniques used with chromatography are equally applicable to nonchromatographic adsorptions. Adsorption using columns or slurries of activated carbon is commonly encountered in bioprocessing, particularly for removing pigments. Useful concepts for scale-up of chromatographic columns have been discussed by Lightfoot and Moscariello [91]. Chromatography is an expensive and low-throughput operation [92]. Consequently, there is continuing interest in separation techniques that might match its resolving power without compromising on throughput.

2.7.8
Separation of Optical Isomers

Pure enantiomers can be produced by enantioselective synthesis with enzymes. Conventional synthetic processes typically produce mixtures of enantiomers. The ability to separate enantiomers is becoming increasingly important in biotechnology processes. Enzymatic resolution has been traditionally used for separating enantiomers. In addition, electrophoretic and chromatographic methods have been developed for this purpose [22, 84]. Chromatographic separation of enantiomers makes use of chiral stationary phases that are available in increasing variety [22]. New technology relying on membranes for separating enantiomers from racemic mixtures has been demonstrated [84].

2.8
Recombinant and Other Proteins

Many of the newer recombinant biotechnology products are proteins [52, 93]. While the general features of a bioseparation scheme for these products are the same as for other proteins, there are some unique constraints. Genetically modified microorganisms and cells of higher life forms are often more fragile than the corresponding wild strains [94–97]. This has implications for the design of cell–liquid separation stages. Also, recombinant proteins formed in bacteria and yeasts frequently precipitate inside the cell as dense, insoluble, denatured inclusion bodies. In this form, proteins which may otherwise be toxic to the cell may be overproduced and remain protected against proteolytic activity within the cell.

Most bacteria and fungi used in producing recombinant proteins also produce a variety of proteases that may degrade some of the desired protein within the cell and during recovery. Soluble, noninclusion body proteins being particularly susceptible to degradation. Degradation by acid proteases with a pH optimum of 2–4 may be minimized by processing at higher pH and low temperatures. Neutral proteases are not particularly thermostable and may be inactivated by heating to 60–70°C for 10 min [37]. Many proteases are metalloproteins and require a divalent metal ion for proteolytic activity; chelating agents such as EDTA or citric acid may be used to inactivate such proteases by binding the metal ions. Alkaline proteases of *Bacilli*, such as subtilisin, contain serine at the active site and are not affected by EDTA, but are inhibited by diisopropylfluorophosphate. The short-lived reagent phenylmethylsulfonyl fluoride protects against serine proteases. Antioxidants such as vitamin E and ascorbic acid protect against oxidation [37].

Proteins tend to be more stable in concentrated solutions. Addition of PEGs and other proteins such as albumins may have a stabilizing effect. Glycerol, sucrose, glucose, lactose and sorbitol are often used as stabilizers in concentrations of 1–30%. Enzyme substrates usually have a stabilizing effect as do high concentration of salts such as ammonium sulfate and potassium phosphate. Metalloproteins may be stabilized by addition of metal salts. Divalent metal ions such as Ca^{2+}, Cd^{2+}, Mn^{2+} and Zn^{2+} stabilize various enzymes [37].

Some commonly used sequences of protein purification methods have been outlined by Bonnerjea and coworkers [76] and by Wheelwright [5]. Chromatographic procedures are indispensable for producing high-purity proteins. Typically, the mean recovery or yield of separation steps such as those listed in Tab. 2.1 is about 60–80% [76]. Average and high values of purification factors associated with some protein purification operations are shown in Tab. 2.8, which is based on data compiled by Bonnerjea and coworkers [76]. Clearly, affinity chromatography far outperforms other methods, but compared to operations such as ion-exchange chromatography, the scope for further improving performance is small because many affinity separations already operate close to their theoretical maximum performance [76].

Changes in processing volume, product yield, and total and specific activities occur during processing as illustrated in Tab. 2.9 for a relatively simple purification of brain tumor plasminogen activator (PA) from supernatants of cultured, anchorage-dependent rat cells [98]. The purification in Tab. 2.9 was performed at 4°C. The serum-free conditioned medium used for recovery had an initial PA activity of only $9 IU mL^{-1}$ [98]. Zinc chelate–agarose chromatography was used as the first concentration/purification step. The culture fluid (6 L) was applied to the column (5×8 cm) at a flow rate of $200 mL h^{-1}$. The column was washed with Tris–HCl

Tab. 2.8 Approximate values of purification factors observed during protein purifications (based on Bonnerjea and coworkers [76])

Operation	Purification Factor	
	Average	**High**
Affinity chromatography	100	3000
Dye–ligand affinity	17	–
Inorganic adsorption	12	100
Size-exclusion chromatography	6	100
Hydrophobic-interaction chromatography	15	60
Ion-exchange chromatography	8	50
Detergent extraction	4	12
Precipitation	3	12

Tab. 2.9 Purification of tumor PA [98]

Process stage	Stream attributes						
	Volume (mL)	Total protein (mg)	Total activity (IU)	Volumetric activity (IU mL^{-1})	Specific activity (IU mg^{-1})	Yield (%)	Purification factor
Clarified medium	6000	270	53000	8.8	196	100	1
Zinc chelate–agarose	100	138	50000	500	362	94	1.9
Concanavalin A–agarose	52	2.4	19400	373	22750	37	116
Gel filtration	7.5	0.53	20800	2773	39000	39	199

buffer (0.02 M, pH 7.5, 1 L) that contained 1 M sodium chloride, aprotinin and Tween-80 [poly(oxyethylene sorbitane monooleate) (0.01% v/v)]. Aprotinin, a protease inhibitor, and Tween-80, a nonionic surfactant, are generally added at all stages of PA processing to, respectively, suppress proteolysis and overcome the surface adherent tendency of PAs [52]. After the wash, the column was eluted with a linear gradient of imidazole (0–0.05 M) in the wash buffer (1 L, 120 mL h^{-1}). Pooled PA fractions were further purified on a concanavalin A–agarose-affinity chromatography column. Dialysis was used to concentrate the pooled fractions, and a final gel-filtration step (Sephadex G-150 superfine) was employed. The overall yield was 39% [98]. This figure is fairly typical of large-scale protein recovery. For example, overall recoveries of 23–47% were noted for a variety of processes (e.g. recombinant bovine somatotropin, recombinant human α-interferon, L-leucine dehydrogenase for use in chiral syntheses) reviewed by Wheelwright [5]. One exception was a somewhat impractical process for tissue-type PA (tPA) for which the overall yield was only 6% [5]. Other methods for large-scale tPA recovery have been presented by Rouf and coworkers [52].

2.8.1
Inclusion Body Proteins

When possible, production of recombinant proteins as inclusion bodies has important advantages. Some proteins that form inclusion bodies are listed in Tab. 2.10. Inclusion bodies are easy to isolate, highly concentrated forms of the desired recombinant protein. Typically, inclusion bodies are spheroidal particles, 0.2–2.0×10^{-6} m in diameter and 1100–1300 kg m^{-3} density. The sequence of steps in recovery of inclusion body proteins is cell disruption, centrifugal separation of the inclusion body, washing, solubilization of the protein and renaturation [37, 99]. Cell disruption by homogenization is the preferred technique in large-scale processing. Disruption by high-pressure homogenization has been detailed by

Tab. 2.10 Some proteins produced as inclusion bodies [37]

Bovine pancreatic ribonuclease	Human interleukin-2	Lysozyme
Bovine somatotropin	Human interleukin-4	Porcine phospholipase
Epidermal growth factor	Human macrophage colony-stimulating factor	Pro-chymosin
Human insulin	Human serum albumin	Pro-urokinase
Human γ-interferon	Immunoglobulins	tPA

Chisti and Moo-Young [24]. Inclusion bodies are not affected by homogenization. Cell homogenates are centrifuged to sediment the dense inclusion body fraction. Centrifugation at 1000–12000 g for 3–5 min is sufficient. Sedimentation of cell debris can be minimized by increasing the density and viscosity of the homogenate with additives such as 30% sucrose or 50% glycerol. The inclusion body fraction is washed with buffers containing 1 M sucrose, 1–5% Triton-100 surfactant [100], and, in some cases, low concentrations of proteolytic enzymes and denaturants. The wash steps remove soluble contaminants, membrane proteins, lipids and nucleic acids. At this stage the remaining solids fraction is above 90% recombinant protein. The protein solids are solubilized in highly denaturing chaotropic media. Typically, 6–8 M guanidine hydrochloride or 8 M urea are used for solubilization at pH 8–9, 25–37°C for 1–2 h [100]. Reducing agents are added to the solubilization media to break any inter- and intra-molecular disulfide bonds to fully solubilize the protein. Some reducing agents are 2-mercaptoethanol, dithiothreitol, dithioerythritol, glutathione and 3-mercaptopropionate. Some typical concentrations are 0.1 M 2-mercaptoethanol or 10 mM dithiothreitol [100]. The latter has a shorter half-life than 2-mercaptoethanol, but does not have the odor of 2-mercaptoethanol. Stability of thiol compounds in solution is dependent on pH, temperature and the presence of metal ions such as Cu^{2+}, which lower stability, and of stability enhancers such as EDTA. Good yields of some proteins can be obtained by solubilization without the reducing reagents, but for others reducing agents are essential. Of the denaturants, guanidine hydrochloride is preferable to urea, which may contain cyanate causing carbamylation of the free amino groups on the protein, particularly during long incubation periods in alkaline environments. For some proteins, use of one denaturant may produce significantly higher overall yield than if solubilization with the other is used [37]. Performance has to be empirically evaluated.

For refolding of solubilized protein into active entities, concentrations of the denaturant and the reducing agent are reduced by dilution with a refolding buffer. Denaturants can be completely removed by ultrafiltration with addition of renaturing buffer, dialysis or gel filtration. Renaturation from concentrated protein solutions produces lower yields of the active protein because of intermolecular aggregation in these solutions. Thus, renaturation is done at low protein concentrations, typically $1–20 \times 10^{-3} \text{kg m}^{-3}$ protein [100]. Yield of the active protein is enhanced by refolding in the presence of small, nondenaturing amounts (1–2 M) of urea or guanidine hydrochloride [100]. Presence of high-molecular-weight polymers such as PEG may also improve yield [37].

During refolding, formation of the disulfide bonds is achieved by one of three ways. The air oxidation method uses dissolved oxygen for oxidation of the cystine residues. The refolding buffer containing solubilized protein is aerated or exposed to atmosphere. Oxidation is accelerated by Cu^{2+} ions at approximately 10^{-6} M. Typical reaction conditions are pH 8–9, 4–37°C for up to 24 h [100]. Traces of 2-mercaptoethanol may enhance yield. Air oxidation is difficult to control [37].

The glutathione reoxidation method typically uses a 10:1 mixture of reduced and oxidized forms of glutathione at a concentration of 10^{-3} M reduced glutathione [100]. Air oxidation is suppressed by using deaerated buffers held under a nitrogen atmosphere. The ratio of the reduced and oxidized forms of glutathione, the ratio of the glutathione and the cystine residues on the protein, the reoxidation temperature (4–37°C), and the time (1–150 h) provide flexibility to this method [100]. Low-molecular-weight thiols other than glutathione may also be used. The third method of disulfide bond formation, the mixed disulfide interchange technique, has been detailed by Fischer [100].

Methods have been developed for solid-phase-assisted refolding of proteins in chromatography columns. Size-exclusion, ion-exchange and affinity chromatography are used for this purpose [101, 102]. Matrix-assisted refolding is potentially suited to proteins with slow kinetics of refolding and proteins that have a high tendency to aggregate. Natural and artificial chaperons have been used to assist in refolding. How solid matrices contribute to refolding is not entirely clear.

For optimal processing, the inclusion body production stage should be optimized to rapidly form relatively pure, large and dense inclusion bodies that are easy to recover and solubilize. Production of proteolytic activity should be suppressed as far as possible. Purification and concentration are greatly simplified because of the already high starting protein concentration and purity in the inclusion bodies which are easy to separate from the bulk of the soluble proteins by centrifugation. The recovery of active protein from inclusion bodies is variable, but can approach 100%. In general, smaller polypeptides are easier to refold into active forms [37]. Due to the added processing and the need to refold in dilute solutions, proteins produced as inclusion bodies tend to be expensive. With certain proteins such as tPA, production as an inclusion body in bacteria is technically feasible, but it is not competitive with animal cell culture-derived product [52], even though the latter is a fairly expensive production method. Protein refolding has been further reviewed by Mukhopadhyay [99], Bernardez Clark [103] and Middelberg [101].

2.9
Conclusions

The variety of bioseparations is tremendous, but usually a small selection of the available methods is sufficient to achieve the requisite purity. The aim always is to employ the fewest possible process steps consistent with the product quality specifications. In-depth knowledge of individual separations must be combined with insight into the bioreaction step to design an efficient, consistent and integrated overall production process. Whereas the scientific understanding of bioseparations continues to improve and several new capable separations have been introduced, downstream processing of biologicals remains an empirical art. Invariably, experimentation must be relied upon to aid process selection, implementation and scale-up.

Although many high-resolution bioseparations have been developed, they apply generally to purification of relatively small quantities of high-value products. Increasing need exists for inexpensive bioseparation processes that could be applied to recovery of bulk chemicals, fuels and materials [104] from novel biotechnology-based production methods that are being introduced in a quest for sustainable manufacturing [105].

References

1 Houdebine, L. M., Transgenic animal bioreactors. *Transgen Res* **2000**, *9*, 305–320.

2 Nikolov, Z. L., Woodard, S. L., Downstream processing of recombinant proteins from transgenic feedstock. *Curr Opin Biotechnol* **2004**, *15*, 479–486.

3 Chisti, Y., Moo-Young, M., Fermentation technology, bioprocessing, scale-up and manufacture. In *Biotechnology: The Science and the Business*, 2nd edn. , Moses, V., Cape, R. E., Springham, D. G. (Eds.), Harwood Academic, New York, **1999**, pp. 177–222.

4 Belter, P. A., Cussler, E. L., Hu, W.-S., *Bioseparations: Downstream Processing for Biotechnology*, Wiley, New York, **1988**.

5 Wheelwright, S. M., *Protein Purification: Design and Scale up of Downstream Processing*, Hanser, New York, **1991**.

6 Jornitz, M. W. (Eds.), Meltzer, T. H., *Filtration in the Biopharmaceutical Industry*, Dekker, New York, **1998**.

7 Christian, J. B., Improve clarifier and thickener design and operation. *Chem Eng Prog* **1994**, *90(7)*, 50–56.

8 Tiller, F. M., Tarng, D., Try deep thickeners and clarifiers. *Chem Eng Prog* **1995**, *91(3)*, 75–80.

9 Gyure, D. C., Set realistic goals for cross-flow filtration. *Chem Eng Prog* **1992**, *88(11)*, 60–66.

10 Fleming, H. L., Consider membrane pervaporation. *Chem Eng Prog* **1992**, *88(7)*, 46–52.

11 Lipnizki, F., Hausmanns, S., Laufenberg, G., Field, R., Kunz, B., Use of pervaporation–bioreactor hybrid processes in biotechnology. *Chem Eng Technol* **2000**, *23*, 569–577.

12 Vane, L. M., A review of pervaporation for product recovery from biomass fermentation processes. *J Chem Technol Biotechnol* **2005**, *80*, 603–629.

13 Abbott, N. L., Hatton, T. A., Liquid–liquid extraction for protein separations. *Chem Eng Prog* **1988**, *84(8)*, 31–41.

14 Raghavarao, K. S. M. S., Rastogi, N. K., Gowthaman, M. K., Karanth, N. G., Aqueous two-phase extraction for downstream processing of enzymes/proteins. *Adv Appl Microbiol* **1995**, *41*, 97–171.

15 Chisti, Y., Solid substrate fermentations, enzyme production, food enrichment. In *Encyclopedia of Bioprocess Technology: Fermentation, Biocatalysis and Bioseparation*, Flickinger, M. C., Drew, S. W. (Eds.), Wiley, New York, **1999**, vol. 5, pp. 2446–2462.

16 Patnaik, P. R., Liquid emulsion membranes: principles, problems and applications in fermentation processes. *Biotechnol Adv* **1995**, *13*, 175–208.

17 Pyle, D. L., Protein separation using reverse micelles. *J Chem Technol Biotechnol* **1994**, *59*, 107–108.

18 Schügerl, K., *Solvent Extraction in Biotechnology*, Springer, New York, **1994**.

19 Jarvis, A. P., Morgan, D., Isolation of plant products by supercritical-fluid extraction. *Phytochem Anal* **1997**, *8*, 217–222.

20 Reverchon, E., Supercritical fluid extraction and fractionation of essential oils and related products. *J Supercrit Fluids* **1997**, *10*, 1–37.

21 Chisti, Y., Moo-Young, M., Large scale protein separations: engineering aspects of chromatography. *Biotechnol Adv* **1990**, *8*, 699–708.

22 Juza, M., Mazzotti, M., Morbidelli, M., Simulated moving-bed chromatography and its application to chirotechnology. *Trends Biotechnol* **2000**, *18*, 108–118.

23 Snowman, J. W., Lyophilization techniques, equipment, and practice. *Adv Biotechnol Process* **1988**, *8*, 315–351.

24 Chisti, Y., Moo-Young, M., Disruption of microbial cells for intracellular products. *Enzyme Microb Technol* **1986**, *8*, 194–204.

25 Harrison, S. T. L., Bacterial cell disruption: a key unit operation in the recovery of intracellular products. *Biotechnol Adv* **1991**, *9*, 217–240.

26 Middelberg, A. P. J., Process-scale disruption of microorganisms. *Biotechnol Adv* **1995**, *13*, 491–551.

27 Karumanchi, R. S. M. S., Doddamane, S. N., Sampangi, C., Todd, P. W., Field-assisted extraction of cells, particles and macromolecules. *Trends Biotechnol* **2002**, *20*, 72–78.

28 Chisti, Y., Sonobioreactors: using ultrasound for enhanced microbial productivity. *Trends Biotechnol* **2003**, *21*, 89–93.

29 Asenjo, J. A. (Ed.), *Separation Processes in Biotechnology*, Dekker, New York, **1990**.

30 Verrall, M. (Ed.), *Downstream Processing of Natural Products: A Practical Handbook*, Wiley, New York, **1996**.

31 Goldberg, E. (Ed.), *Handbook of Downstream Processing*, Blackie Academic, London, **1997**.

32 Drew, S. W. (Eds.), Flickinger, M. C., *Encyclopedia of Bioprocess Technology: Fermentation, Biocatalysis, and Bioseparation*, Wiley, New York, **1999**, vols 1–5.

33 Ahuja, S. (Ed.), *Handbook of Bioseparations*, Academic Press, San Diego, **2000**.

34 Tamer, I. M., Chisti, Y., Production and recovery of recombinant protease inhibitor α_1-antitrypsin. *Enzyme Microb Technol* **2001**, *29*, 611–620.

35 Molina Grima, E., Belarbi, E.-H., Acién Fernández, F. G., Robles Medina A., Chisti, Y., Recovery of microalgal biomass and metabolites: process options and economics. *Biotechnol Adv* **2003**, *20*, 491–515.

36 Zhang, Z., Chisti, Y., Moo-Young, M., Isolation of a recombinant intracellular β-galactosidase by ammonium sulfate fractionation of cell homogenates. *Bioseparation* **1995**, *5*, 329–337.

37 Chisti, Y., Moo-Young, M., Separation techniques in industrial bioprocessing. *I Chem E Symp Ser* **1994**, *137*, 135–146.

38 Fish, N. M., Lilly, M. D., The interactions between fermentation and protein recovery. *Biotechnology* **1984**, *2*, 623–627.

39 Dunnill, P., Trends in downstream processing of proteins and enzymes. *Process Biochem* **1983**, *18(5)*, 9–13.

40 Chisti, Y., Build better industrial bioreactors. *Chem Eng Prog* **1992**, *88(1)*, 55–58.

41 Chisti, Y., Assure bioreactor sterility. *Chem Eng Prog* **1992**, *88(9)*, 80–85.

42 Chisti, Y., Moo-Young, M., Clean-in-place systems for industrial bioreactors: Design, validation and operation. *J Ind Microbiol* **1994**, *13*, 201–207.

43 Flickinger, M. C., Sansone, E. B., Pilot- and production-scale containment of cytotoxic and oncogenic fermentation processes. *Biotechnol Bioeng* **1984**, *26*, 860–870.

44 Lubiniecki, A. S., Wiebe, M. E., Builder, S. E., Process validation for cell culture-derived pharmaceutical proteins. In *Large-Scale Mammalian Cell Culture Technology*, Lubiniecki, A. S. (Ed.), Dekker, New York, **1990**, pp. 515–541.

45 Willig, S. H., *Good Manufacturing Practices for Pharmaceuticals: A Plan for Total Quality Control*, 5th edn., Dekker, New York, **2000**.

46 Garg, V. K., Costello, M. A. C., Czuba, B. A., Purification and production of therapeutic grade proteins. In *Purification and Analysis of Recombinant Proteins*, Seetharam, S., Sharma, S. K. (Eds.), Dekker, New York, **1991**, pp. 29–54.

47 Anicetti, V. R., Keyt, B. A., Hancock, W. S., Purity analysis of protein pharmaceuticals produced by recombinant DNA technology. *Trends Biotechnol* **1989**, *7*, 342–349.

48 Taverna, M., Tran, N. T., Merry, T., Horvath, E., Ferrier, D., Electrophoretic methods for process monitoring and the quality assessment of recombinant glycoproteins. *Electrophoresis* **1998**, *19*, 2572–2594.

49 Pelton, J. T., McLean, L. R., Spectroscopic methods for analysis of protein secondary structure. *Anal Biochem* **2000**, *277*, 167–176.

50 Maggioni, C., Braakman, I., Synthesis and quality control of viral membrane proteins. *Curr Topics Microbiol Immun* **2004**, *285*, 175–198.

51 Raetz, C. R. H., Whitfield, C., Lipopolysaccharide endotoxins. *Annu Rev Biochem* **2002**, *71*, 635–700.

52 Rouf, S. A., Moo-Young, M., Chisti, Y., Tissue-type plasminogen activator: characteristics, applications and production technology. *Biotechnol Adv* **1996**, *14*, 239–266.

53 Prazeres, D. M. F., Ferreira, G. N. M., Monteiro, G. A., Cooney, C. L., Cabral, J. M. S., Large-scale production of pharmaceutical-grade plasmid DNA for gene therapy: problems and bottlenecks. *Trends Biotechnol* **1999**, *17*, 169–174.

54 Ferreira, G. N. M., Monteiro, G. A., Prazeres D. M. F., Cabral J. M. S., Downstream processing of plasmid DNA for gene therapy and DNA vaccine applications. *Trends Biotechnol* **2000**, *18*, 380–388.

55 Burnouf, T., Radosevich, M., Nanofiltration of plasma-derived biopharmaceutical products. *Haemophilia* **2003**, *9*, 24–37.

56 Morgenthaler, J. J., Securing viral safety for plasma derivatives. *Trans Med Rev* **2001**, *15*, 224–233.

57 Roberts, P., Virus safety in bioproducts. *J Chem Technol Biotechnol* **1994**, *59*, 110–111.

58 Darling, A., Validation of biopharmaceutical purification processes for virus clearance evaluation. *Mol Biotechnol* **2002**, *21*, 57–83.

59 Watts, K. T., Mijts, B. N., Schmidt-Dannert, C., Current and emerging approaches for natural product biosynthesis in microbial cells. *Adv Synth Catal* **2005**, *347*, 927–940.

60 Otten, L. G., Quax, W. J., Directed evolution: selecting today's biocatalysts. *Biomol Eng* **2005**, *22*, 1–9.

61 Chisti, Y., Moo-Young, M., Bioprocess intensification through bioreactor engineering. *Trans IChemE* **1996**, *74A*, 575–583.

62 Garrido, F., Banerjee, U. C., Chisti, Y., Moo-Young, M., Disruption of a recombinant yeast for the release of β-galactosidase. *Bioseparation* **1994**, *4*, 319–328.

63 Hochuli, E., Purification techniques for biological products. *Pure Appl Chem* **1992**, *64*, 169–184.

64 Beitle, R. R., Ataai, M. M., One-step purification of a model periplasmic protein from inclusion bodies by its fusion to an effective metal-binding peptide. *Biotechnol Prog* **1993**, *9*, 64–69.

65 Fexby, S., Bzyxsevenninexyzlow, L., Hydrophobic peptide tags as tools in bioseparation. *Trends Biotechnol* **2004**, *22*, 511–516.

66 French, C., Ward, J. M., Production and release of recombinant periplasmic enzymes from *Escherichia coli* fermentations. *J Chem Technol Biotechnol* **1992**, *54*, 301.

67 Tamer, I. M., Chisti, Y., Moo-Young, M., Disruption of *Alcaligenes latus* for recovery of poly- β-hydroxybutyrate: comparison of high-pressure homogenization, bead milling, and chemically induced lysis. *Ind Eng Chem Research* **1998**, *37*, 1807–1814.

68 Dwyer, J. L., Scaling up bioproduct separation with high performance liquid chromatography. *Biotechnology* **1984**, *2*, 957–964.

69 Baker, N. V., Segregation and sedimentation of red blood cells in ultrasonic standing waves. *Nature* **1972**, *239*, 398–399.

70 Kilburn, D. G., Clarke, D. J., Coakley, W. T., Bardsley, D. W., Enhanced sedimentation of mammalian cells following acoustic aggregation. *Biotechnol Bioeng* **1989**, *34*, 559–562.

71 Whitworth, G., Grundy, M. A., Coakley, W. T., Transport and harvesting of suspended particles using modulated ultrasound. *Ultrasonics* **1991**, *29*, 439–444.

72 Mandralis, Z. I., Feke, D. L., Fractionation of suspensions using synchronized ultrasonic and flow fields. *AIChE J* **1993**, *39*, 197–206.

73 Weiser, M. A. H., Apfel, R. E., Interparticle forces on red cells in a standing wave field. *Acustica* **1984**, *56*, 114–119.

74 De Loggio, T., Letki, A., New directions in centrifuging. *Chem Eng* **1994**, *101(1)*, 70–76.

75 Tebbe, H., Lutkemeyer, D., Gudermann, F., Heidemann, R., Lehmann, J., Lysis-free separation of hybridoma cells by continuous disc stack centrifugation. *Cytotechnology* **1996**, *22*, 119–127.

76 Bonnerjea, J., Oh, S., Hoare, M., Dunnill, P., Protein purification: the right step at the right time. *Biotechnology* **1986**, *4*, 954–958.

77 Dörnenburg, H., Knorr, D., Release of intracellularly stored anthraquinones by enzymatic permeabilization of viable plant cells. *Process Biochem* **1992**, *27*, 161–166.

78 Lockwood, C. E., Bummer, P. M., Jay, M., Purification of proteins using foam fractionation. *Pharm Res* **1997**, *14*, 1511–1515.

79 Chisti, Y., Moo-Young, M., Improve the performance of airlift reactors. *Chem Eng Prog* **1993**, *89(6)*, 38–45.

80 Brennecke, J. F., New applications of supercritical fluids. *Chem Ind (Lond)* **1995**, *21*, 831–834.

81 Chisti, Y., Animal cell culture in stirred bioreactors: observations on scale-up. *Bioproc Eng* **1993**, *9*, 191–196.

82 Kroner, K. H., Nissinen, V., Ziegler, H., Improved dynamic filtration of microbial suspensions. *Biotechnology* **1987**, *5*, 921–926.

83 Lee, S. S., Burt, A., Russotti, G., Buckland, B., Microfiltration of recombinant yeast cells using a rotating disk dynamic filtration system. *Biotechnol Bioeng* **1995**, *48*, 386–400.

84 Jirage, K. B., Martin, C. R., New developments in membrane-based separations. *Trends Biotechnol* **1999**, *17*, 197–200.

85 Reschiglian, P., Zattoni, A., Roda, B., Michelini, E., Roda, A., Field-flow fractionation and biotechnology. *Trends Biotechnol* **2005**, *23*, 475–483.

86 Brors, A., Kroner, K. H., Electrically enhanced cross-flow filtration of biosuspensions. In *Harnessing Biotechnology for the 21st Century*, Ladisch, M. R., Bose, A. (Eds.), American Chemical Society, Washington, DC, **1992**, pp. 254–257.

87 Rudge, S. R., Todd, P., Applied electric fields for downstream processing. *ACS Symp Ser* **1990**, *427*, 244–270.

88 Shanley, A., Parkinson, G., Fouhy, K., Biotech in the scaleup era. *Chem Eng* **1993**, *100(1)*, 28–33.

89 Kim, I., Biotech's new mandate: more, cheaper, and faster. *Chem Eng* **1997**, *104(1)*, 28–33.

90 Brandt, S., Goffe, R. A., Kessler, S. B., O'Connor, J. L., Zale, S. E., Membrane-based affinity technology for commercial scale purifications. *Biotechnology* **1988**, *6*, 779–782.

91 Lightfoot, E. N., Moscariello, J. S., Bioseparations. *Biotechnol Bioeng* **2004**, *87*, 259–273.

92 Przybycien, T. M., Pujar, N. S., Steele, L. M., Alternative bioseparation operations: life beyond packed-bed chromatography. *Curr Opin Biotechnol* **2004**, *15*, 469–478.

93 Zhang, Z., Moo-Young, M., Chisti, Y., Plasmid stability in recombinant *Saccharomyces cerevisiae*. *Biotechnol Adv* **1996**, *14*, 401–435.

94 Dunnill, P., Biochemical engineering and biotechnology. *Chem Eng Res Des* **1987**, *65*, 211–217.

95 Moo-Young, M., Chisti, Y., Considerations for designing bioreactors for shear-sensitive culture. *Biotechnology* **1988**, *6*, 1291–1296.

96 Chisti, Y., Animal-cell damage in sparged bioreactors. *Trends Biotechnol* **2000**, *18*, 420–432.

97 Chisti, Y., Hydrodynamic damage to animal cells. *Crit Rev Biotechnol* **2001**, *21*, 67–110.

98 Bykowska, K., Rijken, D. C., Collen, D., Purification and characterization of the plasminogen activator secreted by a rat brain tumor cell line in culture. *Thromb Haemostas* **1981**, *46*, 642–644.

99 Mukhopadhyay, A., Inclusion bodies and purification of proteins in biologically active forms. *Adv Biochem Eng Biotechnol* **1997**, *56*, 61–109.

100 Fischer, B. E., Renaturation of recombinant proteins produced as inclusion bodies. *Biotechnol Adv* **1994**, *12*, 89–101.

101 Middelberg, A. P. J., Preparative protein refolding. *Trends Biotechnol* **2002**, *20*, 437–443.

102 Jungbauer, A., Kaar, W., Schlegl, R., Folding and refolding of proteins in chromatographic beds. *Curr Opin Biotechnol* **2004**, *15*, 487–494.

103 Bernardez Clark, E. D., Protein refolding for industrial processes. *Curr Opin Biotechnol* **2001**, *12*, 202–207.

104 Keller, K., Friedmann, T., Boxman, A., The bioseparation needs for tomorrow. *Trends Biotechnol* **2001**, *19*, 438–441.

105 Gavrilescu, M., Chisti, Y., Biotechnology – a sustainable alternative for chemical industry. *Biotechnol Adv* **2005**, *23*, 471–499.

Part II   Bioprocess and Early DSP

3

Processes Development and Optimization for Biotechnology Production – Monoclonal Antibodies

Jochen Strube, Sven Sommerfeld, and Martin Lohrmann

3.1
Introduction

In manufacturing biological drugs, product quality is defined by the process (e.g. equipment, sequence of unit operations, operation parameters) because no complete analysis of these complex molecules is possible. Therefore, the process is unavoidably fixed after the first clinical lot production in a pilot plant. There is no further process optimization option parallel to production that, in the case of small-molecule production, allows further process optimization. Process development times will not increase in future due to increasing pressure on "time to market". In addition, no change in paradigm seems possible, as complete analysis of complex biomolecules comparable to small synthetic drugs is not foreseeable in the near future. As a consequence, the challenge is to establish generic processes for different drug classes and to find consistent process development methods which allow reliable prediction of large-scale production. Generic in this sense is not understood as a fixed sequence of unit operations with a certain set of generic process parameters. Here, generic means that a typical arrangement of unit operations is set up in an efficient sequence to fulfill the separation task [1].

Monoclonal antibodies came of age as therapeutic products in the early 1990s [2]. Since then, the requirements for purified monoclonal antibodies have advanced exponentially in order to meet the ever-growing clinical and commercial demand [3]. Original monoclonal antibody preparations were the result of *in vivo* production [4], yielding milligram to gram quantities. Today, large-scale cell culture techniques can yield many kilograms of monoclonal antibodies per batch. The challenge is thus to develop and scale purification processes that are capable of yielding kilogram quantities of purified monoclonal antibody at the lowest possible cost.

Process development and design differences between chemical and biotechnological processes are shown in Fig. 3.1.

Performance considers process economy, safety, operability, controllability, etc., while *total effort* is equivalent to time multiplied by the concurrent effort.

Bioseparation and Bioprocessing. Edited by G. Subramanian
Copyright © 2007 WILEY-VCH Verlag GmbH & Co. KGaA, Weinheim
ISBN: 978-3-527-31585-7

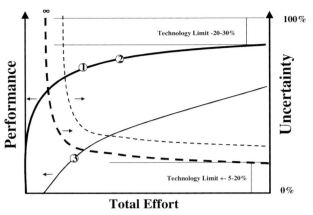

Fig. 3.1 The dependence of process performance (solid lines) and uncertainty (dashed lines) on the total amount of effort invested in project development in chemical (bold lines) and biotechnological (thin lines) processes.

Performance and the reduction of *uncertainty* are limited by the technology limit, which in the case of performance is a reduction of 20–30% (equipment, valves, pumps, etc.) and a remaining uncertainty of 5–20% due to a lack of information concerning methods, amounts of side-components, etc. These values are valid for chemical processes.

For process development of chemical processes, in many cases historical data and process development methods are available so that a certain performance is available after a very short development time with marginal effort. Thus, uncertainty can be reduced very quickly. A typical operating point (point 1) after process development includes a safety margin of about 20–30%. Further process optimization is possible parallel to the production process, raising process performance (point 2) and reducing uncertainty.

In contrast, for the development of biotechnological processes, the technology limit (40–50%) concerning uncertainty is much higher due to a lack of basic fundamentals like biothermodynamics that are still to be examined. In addition, there is still the need for process development methods that are rarely available. This results in a much lower reachable process performance, a lower performance increase and 30–40% more effort. Due to the pressure to reduce "time to market", process development times become shorter. Therefore, only 20–40% of the possible performance can be reached during process development (point 3). As the product is defined by the process (although not for well-characterized drugs), further process optimization is impossible after the first clinical lot has been produced. Moreover, feasibility and reliability in large-scale processes is at much higher uncertainty due to the lack of fundamentals.

In this chapter, monoclonal antibodies are chosen as synonymous for all complex biological drugs in general. Six monoclonal antibody production process concepts are analyzed by literature research. Based on the data, a generic process scheme is proposed. A representative example of monoclonal antibody production is modeled and simulated so that the optimization potential can be identified and even quantified by a sensitivity analysis of influencing parameters and definition of benchmark criteria.

First, emphasis is placed on selective separations, e.g. affinity and ion-exchange chromatography, as well as on membrane ultrafiltration, which are the state-of-the-art monoclonal antibody production processes. The method has proved to be valuable and of great benefit; therefore, further analyses followed which took into account other relevant unit operations and yield losses.

3.2
Monoclonal Antibody Production

3.2.1
Fundamentals

Antibodies are "Y"-shaped molecules consisting of four chains – two light chains and two heavy chains – that are connected by disulfide bonds (Fig. 3.2). Each light chain has a molecular weight of approximately 25 kDa and is composed of two domains – one variable domain (V_L) and one constant domain (C_L).

Each heavy chain has a molecular weight of approximately 50 kDa, and consists of a constant and a variable region. The heavy chain contains one variable domain (V_H) and either three or four constant domains (C_H1, C_H2, C_H3 and C_H4) depend-

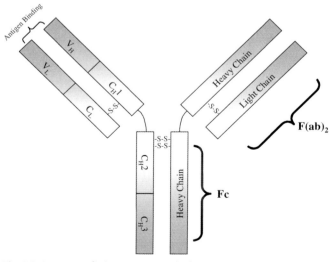

Fig. 3.2 Structure of a human IgG antibody.

ing on the class of the antibody. The variable regions are responsible for antigen binding. They contain the so-called complementarity-determining region (CDR), whereas the constant regions are responsible for biological functions, such as binding to cell walls. This part of the antibody binds to the cell wall Proteins A or G which are used for purification.

There are five classes of antibodies, i.e. IgG, IgA, IgM, IgE and IgD, which are distinguished by their heavy chains γ, α, μ, ε and δ.

The IgG, IgE and IgD antibody classes are each made of a single structural unit, whereas IgA antibodies may contain one or two units and IgM antibodies contain five disulfide-linked structural units. Furthermore, IgG antibodies are divided into four subclasses (isotypes) and IgA into two subclasses, each of which differs slightly in molecular weight and other physical properties [5].

3.2.2
Market/Potential

The 28 major protein-based therapeutics generated more than $13 billion of sales in 2000 with annual growth rates of slightly more than 25% per year [6]. The market for therapeutic monoclonal antibodies is forecasted to reach nearly $9 billion by 2005 [7].

3.2.2.1 Antibodies on the Market
Currently, there are 20 therapeutic antibodies and fusion proteins on the market targeting different diseases [8]. The majority of antibody products are used against different kinds of cancer (see Tab. 3.1). Other major target indications include transplantation, and respiratory, skin and neurological disorders, as well as auto-immune conditions.

Fusion proteins like Immunex's Embrel™ are similar to monoclonal antibodies in terms of fermentation capacities and purification [9]. These protein molecules consist of the Fc region of an antibody, while the Fab region is replaced by a cytokine (interleukin, human growth factor, etc.).

The first antibody which was approved by the US Food and Drug Administration (FDA) was OKT3™ in 1986. OKT3 belongs to the first generation of monoclonal antibodies and has a completely murine structure, forcing the human immune system to build human anti-murine antibodies (antibodies that are directed against the murine antibodies). The first chimeric antibody was ReoPro™ in 1993 marketed by Johnson & Johnson. Due to the partly human structure of chimeric antibodies (only the light chains are of murine origin), the immune response was less strong as compared with the case of OKT3™.

The next step to reduce immunogenicity was Genentech's first humanized antibody, Herceptin™ against breast cancer, that was marketed in 1998. In January 2003, Abbott's antibody Humira™ was approved by the FDA as the first human antibody created by phage-display technology [10]. This antibody has a completely

Tab. 3.1 List of approved therapeutic antibodies and their indication

Product	Company	Indication	Category	Year
OKT-3™	Johnson & Johnson	transplant rejection	mMAB	1986
ReoPro™	Johnson & Johnson	percutaneous transluminal coronary angioplasty (PCTA)	cMAB	1993
Rituxan™	IDEC/Genentech	B-cell lymphoma	mMAB	1997
Herceptin™	Genentech/Roche	breast cancer	hzMAB	1998
Remicade™	Johnson & Johnson	rheumatoid arthritis; Crohn's disease	cMAB	1998
Simulect™	Novartis	transplant rejection	cMAB	1998
Synagis™	Abbott/MedImmune	respiratory syncytial virus	hzMAB	1998
Zenapax™	Roche/PDL	transplant rejection	hzMAB	1998
Mylotarg™	Celltech/AHP	acute myeloid leukemia	hzMAB	2000
Campath™	Millenium/Ilex/Baxter	chronic lymphocytic leukemia	hzMAB	2001
Zevalin™	IDEC	non-Hodgkin's leukemia	cMAB	2002
Humira™	Abbott	rheumatoid arthritis	humMAB	2003
Xolair™	Genentech/Novartis	allergic asthma	humMAB	2003
Bexxar™	Corixa/GlaxoSmithKline	non-Hodgkin's leukemia	mMAB	2003
Raptiva™	Genentech/Xoma	moderate to severe psoriasis	hzMAB	2003
Erbitux™	ImClone/BMS/Merck	various cancers	cMAB	2004
Avastin™	Genentech/Roche	various cancers	hzMAB	2004
Tysabri™	Biogen Crop.	multiple sclerosis	hzMAB	2004
Enbrel™	Immunex	rheumatoid arthritis; active ankylosing spondylitis	fusion protein	1998
Amevive™	Biogen Corp.	chronic plaque psoriasis	fusion protein	2003

MAB, monoclonal antibody; m, mouse; c, chimeric; hz, humanized; hum, human.

human protein structure, making it essentially indistinguishable from antibodies present in the human body so that the immunogenicity is very low.

Several of these antibody products are expected to reach blockbuster status (sales over $1 billion) or have reached this status already. The average compound annual growth rate for antibody-based therapeutics is more than 30%. For other recombinant proteins the predicted average growth rate is only 15%; for the global market of pharmaceuticals it is about 10%. Thus, the market for antibodies has a huge growth potential [6, 7].

Within the next few years the number of marketed antibody products is expected to double. Ginsberg and coworkers report 21 monoclonal antibodies, fusion proteins or antibody fragments in phase III and 39 in phase II development (data from 2001). Assuming a success rate of 50% for phase III and 25% for phase II candidates, there will be 21 new antibody products in 2005. Additionally, there will be 19 nonantibody protein therapeutics so that in 2005 approximately 40 new protein therapeutics will be on the market [9].

These 40 new protein drugs demand additional manufacturing capacities. Based on current capacities of approximately 500 000 L, there will be a shortage of 400% in 2006. To overcome this problem many companies are building new production facilities so that an additional 1 000 000 L for mammalian cell culture will be available. Even with these additional capacities, Molowa and coworkers expect a shortage of 200% unless major process improvements will be made [11].

A further problem of capacity shortage is that a certain part of production capacities is used for process development, scale-up, maintenance and clinical trial material, so that the actual production capacity is much lower than the theoretical maximum. In particular, strain and process development take many months, and tie up valuable production capacity [9].

3.2.2.2 **Product Demand**

Product demand depends greatly on the patient population so that the demand is higher for "popular" diseases (e.g. rheumatoid arthritis). The typical dose for a monoclonal antibody therapeutic is assumed to be 2–5 g per patient per year. If one assumes an average of 80 000–100 000 patients per year taking this drug, this works out to approximately 200 kg product per year [3, 11].

Another problem might be the high product costs so that treatment with that kind of product is very expensive and may not be affordable for everyone. From this point of view it is imaginable that within the next years the health authorities will declare maximum prices [6, 11].

Small-molecule therapeutics that are produced through classical chemical synthesis can be produced for less than $5 g^{-1}, while proteins produced in living cells incur production costs of $100–1000 g^{-1}. These high costs make products dosed at levels higher than 5 mg kg^{-1} body weight per week not commercially viable unless major advances in biologics manufacturing processes are made [11].

In Fig. 3.3, the high degree of the technology jump for required for monoclonal antibody production is shown in relation to other life science manufacturing routes [12]. This points out the extraordinary demand for production improvements for this product class.

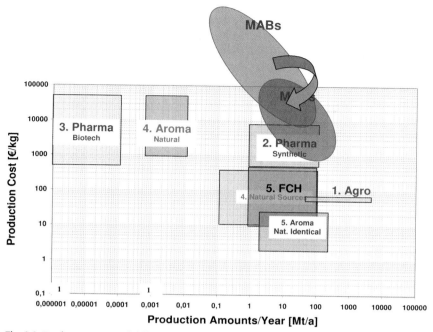

Fig. 3.3 Production costs of different life science product lines.

3.3
Process Development and Optimization

3.3.1
Regulations and Quality Assurance

The Process Analytical Technology (PAT) initiative of the FDA started in 2005 is a major change in not only the allowed, but also the demanded, process development philosophy of recombinant drugs [13].

The PAT initiative has observed that causes for low manufacturing performance and early built-in weaknesses are related to reasons for batch failure like nonrobust processes, mechanical errors, operator errors, automation errors, bed integrity test failure, cleanliness and aseptic status issues, utilities, contaminated water and buffer, etc.

Moreover, transfer from preclinical into clinical phases is crucial. Whereas in preclinical development the cell line is developed, clinical process development causes the transfer of upstream, downstream, formulation and analytics into commercial-scale processes, and has to be finalized with process validation by clinical batch runs. The urgent demand is for these steps to get the drug into humans as soon as possible. It can be clearly seen that fast and reliable process development tools are needed to reduce time-to-process.

The critical parameters of operation and the root course of any decision have to be identified and quantified. Changes in any production parameter are no longer forbidden after the clinical lot production to validate the process. The product quality is no longer only defined by the process which could not be changed to avoid product quality changes. Thus, the FDA has opened up the transfer of well-established methods from small-molecule production to large-molecule production like online process analytics, process modeling and simulation to quantify parameter influences and to optimize process performance. This will be a major contribution to the necessary cost and manufacturing robustness improvements.

3.3.2
Analytical and Experimental Complexity

Any benefit of process modeling is directly linked to the accuracy and efforts of experimental model parameter determination at the laboratory scale and the final model validation by mini-plant technology. From small-molecule production it is well know that in addition to detailed process comprehension and proper process optimization, any contribution of process modeling to a higher efficiency – like shorter development times, lower laboratory-scale efforts and higher manufacturing process robustness – is a direct function of company organization – the scientists who work in the laboratory and mini-plant technique have to be the ones simulating as well, in order to reduce experimental efforts, and to evaluate the required degree of model depth and parameter accuracy. Apart from obstacles in company organization to educate such broad in methodology but highly product specific educated scientist is in biotechnology quite a challenge which needs to be organized and followed up over many years. Nevertheless, companies urgently need to have the opportunity to gain a lead over their competitors as methods used in drug searches like high-throughput screening, combinatory chemistry and molecular modeling, etc., have already almost equaled competition over the last 5 years.

To achieve the optimal benefits of process simulation, the analytical methods to determine model parameters at the laboratory scale have to be improved. The analytical complexity is extreme in biotechnology. Product is highly diluted between 0.1 and about $10 \, g \, L^{-1}$ and has to be measured within ±5% accuracy in the presence of a great number of different side-components and contaminants – some of these (e.g. human/bovine serum albumin) with concentrations as high $1–10 \, g \, L^{-1}$. Product-related impurities are glycosylation, dimer formation and aggregation, incorrect folding, unfolding, proteocyclic cleavage, hydrolysis of peptide bonds, deamination, truncation, disulfide exchange, oxidation (Met, Try, Trp), and ester formation (Glu, Asp). Process-related impurities are host cell proteins, host cell DNA, Protein A leakage of affinity adsorbents, media components, proteases, glycosidases and viruses.

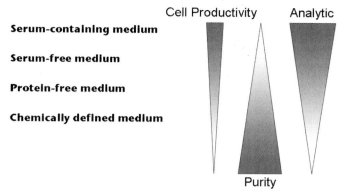

Fig. 3.4 Cell culture medium.

Figure 3.4 shows the interference of maximal cell productivity and lower purity which causes maximal analytical efforts for the four different types of cell culture media.

The classical laboratory-scale analytical methods like sodium dodecylsulfate–polyacrylamide gel electrophoresis (SDS–PAGE), enzyme-linked immunosorbent assay (ELISA), protein chips, the Bradford assay, chromatography and ultraviolet (UV) detection have to be automated by robotics to eliminate manual inaccuracies and to be able to analyze about 20–30 fractions of each of one to three experimental runs per day for about four to six components within ±5% accuracy. This is a representative number of fractions to be able to quantify purity–yield relations by short-cut models or to determine side-component positions in elution profiles for detailed modeling.

3.3.3
Fermentation – Upstream

3.3.3.1 **Cell Types**
Different mammalian cell lines are used for the production of monoclonal antibodies. Originally, monoclonal antibodies were produced in hybridoma cells [4]. For modern *in vitro* production processes, the following cell lines are used:
- Chinese hamster ovary (CHO).
- Mouse myeloma (NSO, Sp2/0).
- Monkey kidney (COS).
- Baby hamster kidney (BHK).

Various criteria have to be considered in the choice of the cell line. First, the cell line must provide a close human glycosylation pattern. Another important point is that the cell line must have a stable expression system to produce the desired protein. This is necessary because the product is only produced when the cells accept the foreign DNA part in which the antibody information is stored. Furthermore, the specific production rate of the cells should be as high as possible. To achieve this, different amplification systems are used. One of the most common systems is the glutamine synthetase system, which is used for some monoclonal antibodies and which is licensed to Lonza [14].

The previously described criteria are best met by CHO and mouse myeloma cells since they provide a nearly human glycosylation pattern as well as the chance to establish a stable expression system. Furthermore, these cell lines are very well known so that there is extensive experience in culturing and modifying them [15].

A further advantage is that product concentrations above 1 g L^{-1} can be reached [16] and that those cell lines can grow in suspension culture in a serum-free medium. Therefore, the study can be extended to 10 g L^{-1} titer concentration. Normally, these cells require a surface on which to attach.

3.3.3.2 Serum
Cell culture medium has a very complex composition so that the cells can be permanently provided with essential substances. An often used substance is animal-derived serum that is added to the medium in different concentrations depending on the individual cell line.

Serum is used to supply growth and attachment factors to the cells. The most often used serum types are bovine serum or fetal calf serum. The advantage of the latter is the very small γ-globulin fraction. This fraction comprises bovine antibodies so that the concentration of unwanted antibodies is very small.

The use of animal-derived serum has a number of disadvantages. For example, serum might contain viruses or prions (e.g. in bovine spongiform encephalopathy) so that there is a risk of viral contamination. A further disadvantage is the lot-to-lot variability so that no consistent composition is guaranteed. The major disadvantage is the high protein content of this complex mixture so that purification becomes more complex. For these reasons it is desirable to avoid the use of animal serum in the production of recombinant proteins like human antibodies.

Instead of animal serum, a chemically well-defined serum-free medium, consisting of essential amino acids, trace elements, proteins, etc., should be used in fermentation processes [17]. This requires adaptation of the cells to the new conditions, which is a very time-intensive process. Furthermore, adaptation to serum-free conditions is not always possible or the specific production rate of the cells is significantly impacted. A further improvement is the use of completely protein-free medium [18].

The use of chemically defined media, free of animal-derived substances, results in smaller expenditure on downstream costs. Purity of monoclonal antibodies at harvest (after fermentation) is less than 30% in an optimized protein-containing culture, whereas it is 60–75% in an optimized protein-free culture [19].

For future antibody products it is expected that cell lines will be established that are capable of being cultured in serum-free medium with only a small protein content; thus, this chapter will focus on serum-free cell culture medium.

3.3.3.3 Contaminants

Typical substance classes that have to be separated in the downstream section are shown in Fig. 3.5. Requirements for the fermentation process are, amongst others, a cell culture medium, eventually additional supplements and the cells themselves. During the fermentation process the cells secret product into the fermentation broth so that the product has to be separated from a complex mixture of different substances. Cell culture medium comprises amino acids, inorganic salts, vitamins, glucose and other organic substances. Moreover, as mentioned above, certain proteins (e.g. from bovine or fetal calf serum) are added as supplements. These proteins make the purification of the product protein more difficult.

Other problems arise from the cells since DNA, viruses and lipids can get into the fermentation broth when the cells are destroyed or cell lysis occurs during the fermentation process. Moreover, certain host cell proteins are secreted and have to be separated from the product protein.

The intention of this analysis of possible contaminants is to establish certain substance classes and to measure typical physical properties so that downstream unit operations can be designed more effectively.

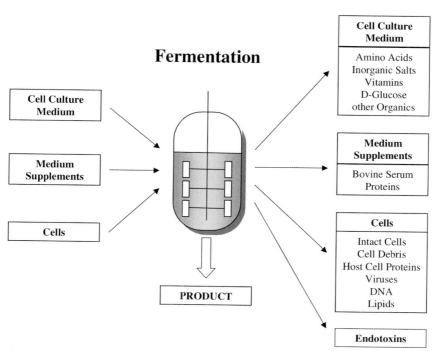

Fig. 3.5 Downstream processing task.

3.3.4
Downstream

After fermentation, monoclonal antibodies are further processed in the downstream section to meet purity and quality requirements. The downstream section can be divided into three parts – capture, purification and polishing.

In the capture step, the product is isolated and concentrated before it is further processed. In the purification section, bulk impurities are removed. The final purity is achieved in the polishing section. For this purpose several unit operations can be used that have different separation mechanisms and capacities. Chromatographic and membrane filtration units are mainly used in downstream processing of biological products.

A further important aspect of downstream processing is the clearance of any viral contamination to provide product safety.

3.3.4.1 Chromatography

Different chromatographic techniques are used for the purification of monoclonal antibodies.

Affinity chromatography can provide very high enrichment factors in one step due to a very high selectivity. The most common types of affinity ligands for the purification of monoclonal antibodies are immobilized bacterial cell wall proteins, e.g. staphylococcal Protein A or streptococcal Protein G.

The interaction of these two proteins occurs primarily through the Fc region (Fig. 3.2) of the monoclonal antibodies. Both Protein A and Protein G are available in recombinant forms in which nonessential regions (e.g. albumin binding site) have been removed, leaving four binding sites for IgG intact [20].

A disadvantage of Protein A or Protein G usage is that there is always a small degree of leakage of the protein ligand so that additional purification steps are required for therapeutic products. Moreover, there are nonspecific interactions resulting in contamination of the target molecule with impurities that are retained by the affinity matrix due to hydrophobic or ion-exchange effects and then coeluted with the affinity bound products [21]. Unspecific interactions take place between the support and other matrix molecules that interfere with the selectivity, and, more importantly, can block the access of the target proteins to the affinity ligands and result in significant loss of capacity of the affinity matrix.

Ion-exchange chromatography is used to remove these remaining impurities. The basis for ion-exchange chromatography is the competition of the protein and a salt for the binding sites on the surface of the ion exchanger. The strength of binding is proportional to the charge of the protein for an oppositely charged ion exchanger. The charge of the protein depends on the pH value of the solution. The pH value at which the net charge of the protein is zero is called the isoelectric point (pI value). Ion exchangers with negatively charged groups are cation exchangers because they bind positively charged proteins (cations) when the mobile-phase pH value is below the pI value of the protein. Conversely, anion exchangers are positively charged [22].

Whether the protein and stationary phase are charged oppositely is determined by the pH value of the mobile phase. The salt concentration of the mobile phase will determine if protein binding occurs as both the salt and protein compete for the charged sites of the stationary phase.

The strength of binding of the individual protein depends on its pI value so that different proteins can be separated by their different pI values.

Moreover, hydrophobic-interaction chromatography, which separates proteins on the basis of differences in their surface hydrophobicity, can be used since proteins contain hydrophobic (i.e. nonpolar) aliphatic as well as aromatic side-chains. The hydrophobic side-chains are mostly hidden within the molecule, whereas the hydrophilic side-chains are preferentially located on the outer side of the molecule's surface. Nevertheless, some hydrophobic amino acids are located at the protein's surface, which will associate with hydrophobic ligands attached to a supporting chromatographic matrix [20].

Size-exclusion chromatography can be used to separate impurities according to their different molecular weight. This technique is often used as a polishing step to remove antibody aggregates or fragments which differ significantly in size (by a factor 5–10) [21].

Each chromatography process itself comprises several steps. Before the column can be used it must be equilibrated to ensure that the same conditions with the lowest elution strength can be found everywhere within the column. The same buffer as used for loading the column is used. After equilibration, the column is loaded with the substance mixture to be separated. The column is then washed to remove some impurities that do not adsorb as strong as the product before the desired product is eluted from the column by a stepwise or continuous change of the elution strength of the mobile phase. Afterwards, the column is washed again to remove the remaining impurities. The column has to be regenerated and sanitized before the next batch can be processed.

3.3.4.2 Filtration

Different filtration techniques are used in addition to chromatography unit operations. Particles such as cells and cell debris are separated by microfiltration with pore sizes between 0.01 and 10 µm. Microfiltration is often used as a first step in downstream processing to clarify the fermentation broth.

Ultrafiltration is used to separate proteins from buffer components for buffer exchange or to concentrate the product. Ultrafiltration membranes are able to retain molecules in a molecular weight range of 300–500 000 Da (1 Da = 1 g mol^{-1}). The membranes are classified based on the nominal molecular weigh cut-off, which is defined as the smallest molecular weight species for which the membrane has more than 90% rejection. In the case of antibodies (150 kDa), pore sizes vary in the range of 30–500 kDa depending on whether the antibody shall be retained or not.

Diafiltration is used to enhance either product purity or yield. During diafiltration, a buffer is introduced into the recycle tank while the filtrate is removed from the unit operation and the retentate is recycled. In processes where the product is the retentate, diafiltration washes components out of the product pool into the filtrate. In this way the buffer can be exchanged and the concentration of undesired components is reduced, and thus purity is increased. On the other hand, when the product is in the filtrate, diafiltration washes it through the membrane into a collecting storage tank, increasing product yield [23].

3.3.4.3 Virus Clearance

Virus clearance consists of virus removal and virus inactivation. For biological products, the FDA demands at least two different steps for virus reduction to guarantee product safety and efficacy. These steps must provide evidence that they are capable of inactivating or removing viruses that could potentially be transmitted by the product [24]. Generally, it is considered prudent to use two orthogonal virus clearance steps that employ complementary mechanisms, so that viruses resistant to the first mechanism may be cleared by the second.

Viruses might get into the fermentation broth on different ways. Many cell lines, especially hybridoma cell lines, carry retrovirus or retrovirus-like particles. Furthermore, the presence of unknown, but potentially harmful, viruses cannot be excluded [25]. In addition, any material used in the fermentation may contain virus contamination. This applies especially to all animal-derived materials such as serum. Therefore, the use of animal serum should be avoided, making it essential to adapt cells to serum-free conditions.

In antibody processes that use serum-free medium, only two different virus clearance steps might be necessary. On the other hand, when animal serum is used, the risk of viral contamination is much higher so that additional virus clearance steps are necessary.

Virus inactivation is preferably performed using very harsh physical/chemical measures like:

- Acid treatment (destruction of the core protein (nucleocapsid) and genome).
- Urea treatment (disintegration of nucleocapsid).
- Solvent/detergent treatment (limited to dissolution of the envelope of enveloped viruses).
- Heat treatment.
- UV radiation.

Using one of these methods may conflict with the stability of the protein product. Therefore, the best choice is a method which inactivates viruses, but does not harm the product.

There are several methods widely used for the mechanical removal of viruses. A separation of virus from product can be achieved by adsorption of protein onto a chromatographic matrix, while leaving virus in the flowthrough or *vice versa*. Another approach is the use of membrane filters – either depth filters (20–40 nm pores) in dead-end mode or ultrafiltration membranes (cut-off below 300 kDa) in a tangential-flow mode.

3.4
Production Processes

Existing processes for the production of marketed monoclonal antibodies are evaluated to generate a generic flowsheet. Here, a generic process is not understood as a fixed sequence of unit operations with fixed generic process parameters. Generic means that a certain set of unit operations is set up in an efficient sequence to fulfill the separation task. Necessary data can be found in the FDA's Summary Basis of Approvals and European Agency for the Evaluation of Medicinal Products' scientific discussions databases.

3.4.1
Examples

The production processes for Herceptin™, Rituxan™, MabCampath™, Synagis™, Remicade™ and Simulect™ are analyzed. The different steps of each process can be seen in Tab. 3.2. The numbers indicate the position of the step within the downstream processing. The first step in all cases is cell removal. This is done by either centrifugation or microfiltration. The next step in almost all cases (except of Synagis) is a Protein A-affinity chromatography step as a first capture step. Since the elution will be at very low pH, virus inactivation follows at low pH. The next chromatography steps are usually one or more ion-exchange steps to remove DNA fragments or leached Protein A. If the final purity cannot be achieved within these steps, additional chromatography units (e.g. size-exclusion or hydrophobic-interaction) have to be considered.

This sequence is typical for monoclonal antibody processes.

Tab. 3.2 Analysis of different production processes

	Herceptin	Rituxan	MabCampath	Synagis	Remicade	Simulect
Cell removal	1	1	1	1	1	1
Affinity chromatography	2	2	2		2	2
Virus inactivation	3	3	3	4	3	3
Cation exchange	4	5	4	3	4	5
Anion exchange	5	4		2, 6	6, 7	4
Hydrophobic interaction	6					
Size exclusion			5	8		
Virus clearance		6	6	5, 7	5	6
Sterile filtration	7	7	7	9	8	7

3.4.2
Generic Process

Comparing all these different flowsheets it becomes obvious that certain unit operations are present in each purification process.

Therefore, the production process in Fig. 3.6 is proposed as a generic process. The process can be divided into the upstream section, in which the fermentation and the media preparation are the most important tasks, and the downstream section, in which purification and buffer handling take place.

Before the production of monoclonal antibodies can start, cells must be grown step by step. The original cells are frozen in one master cell bank and some working cell banks. The cells from one of these working cell banks are defrosted and cultured in small T-Flasks. Over a period of 2 weeks cells are progressively grown in larger and larger volumes to provide a seed culture (inoculum) for the large fermentation tanks (10 000–20 000 L). This gradual step up in volume allows for the most rapid growth of a large volume of cells. When the cells are added to the main fermenter, they are grown to an optimal density. This takes between 1 and 2 weeks for fed-batch cultures depending on the tank volume. Several fermenter trains are often used in the case of large-scale production of monoclonal antibodies. Since the cells secrete product into culture fluid, cells have to be removed by centrifugation and microfiltration, whereas culture medium including product

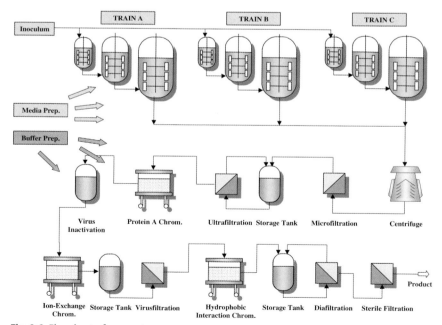

Fig. 3.6 Flowsheet of a generic process.

is concentrated by ultrafiltration and captured in a storage tank before it is further processed in the downstream section.

As described above, at least two chromatography steps are necessary. Affinity chromatography is used as a first capture and purification step. Due to its specificity, the antibody may reach a purity of about 90%. Ion-exchange chromatography is used to remove other protein impurities that are still present after Protein A chromatography. Anion exchangers can be found in nearly all processes since most of the contaminants have an electronegative character and bind very strongly to the anion exchanger, while the antibody remains in the flowthrough [26]. If these two unit operations cannot guaranty the required purity, additional chromatographic unit operations have to be used, e.g. cation-exchange chromatography can be used to remove leached Protein A [26].

Apart from chromatography, membrane filtration plays an important role in downstream processing of monoclonal antibodies. Necessary units are:
- Microfiltration for cell removal.
- Ultrafiltration for concentration.
- Diafiltration for buffer exchange.

Potential viral contaminants have to be removed to guarantee product safety. At least one storage tank is used to provide the necessary residence time for virus inactivation (solvent/detergent or pH treatment).

Based on these data, unit operations for a minimal step process with a serum-free cell culture medium can be defined. The process comprises two chromatography units, the different membrane filtration units and a tank for the virus inactivation.

3.5
Process Analysis: Optimization Potential

Current processes for the production of monoclonal antibodies have not yet reached an optimum so that there is still a large potential for optimizing these processes. This potential needs to be quantified to have a benchmark of the state-of-the-art for further examination. The efficiency of future developments can be evaluated based on these data. Currently, the cost distribution between upstream and downstream process is assumed to be 50–80% for downstream costs [27].

3.5.1
Needs

Further optimization is necessary since current production costs are in the range of Euro1000 g^{-1} product, which is significantly higher than for typical small molecules. Future developments must therefore reduce unit production cost by a factor of 100–1000. This is only possible if upstream as well as downstream processing are optimized.

Table 3.3 shows the sales prices for some typical antibody therapeutics with their corresponding dosage. These data were obtained by various online pharmacies and point out the high level of prices per unit, which should provide an indication of high unit production cost. Moreover, Tab. 3.3 gives a proof of principle for the following simulation results which are of a similar order of magnitude.

3.5.2
Upstream/Downstream

Within the upstream section the major optimization potential is to increase the product titer. For monoclonal antibody-secreting cell lines the product concentration reaches values up to 1 g L^{-1}. A theoretical limit for the concentration is expected to be around 10 g L^{-1} since mammalian cells are unable to grow in such high densities as *Escherichia coli* cells. For *E. coli*, the theoretical limit is assumed to be approximately about 40 g L^{-1}.

The product concentration can be either raised by increasing the specific production rate of the cells or by having a high cell number over a long fermentation period (high cell space–time yield).

(i) The specific production rate can be raised by manipulating the genotype (genetic constitution of an organism) and phenotype (observable characteristics of an organism produced by the organism's genotype interacting with the environment). The manipulation of the genotype should result in a more stable expression system. The selection procedure is quite important in order to obtain a cell clone with a very high specific production rate and the necessary stability. The phenotype of cells can be manipulated by changes of the environmental conditions. In particular, the fermentation conditions are very important for a high specific production rate [16].

(ii) Cell space–time yield depends on the cultivation time and cell density. Both are influenced by bioreactor design, process control, medium composition and feeding strategy. Cells can reach much higher cell densities in perfusion reactors than in conventional batch or fed-batch reactors, so that it can be assumed that future production processes will make use of this technology as far as possible.

Even if the product concentration can be raised by a factor of 10–100, unit production costs are still far away from an acceptable price. Thus, it is necessary to optimize downstream processing.

As mentioned above, it can be assumed that within the next 5 years product concentration will increase by a factor of 3 and possibly within the next 10 years by a factor of 10 [28]. This will have a great influence on purification costs since 3-fold, respectively, 10- to 100-fold amounts of product have to be purified. (The amount

Tab. 3.3 Prices per unit for different antibody products

Year on market	Company	Indication	Antibody	Dose (mg)	Price ($)	Price per unit ($/g)	Source
1993	J & J/Eli Lilly	PCTA	ReoPro™	10	583	58300	www.drugsdepot.com
1997	IDEC/Genentech	B-cell lymphoma	Rituxan™	100	518	5180	www.drugsdepot.com
1998	Abbott	respiratory syncytial virus	Synagis™	100	1.165	11650	www.savrx.com
1998	Amgen	rheumatoid arthritis	Enbrel™	25	540	21600	www.pillbot.com
1998	Genentech/Roche	breast cancer	Herceptin™	440	2640	6000	www.drugsdepot.com
1998	J & J	rheumatoid arthritis; Crohn's disease	Remicade™	100	1100	11000	www.pillbot.com
1998	Novartis	transplant rejection	Simulect™	20	1750	87500	www.drugsdepot.com
1998	Roche	transplant rejection	Zenapax™	25	510	20400	www.walgreens.com
2000	Wyeth	acute myeloid leukemia	Mylotarg™	5	2025	405000	www.rxusa.com
2001	Berlex/ILEX	chronic lymphocytic leukemia	Campath™	90	5486	60956	www.drugsdepot.com
2003	Abbott	rheumatoid arthritis	Humira™	40	715	17875	www.farmamondo.com
2003	Biogen Corp.	chronic plaque psoriasis	Amevive™	30	2750	91667	www.rxusa.com

to be purified is not so much determined by the titer of the fermentation step, but by the demand for the finished drug and its growth potential. This growth potential depends to a large degree on the production cost and by that on the sales potential as a function of market price. The higher concentration of the titer means in the first stage only that the required amount of drug can be recovered from a smaller initial volume, thus dealing with higher concentrated, lower process volumes.)

There are two priorities to optimize downstream processing:

   (i) Reduction of the number of unit operations (integration of unit operations, higher selectivity).

  (ii) Reduction of yield losses in each unit operation.

3.5.2.1 Cost Distribution between Upstream and Downstream

To optimize the process most effectively, it is important to know the cost distribution between upstream and downstream section. As mentioned above, downstream costs currently account for nearly 50% of total cost.

With increasing product concentration, the specific costs for the downstream as well as for the upstream section decline (see Fig. 3.7). Specific costs of the upstream section are inversely proportional to the increase of product concentration. This is valid if it can be assumed that the culture medium for the high-level secreting cell line is the same as for the low-level secreting cell line and thus costs are equal. Moreover, culture length is the same in all three cases so that costs for labor, utilities, etc., are identical.

In the downstream section, costs do not decline inversely proportional since no economy of scale can be achieved, so that its portion of total specific costs increases up to 90%.

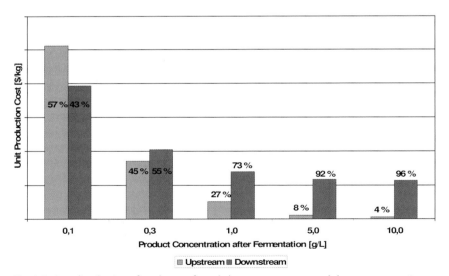

Fig. 3.7 Cost distribution of total cost of goods between upstream and downstream section.

With increasing the titer on the upstream side by a factor of 10–100 or more, the cost for downstream processing will be the major cost factor and therefore offer the largest optimization potential.

3.5.2.2 Influence of Production Rate on Specific Downstream Costs

The specific downstream costs can be more detailed if one differentiates between costs for:

- Equipment (depreciation time 10 years).
- Consumables (regularly replaced chromatography materials and membranes).
- Labor.
- Waste treatment and disposal.
- Raw materials.

The equipment-dependent costs decline nearly inversely in proportion to the amount of product. For example, the volume of the affinity column has to be increased by a factor of 10 to be able to process the 10-fold amount of product (capacity is assumed to be constant), while the equipment-dependent costs increase only by a factor 1.5. A reduction of equipment costs is necessary to effectively reduce production costs, since equipment and consumable-dependent costs account for the major portion of total downstream costs.

The unit production costs for consumables remain nearly constant, e.g. 10-fold amount of stationary phase is necessary for the 10-fold amount of product (capacity = constant).

The membrane area depends on the processed volume since filtration time is fixed in this study. For the first diafiltration and ultrafiltration in downstream process, volume is independent of product concentration, since the fermentation volume is not changed. Volume is increased by a factor of 10 after the affinity chromatography because 10-fold volume (column volumes) is used for the elution of monoclonal antibodies, so that 10-fold membrane area is required for the following filtration steps.

Since costs for consumables remain nearly constant (see Fig. 3.8) in comparison to the equipment- or labor-dependent costs, their portion of total downstream cost increases from 17 to nearly 50%, and thus the development of new and better materials (higher capacity, etc.) is important to reduce unit production costs of the total downstream section.

3.5.2.3 Influence of Process Step Yield on Downstream Costs

Figure 3.9 shows the influence of the yield per step on total downstream costs. The assumptions for this analysis are:

- Fermentation delivers approximately 2 kg per batch.
- 47 batches are processed per year.
- Every selective unit (chromatography and filtration) has the same yield.

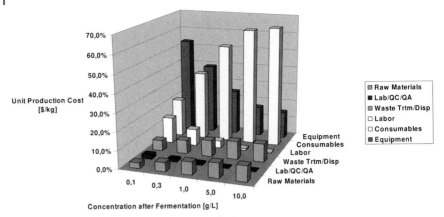

Fig. 3.8 Influence of production rate on specific downstream costs.

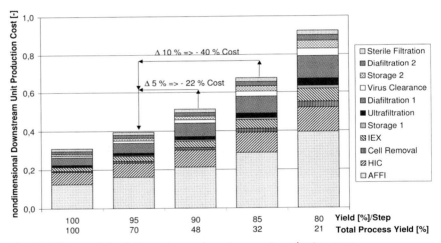

Fig. 3.9 Influence of the yield per step on downstream unit production costs.

The downstream process and the corresponding units are shown in the flowsheet in Fig. 3.6. In this analysis only the yield of the seven selective units (chromatography and filtration) is changed, while storage tanks and virus inactivation are assumed to have no product loss.

A decreasing yield is equivalent to higher product losses in every unit operation so that less product is produced per batch as well as per year. Due to these product losses, unit operations which are located close to the end of the downstream process can be designed smaller or even a parallelism of units is no longer necessary. This results in a reduction of annual downstream costs. However, downstream unit production costs increase since annual costs are allocated to less amount of product.

Current bioprocesses have total process yields of about 50%, meaning in this study an average yield per step of 90%. From Fig. 3.7 it becomes clear that the increase of the average yield per step from 90 to 95% results in a reduction of downstream unit production costs by a factor of 22%.

In cases where the total process yield is only about 30%, representing an average yield per step of 85%, a reduction of 40% seems to be possible for the unit production costs in the case that it is possible to raise the average yield per step from 85–95%, whereby reaching 95% is a reasonable aim.

3.5.3
Process Modeling Tool Concept

Figure 3.10 summarizes the requirements of any modeling tool for use in biotechnology process development and optimization. We can conclude that not one tool alone could fulfill the needed tasks properly. Figure 3.10 shows the resulting concept applied at Bayer Technology Services GmbH. Any batch design tool which is capable of generating and solving mass and energy balances, and running scheduling and cost estimation tasks has to be linked with a flowsheet simulation tool which is capable of detailed and dynamic modeling for any unit operation by any spreadsheet tool. Approaches for rigorous models have been developed in recent years (e.g. Ref. [29]), and short-cut approaches which take into account the relation of purity and yield to influence any parameter are described in detail in Ref. [30], for example.

The proper sequence of unit operations is generated by most companies by a platform of technology which combine different methods like a predefined set of unit operations with the majority of operating parameters templated to reduce the number of parameters to be determined. Through experience in combination with knowledge of physical properties and general knowledge of unit operation process performance, the final sequence of unit operations is gained with a good set of start parameters for final experimental and theoretical optimization [31, 32, 33].

	Batch Design and Scheduling Tool	Flowsheet Simulation Tool
▪ **Short-cut model**	+	+
▪ **Rigorous model**	–	+
▪ **Flowsheeting**	+	–
▪ **Scheduling**	+	–
▪ **Equipment design**	+	–
▪ **Cost calculation**	+	–
▪ **Optimization**	–	+

Fig. 3.10 Modeling tool requirements for biotechnology process development.

3.5.4
Sensitivity Study

A sensitivity analysis is performed to evaluate the influence of certain process parameters on unit production costs as well as on characteristic numbers, such as productivity, eluent demand and concentration factor. These values are extracted from a flowsheet simulation tool based on short-cut models, e.g. AspenCustom-Modeller™, gPROMS™, Super ProDesigner™ and BatchPlus™. The typical bio-process-related chromatographic unit operations (affinity, ion-exchange and hydrophobic-interaction) are considered, while ultrafiltration was examined as an example of membrane unit operations.

This documents the state-of-the-art and gives a proof of principle for the analysis methods applied here. Later, ion-exchange and affinity membranes are analyzed as well to compare similar unit operations and to select the optimal one.

The input data for these sensitivity analyses are generated by the same example flowsheet used for the analysis of the optimization potential. Again, different product concentrations are used in the fermentation (0.1, 0.3 and 1.0 g L^{-1}) to obtain the required feed amount for the subsequent unit operations.

3.5.5
Affinity Chromatography

The parameters used for affinity chromatography are shown below. It is assumed that the feed solution is concentrated 10-fold by the preceding ultrafiltration, resulting in a feed volume of 200 L, and concentrations of 1.0, 3.0 and 10 g L^{-1}.

The following parameters are changed in the sensitivity analysis:
- Resin-binding capacity (mg mL^{-1} column volume).
- Linear velocity (cm h^{-1}).
- Resin lifetime (number of batches before the resin is replaced).
- Buffer costs ($ kg^{-1}).
- Labor cost ($ h^{-1}).
- Number of column volumes for washing, regeneration and equilibration (–).

Exemplarily resin-binding capacity and resin lifetime are shown in this study.
The calculated parameters are defined as follows:
- Operation costs as total separation costs (yield losses are not accounted for!) minus equipment dependent costs (e.g. depreciation).
- Productivity as product quantity in gram per liter column volume and day.
- Concentration factor as volume of product-containing fraction after elution referring to feed volume.

- Eluent demand as cubic meters of buffer solutions,
  including those for elution, washing, regeneration and
  equilibration per processed batch.

3.5.5.1 Resin-binding Capacity

Figure 3.11 shows the influence of resin-binding capacity. A higher binding capacity results in a smaller column volume necessary to process the same amount of
product. Since in this study the bed height is fixed, only the diameter can be
reduced. This results in a decreasing volume flow and thus longer process times
since the linear velocity is kept constant. Other parameters, which are constant,
are shown in the box within each of the figures below.

A higher resin-binding capacity has a major influence on operating costs (dashed
lines) since less consumables, which account for the major portion of operating
costs, are needed. Equipment-dependent costs (dotted line) remain nearly constant
due to the small influence of column volume on equipment-dependent costs.
Resin-binding capacity significantly influences productivity as well as eluent
demand, whereas concentration factor is less affected (Fig. 3.12). The strong
increase in productivity (diamonds) is caused by a decrease of the necessary
column volume. The impact of the increasing process time is only very small,
especially for higher product amounts, and results in the different slopes of the
curves. Eluent demand (triangles) declines drastically with increasing resin-
binding capacities due to the declining column volume. Independent of the
amount of product, eluent demand is only one-third if binding capacity is increased
from 15 to 50 mg mL^{-1}. The concentration factor (squares) rises with increasing
resin-binding capacity since column volume declines and so it is assumed that the
necessary elution volume declines too.

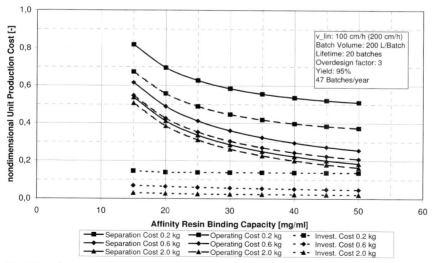

Fig. 3.11 Influence of affinity resin-binding capacity on unit production costs.

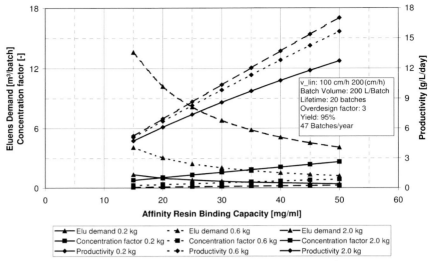

Fig. 3.12 Influence of affinity resin-binding capacity on characteristic numbers.

3.5.5.2 Resin Lifetime

The resin lifetime is the number of batches the chromatography medium can be used for before it is replaced by a new medium. The replacement frequency affects the operating costs as shown in Fig. 3.13. The longer the medium can be used, the lower the operation costs. In particular, for larger amounts of product the costs can be reduced by more than 50% so that it is desirable for future large-scale processes to have longer resin half-lives. The resin lifetime does not affect any characteristic numbers.

3.5.6
Comparison with Ion-exchange and Hydrophobic-interaction Chromatography

The results of affinity chromatography are now compared with those of ion-exchange and hydrophobic-interaction chromatography. In general, total binding capacities in ion exchange are much higher than those in affinity chromatography, while those in hydrophobic-interaction chromatography are comparable with affinity chromatography. Nevertheless, small amounts of salt ions in the feed solution can already significantly lower the resin-binding capacity in ion-exchange chromatography so that in this study a broad range of capacities is covered.

Apart from higher resin-binding capacity, ion-exchange resins are considerably cheaper than affinity resins. This results in a smaller influence of costs for consumables on separation costs. Instead, costs for labor and equipment become more dominating.

Increasing the resin-binding capacity reduces the column volume, as mentioned before. Therefore, separation costs can be lowered. On the other hand, process

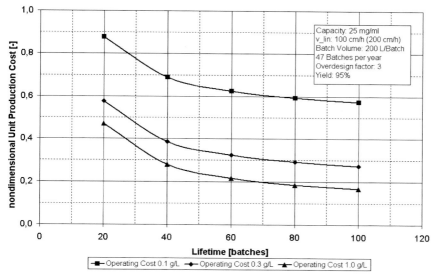

Fig. 3.13 Influence of resin lifetime on costs.

times increase due to the reasons previously described, resulting in higher costs for labor and thus an increase of separation costs. These two phenomena result in a minimum of separation costs (Fig. 3.14), which can be shifted towards higher capacities (smaller columns) by either increasing the feed amount or by reducing bed height.

Hydrophobic-interaction chromatography resins (Fig. 3.15) are normally less expensive than affinity chromatography resins, but more expensive than ion-exchange resins, so that again (compare affinity chromatography), in combination with their low binding capacity, costs for consumables make up the major portion of downstream operating costs. For low resin-binding capacities this results in a strong decrease of operating costs with increasing capacity.

3.5.7
Comparison between Affinity and Ion-exchange Chromatography as Capture Step

Figure 3.16 shows the comparison of affinity and ion-exchange resin-specific separation cost contributions for different titers. Equipment costs are not relevant for both units at higher titers, but labor efforts are. Therefore, any need to minimize investments costs in equipment should be avoided in order to invest in more robust and easy to handle equipment. The affinity-separation costs are always higher than ion-exchange costs and are dominated by the resin costs (consumables). The clear cost benefit of ion-exchange resins at this first affinity step in the process is in most cases neglected for antibody production due to the advantage of having a very clearly determined feedstock for the further following steps which

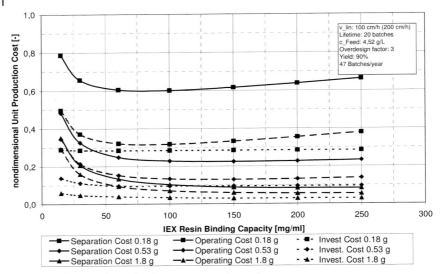

Fig. 3.14 Influence of resin-binding capacity on ion-exchange chromatography unit production costs.

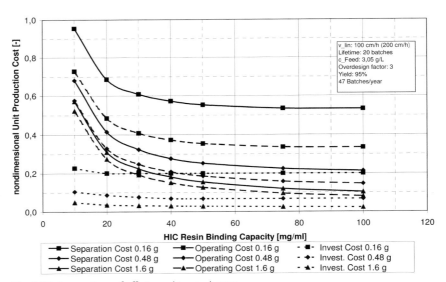

Fig. 3.15 Comparison of affinity and ion-exchange chromatography unit operation production costs.

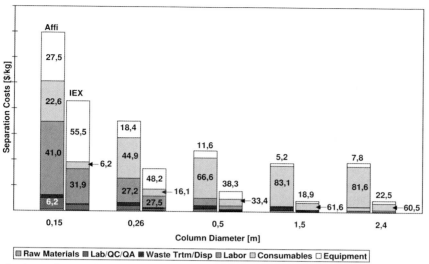

Fig. 3.16 Influence of resin-binding capacity on hydrophobic-interaction chromatography unit production costs.

could be designed according to platform technology given already known parameters which do not have not to determined from scratch.

3.5.8
Ultrafiltration

The parameters used for ultrafiltration are shown below. For the feed concentration it is assumed that the preceding diafiltration in which the cells are removed dilutes the solution, so that a volume of 3000 L, and concentrations of 0.07, 0.21 and 0.7 g L^{-1} have to be processed (fermentation concentration 0.1, 0.3 and 1.0 g L^{-1}).

The filtration process itself proceeds as follows. After concentrating the antibody-containing solution (antibodies remain in the retentate), the membrane has to be flushed with water. After this it is cleaned with a cleaning solution, e.g. sodium hydroxide solution, and flushed with water again. The membrane is sanitized with cleaning solution and flushed once more with water before it can be reused [34]. Extensive integrity tests have to be performed before and after each usage.

The following parameters are changed in the sensitivity analysis:
- Membrane area (m^{-2}).
- Filtrate flux (L m^{-2} h^{-1}).
- Concentration factor (volume retentate per volume feed solution).
- Membrane lifetime (batches).
- Clean-in-place volume (L m^{-2}).
- Sanitization volume (L m^{-2}).
- Flush volume (L m^{-2}).

The influence of these parameters on separation costs and characteristic numbers is considered similar to affinity chromatography. Characteristic numbers are defined as follows:

- Productivity as product amount in gram per square meter membrane area per day.
- Eluent demand as cubic meters of buffer solutions used for flushing, washing and sanitization of the membrane per processed batch.

3.5.8.1 Membrane Area

Figure 3.17 shows the influence of membrane area on costs. Investment costs (dotted lines) increase with increasing membrane area. Operating costs (dashed lines) increase with larger membrane area since more consumables and buffers for cleaning are needed. This results in an increasing eluent demand (Fig. 3.18) and thus increasing costs for raw materials. All these aspects result in higher separation costs.

The influence of a larger membrane area on characteristic numbers is shown in Fig. 3.18. With increasing membrane area, the eluent demand rises proportional since the washing volumes are defined in volume per square meter membrane area.

Productivity declines although process time is shortened with increasing membrane area, since it also depends on the membrane area. The smaller the membrane area, the higher the productivity. Increasing membrane area by a factor of 8 (from 10 to 80 m²) reduces productivity by factor of 2 independent of the initial

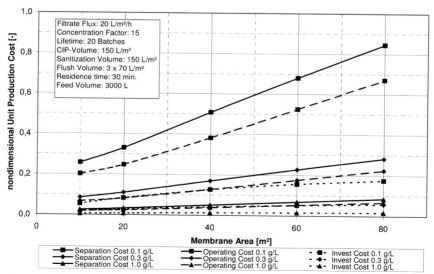

Fig. 3.17 Influence of membrane area on costs.

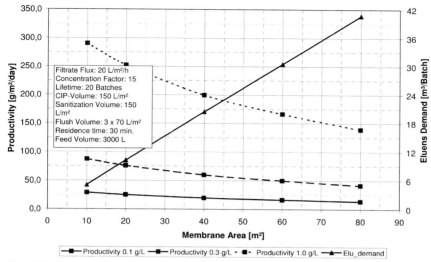

Fig. 3.18 Influence of membrane area on characteristic numbers.

concentration, since process times are the same because in every case a volume of 3000 L has to be processed. The level of productivity only depends on the processed amount of product if feed volume, respectively, process time and membrane area are fixed.

3.5.8.2 Filtrate Flux

Figure 3.19 shows the influence of a higher filtrate flux on productivity, which results in a higher filtrate volume flow. Filtrate flux depends on the applied transmembrane pressure and is limited by the maximum pressure of the membrane since membranes are very pressure sensitive. This pressure limitation is not considered in the current model.

With a higher filtrate flux rate, productivity increases due to the shorter process time. The process time gets shorter since time to process the same volume is shortened at higher flux rates.

3.5.8.3 Single- versus Multi-use Membranes

Ultrafiltration membranes are commonly used several times, making it essential to clean the membranes thoroughly after each use. This requires a huge amount of expensive raw materials such as cleaning buffers and very pure water. In addition, membrane cleaning is a very time-consuming process, since the cleaning procedure can take several hours. The development and validation of this cleaning procedure takes about three person-months and has to be repeated every few years.

From this point of view it might be interesting to utilize single-use membranes that are exchanged after each filtration process. The issue whether to use single- or multi-use membranes depends on the membrane costs.

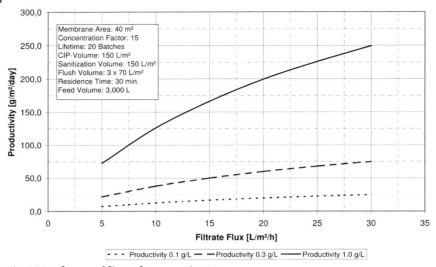

Fig. 3.19 Influence of filtrate flux on productivity.

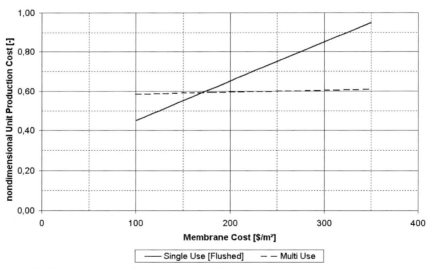

Fig. 3.20 Break-even point of single- versus multi-use membranes.

Figure 3.20 shows the break-even point of single-use membranes. In this case, membranes cost of less than $180 m^{-2} indicate that single-use membranes are the first choice. Moreover, it becomes obvious that the effect on costs for multi-use membranes is only very small since costs for raw materials (cleaning buffers) dominate operating costs. For the single-use membranes, it is assumed that they are only flushed with water before they are removed.

The decision of whether to use disposables could be made with the aid of these simulation tools.

3.6
Conclusions

In this study, a method is proposed for the development of generic processes for the production of monoclonal antibodies and their efficient optimization during process development. As a first step, literature and Internet sources have been screened for relevant process details. Fundamental steps in downstream processing of monoclonal antibodies have been identified. As a result, a generic flowsheet has been developed.

Urgently needed optimization potentials in scale and cost to solve the gap between therapeutic demands and bottlenecks of healthcare systems could be identified in downstream processing. The analysis of optimization potential showed a particularly huge potential in the expensive affinity separations with their high selectivity. Further progress in this field is required, but the proposed analysis and methods show the potential to solve these problems.

The level of operational costs due to cleaning procedures needed in the work-horse membrane separations is astonishingly high. This seems to be the main objective for further investigations and improvements in this area.

Further development of this kind of analysis and of this method will be done, as well as extensions to other monoclonal antibody examples to gain more general results.

Acknowledgments

The authors would like to thank Bayer Technology Services GmbH for financing this work. Special thanks are extended to Dr. Michael Schulte and Dr. Reinhard Ditz (Merck KGaA) for their valuable discussions and fruitful contributions. We are very grateful to Dipl.-Ing. Martina Mutter (Bayer Technology Services GmBH) and Dr. Dieter Melzner, Sartorius for their helpful contributions to membrane technology questions.

References

1 Sommerfeld, S., Strube, J., Challenges in biotechnology production – generic processes and process optimization for monoclonal antibodies, *Chem Eng Proc* **2005**, *44*, 1123–1137.

2 Walsh, G., Biopharmaceutical benchmarks, *Nat Biotechnol* **2000**, *18*, 831–833.

3 Kelly, B. D., Bioprocessing of therapeutic proteins, *Curr Opin Biotechnol* **2001**, *12*, 173–174.

4 Köhler, G., Milstein, C., Continuous cultures of fused cells secreting antibody of predefined specificity, *Nature* **1975**, *256*, 495–497.

5 Seiler, F. R., Gronski, P., Kurrle, R., Lüben, G., Harthus, H. P., Ax, W., Bosselt, K., Schwick, H. G., Monoclonal antibodies: their chemistry, functions and possible uses, *Angew Chem Int Ed* **1985**, *24*, 139–160.

6 Erpel, T., König, B., Schäfer, M. A., 2002, Therapeutische Antikörper – Der Start einer neuen Ära, *McKinsey Health* **2002**, *2*, 54–65.

7 Milroy, D., Auchincloss, C., *Monoclonal Antibodies – On the Crest of a Wave*, Wood Mackenzie, **2003**, 1–6.

8 Rader, R. A., 2003. *BIOPHARMA: Biopharmaceutical Products in the US Market*, 2nd edn., Biotechnology Information Institute, Rockville, MD, **2003**.

9 Ginsberg, P. L., Bhatia, S., McMinn, R. L., *The Road Ahead for Biologics Manufacturing*, US Bancorp Piper Jaffray, Minneapolis, **2002**.

10 Winter, G., Griffiths, A. D., Hawkins, R. E., Hoogenboom, H. R., Making antibodies by phage display technology, *Annu Rev Immunol* **1994**, *12*, 433–455.

11 Molowa, D. T., Shenouda, M. S., Meyers, A. P., *The State of Biologics Manufacturing*. JP Morgan, **2001**, 1–9.

12 Strube, J., Schulte, M., Arlt, W., Chromatographie, in *Fluid-Verfahrenstechnik*, R. Goedecke (Ed.), Wiley-VCH, Weinheim, **2006**, pp. 381–494, .

13 www.fda.gov.

14 Brown, M. E., Renner, G., Field, R. P., Hassell, T., Process development for the production of recombinant antibodies using the glutamine synthetase (GS) system, *Cytotechnology* **1992**, *9*, 231–236.

15 Peakman, T. C., Worden, J., Harris, R. H., Cooper, H., Tite, J., Page, M. J., Gewert, D. R., Bartholemew, M., Crowe, J. S., Brett, S., Comparison of expression of a humanized monoclonal antibody in mouse NSO myeloma cells and Chinese Hamster Ovary cells, *Hum Antibod Hybridomas* **1994**, *5*, 65–74.

16 Racher, A. J., Tong, J. M., Bonnerjea, J., Manufacture of therapeutic antibodies, in *Biotechnology*, 2nd edn., H. J. Rehm (Ed.), Wiley-VCH, Weinheim, **1999**, vol. *5a*, pp. 81–92.

17 Ham, R. G., Clonal growth of mammalian cells in a chemically defined, synthetic medium, *Proc Natl Acad Sci USA* **1965**, *53*, 288–293.

18 Hamilton, W. G., Ham, R. G., Clonal growth of Chinese hamster cell lines in protein-free media, *In Vitro* **1977**, *13*, 537–547.

19 Birch, J., *Lonza: Analyst Event Presentation Part 2*, **2003**, www.lonza.com.

20 Pierce Chemical Technical Library, *Antibody/Protein Purification*, **2003**, www.piercenet.com.

21 Ladisch, M. R., *Bioseparations Engineering, Principles, Practice and Economics*. Wiley-Interscience, New York, **2001**.

22 Scopes, R. K., *Protein Purification: Principles and Practice*, 3rd edn., Springer, Heidelberg, **1996**.

23 Millipore, *Technical Brief: Protein Concentration and Diafiltration by Tangential Flow Filtration – An Overview*, **2003**, www.millipore.com

24 Farshid, M., *Evaluation of Viral Clearance Studies*, **2002**, www.fda.gov.

25 Walter, J. K., Nothelfer, F., Werz, W., Validation of viral safety for pharmaceutical proteins, in *Bioseparation and Bioprocessing: A Handbook*, G. Subramanian (Ed.), Wiley-VCH, Weinheim, **1998**, vol. *1*, pp. 465–496.

26 Gagnon, P., *Purification Tools for Monoclonal Antibodies*, Validated Biosystems, Tucson, **1996**.

27 Lowe, C. R., Lowe, A. R., Gupta, G., New developments in affinity chromatography with potential application in the production of biopharmaceuticals, *J Biochem Biophys Methods* **2001**, *49*, 561–574.

28 Wurm, F., Mammalian cell culture productivity at large scale from grams/liter to tens of grams/liter, soon?, in *Proc 3rd Int Symp on Downstream Processing of Genetically Engineered Antibodies and Related Molecules*, Nice, **2004**, pp. 1–19.

29 Wiesel, A., Schmidt-Traub, H., Lenz, J., Strube, J., Modeling gradient elution, *J Chromatogr A* **2003**, *1006*, 101–120.

30 Lohrmann, M., Schutte, M., Strube, J., Generic method for systematic phase selection and method development, *J Chromatogr A* **2005**, *1092(1)*, 89–100.

31 Ambrosius, D., platform technology for MAB production presented at *IBC*

Conference on Monoclonal Antibody Production, Berlin, **2005**.

32 Slaff, G., MAB production presented at *IBC Conference on Monoclonal Antibody Production*, Berlin, **2005**.

33 Asenjo, J. A. et al., *J Mol Recognit* **2004**, *17*, 236–247.

34 Millipore, *Maintenance Procedures, Pellicon and Pellicon-2 Cassette Filters*, **2003**, www. millipore.com.

4
Dynamics of Cellular Response to Recombinant Protein Overexpression in *Escherichia coli*

Balaji Balagurunathan and Guhan Jayaraman

4.1
Introduction

Proteins have important applications in therapy, diagnostics and other industries. The production of recombinant proteins has to follow a qualitative rationale, which is dictated by regulatory requirements and, to a lesser extent, by process economics. For the production of technical enzymes or food additives, recombinant DNA technology must provide an approach that has to compete with the mass production of such compounds from traditional sources. As a consequence, production procedures have to be developed that employ highly efficient expression platforms and rely on the use of inexpensive media components in fermentation processes. For the production of therapeutics, the rationale is dominated by safety aspects and a focus on the generation of authentic products. The demand for suitable expression systems is increasing as the emerging tools for high-throughput screening lead to an increasing number of gene targets with various applications (Gellissen et al., 2005).

Several expression systems have been developed for the production of recombinant proteins. Some of them include Gram-positive and Gram-negative prokaryotes, yeasts/filamentous fungi, plant cell, insect cell and mammalian cell cultures as well as *in vitro* expression systems. However, there is no universal expression system for heterologous proteins. All expression systems have some advantages as well as some disadvantages that should be considered in selecting a platform for recombinant protein production. Some features of the various expression systems used for recombinant protein expression are compared in Tab. 4.1 (Rai and Padh, 2001; Ma et al., 2003; Baneyx and Mujacic, 2004; Grengross, 2004; Hellwig et al., 2004; Wurm, 2004; Gellissen et al., 2005).

The enteric bacterium *Escherichia coli* is one of the most extensively used prokaryotic organisms for genetic manipulations and for the industrial production of proteins of therapeutic or commercial interest. Compared with other established and emerging expression systems, *E. coli* offers several advantages, including growth on inexpensive carbon sources, rapid biomass accumulation, amenability

Bioseparation and Bioprocessing. Edited by G. Subramanian
Copyright © 2007 WILEY-VCH Verlag GmbH & Co. KGaA, Weinheim
ISBN: 978-3-527-31585-7

Tab. 4.1 Comparison of Expression Systems for Recombinant Proteins

Characteristics	Bacteria (E. coli)	Yeast	Insect cells	Plant cells	Mammalian cells
Cell growth (doubling time)	rapid (30 min)	rapid (90 min)	slow (18–24 h)	slow (24–36 h)	slow (24 h)
Complexity of growth medium	minimum	minimum	complex	complex	complex
Cost of growth medium	low	low	high	high	high
Expression level	high	low–high	low–high	low–moderate	low–moderate
Extracellular expression	secretion to periplasm/medium	secretion to medium	secretion to medium	secretion to medium	secretion to medium
Use of antibiotics	required	not required	not required	not required	not required
Process developed	industrial scale	industrial scale	pilot scale	pilot scale	industrial scale
Contaminant risks	endotoxins	low risk	low risk (serum free medium)	low risk	viruses, prions and oncogenic DNA
Protein folding	refolding usually required	refolding may be required	proper folding	proper folding	proper folding
Post-translational modifications glycosylation	none	high mannose	simple, no sialic acid	yes, terminal fucose	complex, typically human like
disulfide bond formation	yes, in the periplasmic space	possible	possible	possible	possible
phosphorylation	no	possible	possible	possible	possible

to high cell density fermentations and simple process scale-up. Due to its long history as a model system, E. coli genetics are very well characterized and many tools have been developed for chromosomal engineering and to facilitate cloning and gene expression.

Plasmids are circular, double-stranded, nonchromosomal DNA molecules that are capable of autonomous replication. Plasmids often contain genes or gene cassettes that confer a selective advantage to the bacterium harboring them, e.g. the ability to make the bacterium antibiotic resistant. Plasmids used in genetic engineering are called vectors. The schematic of a prokaryotic expression vector is shown in Fig. 4.1. The salient features of a prokaryotic expression vector are summarized in Tab. 4.2. Numerous guidelines have been formulated for designing the various features of the expression system. The guidelines are extensively reviewed and summarized elsewhere (Makrides, 1996; Baneyx, 1999; Sorenson and Mortensen, 2005a, b).

The characteristics of an optimal expression vector for use in E. coli vary with the kind of protein to be overexpressed. For example, a strong, nonleaky promoter on a high-copy number plasmid could often seem to be an attractive system for high-level expression of recombinant proteins; however, such a strong promoter in a high-copy number plasmid would adversely affect cell physiology and growth. In particular, for secretory expression systems and for toxic proteins, often a low-copy weak expression system will be preferred (Choi and Lee, 2004). However, once the plasmid and expression host has been identified, the emphasis is then on the other aspects that affect the system which includes the host responses to the overexpression of the recombinant protein. Appropriate process strategies need to be evolved at this stage, taking into account the host cell response and optimal conditions for heterologous gene expression.

A number of biological constraints are responsible for limiting the amount of foreign gene that can be produced from the organism. These biological constraints

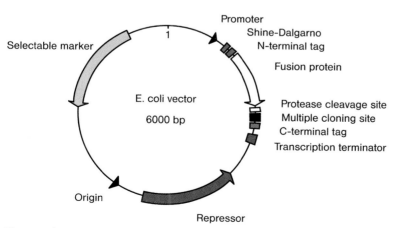

Fig. 4.1 Schematic of the prokaryotic expression vector.

Tab. 4.2 Salient features of a prokaryotic expression system

Feature	Explanation	Examples
Selectable marker	In the absence of selective pressure plasmids are lost from the host, especially in the case of very-high-copy number plasmids and when plasmid-borne genes are toxic to the host or otherwise significantly reduce its growth rate. The simplest way to address this problem is to express from the same plasmid an antibiotic-resistance marker and supplement the medium with the appropriate antibiotic to kill plasmid-free cells.	Some of the commonly used antibiotics include ampicillin, kanamycin, carbenicillin, chloramphenicol, tetracycline and rifampicin.
Promoters	The promoter initiates transcription and is positioned 10–100 nucleotides upstream of the ribosome-binding site. The ideal promoter exhibits several desirable features. (i) It is strong enough to allow product accumulation up to 50% of the total cellular protein. (ii) It has a low basal expression level (i.e. it is tightly regulated to prevent product toxicity). (iii) It is easy to induce using substrate analogues or environmental changes.	Some examples include Lac, LacUV5, tac, trp, T7, T3, T5, araBAD, TetA, λP_L and ProU.
Regulatory gene	Many promoters show leakiness in their expression, i.e. gene products are expressed at low level before the addition of the inducer. This becomes a problem when the gene product is toxic for the host. This can be prevented by the constitutive expression of a repressor protein.	For example, the leaky Lac promoters can be controlled by the insertion of a lac operator sequence downstream of the promoter and the expression of the lac repressor by host strains carry ing the lacIq allele.
Origin of replication	Sequences that allow their autonomous replication within the cell. The origin of replication controls the plasmid copy number.	For example, ColE1 300–500 copies and P15A 10–12 copies.
Transcription terminator	The transcription terminator reduces unwanted transcription and increases plasmid and mRNA stability.	The commonly used ones are the T1 and T2 derived from rrnB rRNA operon in *E. coli.*

Tab. 4.2 (Continued)

Feature	Explanation	Examples
Ribosome-binding site	The Shine–Dalgarno sequence is required for translation initiation and is complementary to the 3'-end of the 16S ribosomal RNA. The efficiency of translation initiation at the start codon depends on the actual sequence.	The consensus sequence is: 5'-TAAGGAGG-3'. It is positioned 4–14 nucleotides upstream the start codon with the optimal spacing being 8 nucleotides.
Start codon	Initiation point of translation of the mRNA.	The most commonly used start codon is AUG.
Tag and Fusion proteins	N- or C-terminal fusions of heterologous proteins to short peptides (tags) or to other proteins (fusion partners) offer several potential advantages like improved expression, solubility, detection and purification.	Some of the commonly used tags include His-Tag, avidin and streptavidin. Maltose-binding protein, green fluorescent protein and various signal peptides are commonly used as fusion proteins.
Protease cleavage sites	Protease cleavage sites are often added to enable the removal of the tag or fusion partner from the fusion protein after expression.	Most commonly used ones are factor Xa, enterokinase, TEV protease and thrombin.
Multiple cloning site	A series of unique restriction sites that would enable to clone the gene of interest into the vector.	
Stop codon	Termination of translation. The efficiency of termination is increased by using two or three stop codons in series.	There are three possible stop codon, but TAA is preferred because it is less prone to read-through.

include the precursor and energy requirements for the foreign gene production, and the competition for the protein synthesis and maintenance machinery of the cell. Understanding these biological constraints and the response of the host cell to such limitations becomes indispensable for designing strategies for high-level expression of recombinant proteins with high biological activity.

This chapter is organized into two parts. The response of the cell to increased requirement of precursor, energy and protein synthesis machinery for foreign gene expression will be discussed in Section 4.3. The cellular response to the competition for protein maintenance machinery in the cell will be discussed in Section 4.5. In general, the response of the cell varies significantly with the properties of the recombinant protein produced. However, the most general/common observations are explained here and specific cases are dealt only as examples.

Since the cellular response of *E. coli* to stress (including recombinant protein production) has been well characterized in the literature, emphasis will be placed only on recombinant *E. coli* cultures. However, parallel analysis can be drawn for different bacterial and eukaryotic systems, although specific mechanisms of stress response will vary from one organism to another. The understanding of cellular response mechanisms in *E. coli* has also allowed for development of strategies to overcome the deleterious effects of stress response. It should be possible to learn from these strategies and develop similar strategies in other organisms undergoing the stress of recombinant protein overexpression.

4.2
Global Analysis of the Cellular Response to Recombinant Protein Overexpression

The molecular events underlying the cellular response to recombinant protein overexpression have been analyzed by monitoring the transcriptional patterns of the host genes during recombinant protein overexpression (Oh and Liao, 2000; Ow et al., 2006). Even though the response was dependent on strain, media and the protein being expressed, there were some common features. A general trend of downregulated biosynthetic/energy metabolism genes, differentially expressed transport genes and upregulated heat shock proteins was observed during the overexpression of the plasmid encoded recombinant proteins. Ow et al. (2006) also observed that the expression ratios of 19 proteins identified from proteomic studies were consistent with these observations. Sections 4.3 and 4.5 will focus on some aspects of metabolic and physiological consequences due to recombinant protein overexpression.

4.3
Metabolic Consequences of Recombinant Protein Overexpression

Metabolism in general relates to the catabolic and anabolic reactions which are responsible for generating the energy and the precursors required by the cell. The microorganism consumes the carbon and the other nutrients in the medium through the catabolic reactions to generate energy required for the synthesis of the building blocks of the cell, i.e. proteins, nucleic acids and structural components, by the process of anabolism. The net effect of these two reactions is reflected as the growth of the organism.

4.3.1
Inhibition of Growth During Recombinant Protein Overexpression

The most general observation in recombinant fermentations is the decline in the specific growth upon induction of the recombinant protein. This is mainly due to the impact of plasmid presence and expression of plasmid-encoded recombinant genes on the host cell. The difference in the growth rate of plasmid-bearing and plasmid-free cells is attributed to the "metabolic burden", defined as the amount of resources (raw material and energy) withdrawn from the host's metabolism for maintenance and expression of the foreign DNA (Bentley et al., 1990; Glick, 1995). In general, the specific growth rate of the plasmid-bearing cells correlates inversely with the recombinant protein synthesis rate and accumulation level of the recombinant proteins, leading to final cessation of cell growth when the recombinant proteins account for approximately 30% of the total cell protein (Dong et al., 1995). For two model proteins analyzed by Dong et al. (1995), the growth rate reduced from an initial value of about 0.84–0.97 h^{-1} in the uninduced phase to 0.14–0.21 h^{-1} in the post-induction phase.

Inhibition of cell growth by recombinant protein production is also evident through lowered biomass, impaired cell division and increased carbon dioxide yields after induction (Schmidt et al., 1999; Hoffmann and Rinas, 2000). The cell number remains nearly constant when E. coli produce a recombinant protein under the control of the very strong T7 promoter. Slow growth after induction and an increase in biomass concentration is not caused by cell proliferation, but by an increasing cell size, as was shown using flow cytometry (Borth et al., 1998; Soriano et al., 1999). In contrast, uninduced cells reveal a decreasing cell size with decreasing growth rate. An unusual elongated shape of cells from producing cultures is often also apparent in electron or light micrographs (Carrio and Villaverde, 2001).

The decrease in the specific growth rate has long been attributed to an increase in the maintenance coefficient in traditional biochemical engineering approaches. However, this decrease could be due to the modification to the catabolism, anabolism or the protein-producing machinery in the cell, as explained in the following sections.

4.3.2
Alteration of the Energy Generation Systems

Protein synthesis is the most energy consuming process in the cell. Under normal growth conditions, 11 523 μmol of NADPH is needed for protein biosynthesis to generate 1 g of E. coli cells. This amount accounts for almost 63% of the total NADPH required for all the biosynthetic reactions in the cell. The ATP consumption for the biosynthesis of 1 g of protein is estimated to be 13.25 mmol and this value is nearly equal to 70% of the energy demands for total biomass formation (18.48 mmol ATP g^{-1} biomass) in normal growing conditions (Flores et al., 2004). Therefore, energy generation may become critical in recombinant protein-overproducing cells. Theoretical considerations show that within the usually achieved range of product concentrations, only minor effects on growth rate and biomass yield are to be expected through the energy and precursor drain towards recombinant protein production. The recombinant protein synthesis rates, however, can exceed the accumulation rates by far if simultaneous degradation of the recombinant protein occurs. Consequently, the high energy demand for recombinant protein production can urge enhanced respiration for ATP regeneration at the cost of biomass formation even with negligible or low accumulation of the product (Schmidt et al., 1999; Hofmann and Rinas, 2001; Weber et al., 2002). In addition, enhanced maintenance requirements have been reported during recombinant protein production, which have been attributed to the higher energy demand for nongrowth-related processes (Bhattacharya and Dubey, 1995). Increased respiratory activities, in response to recombinant protein synthesis, have been documented for various expression systems (Luli and Strohl, 1990; Lin and Neubauer, 2000). Also, recombinant protein production can trigger stress responses resulting in the synthesis of stress proteins at high rates. Some of the stress proteins include ATP-dependent chaperones and proteases, which are required to fold and/or degrade misfolded recombinant proteins. This increase of the synthesis rates of plasmid-encoded and stress proteins is closely correlated with the increase in respiration rates and maintenance energy requirements, indicating that the elevated energy demand for the synthesis and activity of these proteins are the major components of the metabolic burden.

The synthesis of many enzymes of the central catabolic pathways is constitutive, proceeding at a nearly constant rate under a variety of conditions (Fraenkel, 1996). It has been observed that the changes in the catabolic enzyme synthesis rates are much less pronounced in response to recombinant protein production than the changes in stress protein synthesis (Hoffmann and Rinas, 2001; Hoffmann et al., 2002). Accordingly, the mRNA levels of most of the tricarboxylic acid (TCA) cycle enzymes show no significant changes after induction of recombinant protein production (Oh and Liao, 2000). Thus, it could be concluded that the observed changes in the respiration rate during recombinant protein production are due to the changes in the level of enzyme activity and not due to the changes in the level of enzyme synthesis.

When cells experience energy-limiting conditions during recombinant protein synthesis, they may activate alternative pathways for energy generation through substrate-level phosphorylation. This less-efficient way of energy generation is often used under carbon overflow conditions, frequently leading to the accumulation of cell-toxic levels of acetate (Luli and Strohl, 1990; Seeger et al., 1995; Sandén et al., 2003). In addition to acetate, pyruvate excretion has been reported when recombinant proteins are overproduced in rich medium or under conditions of excess carbon (George et al., 1992). Higher carbon flux towards the catabolic pathways has also been observed during temperature-induced recombinant protein overexpression (Hoffmann et al., 2002).

4.3.3
Alteration of the Biosynthetic Machinery (Fig. 4.2)

High fluxes into the energy-generating respiratory pathway during recombinant protein production are coupled to a reduced supply of precursors for biomass

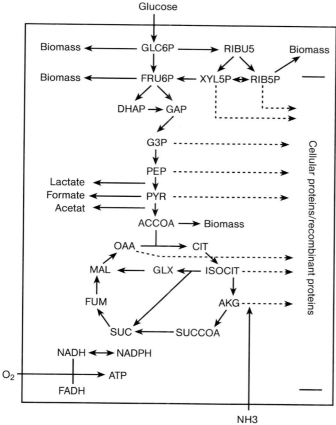

Fig. 4.2 Schematic of the metabolism in *E. coli* is shown. Only the significant members of the pathway are depicted. The dotted lines indicate multistep processes leading finally to cellular protein synthesis. Key: ACCOA, acetyl coenzyme A; AKG, α-ketoglutarate; CIT, citrate; DHAP, dihydroxyacetonephosphate; DTT, dithiothreitol; FRUC6P, fructose-6-phosphate; FUM, fumarate; G3P, 3-phosphoglycerate; GAP, glyceraldehyde-3-phosphate; GFP, green fluorescent protein; GLC6P, glucose-6-phosphate; GLX, glyoxylate; ISOCIT, isocitrate; MAL, malate; MBP, maltose binding protein; OAA, oxaloacetate; PEP, phosphoenolpyruvate; PYR, pyruvate; RIB5P, ribose-5-phosphate; RIBU5P, ribulose-5-phosphate; SUC, succinate; SUCCOA, succinyl CoA; XYL5P, xylulose-5-phosphate. Modified from Hoffmann et al. (2002).

formation. Flux estimations during temperature-induced recombinant protein production revealed that even the fluxes for amino acid biosynthesis are reduced, despite elevated protein synthesis rates, including the additional formation of the recombinant product (Weber et al., 2002). Moreover, the oxidative branch of the pentose phosphate pathway (PPP), which supplies reducing equivalents for biosynthetic purposes and through which one-third of the glucose is processed during unperturbed growth, is completely shut down due to lower NADPH demands and surplus NADPH generation in the TCA cycle (Weber et al., 2002). The nonoxidative branch of the PPP operates in a "backwards" direction for pentose supply. This way all glucose is channeled through the Embden–Meyerhof–Parnas (EMP) pathway towards the TCA cycle. Thus, the "aim" of *E. coli* to maximize growth is severely hampered by the high energy needs during recombinant protein production (Hoffmann and Rinas, 2004).

In wild-type *E. coli* (nonrecombinant strains), the components of the protein producing systems, including the ribosomal proteins and other proteins, are synthesized in balance with the cellular needs, i.e. growth rate-dependent synthesis. During the strong induction of a recombinant protein, even though the protein synthesis rates are very high, the synthesis rate of these protein producing systems are often at a reduced level. The reduction in the synthesis rates of the protein-producing system occurs at different product concentrations, ensuring again that the accumulation level of the recombinant protein is not the major determinant (Dong et al., 1995; Rinas, 1996; Jürgen et al., 2000). However, when the synthesis rate of the recombinant protein is manipulated, e.g. by altering the efficiency of the ribosome binding site (Vind et al., 1993) (without significant effect on the final accumulation level) or by the inducer concentration (Hardsock et al., 1995), higher synthesis rates result in stronger inhibition of the synthesis of components of the protein-producing system and other housekeeping proteins. It has been concluded that the concentration of the free ribosomal subunits decreases after induction, leading to an increased competition among the individual ribosome-binding sites for the ribosomes, thereby reducing the cellular capacity for synthesizing components of the housekeeping machinery. This inhibition can occur transiently directly after induction or can prolong throughout the entire induction period (Hoffmann and Rinas, 2004).

4.4
Strategies to Overcome the Metabolic Consequences of Recombinant Protein Overexpression

It is clear that recombinant protein overexpression affects the host's metabolism in several ways. In general, more resources are diverted towards additional energy generation at the cost of biosynthesis. It is clear that the dynamics of recombinant protein production, i.e. the synthesis rates, and the response of the host cell to the dynamics needs to be understood to tackle such problems. Several processes and biological strategies have been employed by researchers to tackle the situation (Kim and Cha, 2003; Flores et al., 2004). The strategies include the redirection of metabolic flux, knockout/antisense-based downregulation of a metabolic pathway for substrate level phosphorylation and fed-batch strategies to avoid accumulation of toxic metabolites like acetate.

4.4.1
Redirection of Metabolic Flux

Flores et al. (2004) analyzed the effects of increasing the expression of *zwf*, the gene that codes for glucose-6-phosphate dehydrogenase, the enzyme that converts glucose-6-phosphate into 6-phoshogluconolactone and commits carbon flux through the PPP. By redirecting carbon flux through the oxidative branch of the PPP it might be possible to fulfill some of the extra requirements imposed by multicopy plasmid DNA replication and recombinant protein production. By expressing the plasmid encoded *zwf* gene, a growth rate recovery from 0.46 to 0.64 h^{-1} was demonstrated in a wild-type *E. coli* strain. The effect of overexpression of this gene was also studied during the production of recombinant insulin peptide in *E. coli*. In this system, production of pro-insulin strongly reduces growth rates; the coexpression of the *zwf* gene helped to recover from a specific growth rate of 0.1 h^{-1} in the pro-insulin-induced strain to 0.37 h^{-1}. However, there was no significant increase in the recombinant protein expression levels with increased coexpression of the *zwf* gene product.

4.4.2
Antisense Downregulation of Acetate Production

In recombinant *E. coli*, accumulation of acetate, a major metabolite of glucose metabolism, can cause inhibition of growth rate and recombinant protein production, and therefore has been recognized as a serious problem. Some of attempts to tackle this problem include redirection of the acetate excretion pathway, blocking of the pathway by knockout mutations and the controlling carbon sources to reduce the burden on the glycolysis pathway. However, these approaches ignore the fact that the acetate pathway plays an important physiological role in *E. coli* (Kim et al., 2003). It uses the acetate production pathway as a source of ATP formation under anaerobic and even aerobic conditions. Thus, a simple knockout mutation or feed-controlling strategy may not be a good strategy.

Antisense RNA technology has been employed by Kim and Cha (2003) as an attempt to reduce the negative effect of acetate excretion, but still maintaining the useful functions of the pathway. They have observed that exquisite regulation using antisense mRNA could be a better alternative than knockout mutation methodology. The latter is generally not suitable for metabolic control because of fatal effects on the hosts due to the complicated properties of the target pathways. The timed downregulation of the acetate pathway was achieved by employing an antisense for the *ackA* promoter (*ackA* codes for acetate kinase which converts acetyl-CoA to acetate). Although the antisense strategy led to only a marginal (20–30%) reduction in secreted acetate levels, foreign protein production was enhanced by 1.5- to 2-fold. This enhancement of production yield by antisense downregulation of the acetate pathway increased with the culture scale, suggesting that this strategy may be successfully applied to practical large-scale high-cell-density fermentations of recombinant *E. coli*.

4.5
Consequences of Recombinant Protein Overexpression on the Protein Maintenance Machinery in the Cell

As mentioned earlier, the formation of the final three-dimensional structure with high biological activity is indispensable for recombinant proteins with therapeutic applications. However, a large number of eukaryotic proteins fail to attain their proper conformation in the bacterial expression systems, due to limitations of post-translational processing. These proteins normally deposit as insoluble material known as inclusion bodies. Even though production of recombinant proteins as inclusion bodies offers several advantages (Fahnert et al., 2004), they are normally not recommended for therapeutic proteins. However, understanding the cellular protein-folding processes and the modulators of the cellular protein folding could enable one to design strategies for the soluble expression of recombinant proteins.

4.5.1
Cellular Protein Folding

Proteins contain within their complete amino acid sequence all of the information necessary for attaining their functional three-dimensional structure. However, all newly synthesized proteins face challenges in reaching their native state within the crowded environment of the cell. In general, small single-domain host proteins efficiently reach there native conformation due to their fast folding kinetics, whereas large multidomain proteins and overexpressed recombinant proteins often require the assistance of folding modulators. Folding helpers include molecular chaperones, which favor on-pathway folding by shielding the interactive surfaces from each other and from the solvent, and folding catalysts that accelerate the rate-limiting steps, such as the isomerization of the peptidyl–prolyl bonds from an abnormal *cis* to a *trans* conformation, and the formation and reshuffling of disulfide bonds.

Molecular chaperones are ubiquitous and highly conserved proteins that help other polypeptides to reach their native conformation without becoming part of the final structure. They are not true folding catalysts, since they do not accelerate folding rates. Instead, they prevent off-pathway aggregation reactions by transiently binding hydrophobic domains in partially folded or unfolded polypeptides collectively designated as nonnative proteins.

Proteases are the other important members of the protein maintenance machinery in the cell. There are two main physiological functions of proteolytic degradation in *E. coli* – the inactivation of the short-lived regulatory proteins and the degradation of the misfolded proteins. The degradation of misfolded proteins by the host proteases ensures that abnormal polypeptides do not accumulate within the cell and allow amino acid recycling. Targets of degradation include prematurely terminated polypeptides, proteolytically vulnerable folding intermediates that are kinetically trapped off-pathway and partially folded proteins that have failed to reach a native conformation after multiple cycles of interactions with folding modulators.

4.5.2
Cytoplasmic Folding Modulators

Molecular chaperones, especially the cytoplasmic folding modulators, can be divided into three functional subclasses based on their mechanism of action. "Folding" chaperones (e.g. DnaK and GroEL) rely on ATP-driven conformational changes to mediate net refolding/unfolding of their substrates. "Holding" chaperones (e.g. IbpB) maintain partially folded proteins on their surface to await availability of folding chaperones upon stress abatement. Finally, the "disaggregating" chaperone ClpB promotes the solubilization of proteins that have become aggregated as a result of stress. The various chaperones with their classification and substrate specificities are summarized in Tab. 4.3.

In the cytoplasm, proteolytic degradation is initiated by five ATP-dependent heat shock proteases (Lon, ClpYQ/HslVU, ClpAP, ClpXP and FtsH) and completed by peptidases that hydrolyze sequences 2–5 residues in length (Gottesman, 1996, 1999, 2003).

4.5.3
Protein Export

Recombinant proteins can be secreted to the periplasmic space using an appropriate signal peptide. Periplasmic expression has a number of advantages compared to the cytoplasmic production. (i) An authentic N-terminus can be obtained after removal of the signal peptide by leader peptidases. (ii) The periplasm is conducive to disulfide bond formation. (iii) There are few proteases in the periplasm compared to the cytoplasm and many have specific substrates. (iv) As the periplasm contains fewer proteins, purification of the target protein has been facilitated. In E. coli, the vast majorities of proteins designed for export are secreted by the Sec-dependent pathway and are synthesized with an N-terminal signal sequence, 20–30 amino acids in length that consists of a hydrophobic core followed by a proteolytic cleavage site (Xu et al., 2000). Efficient export of the resulting preprotein requires targeting to the membrane-associated translocation apparatus in an extended conformation. A subset of proteins is exported via the signal recognition particle (SRP)-dependent pathway (Harms et al., 2001). Because of its ability to bind nascent proteins, trigger factor (TF) also plays a major role in both Sec- and SRP-dependent protein secretion. The twin arginine-dependent secretion pathway (Delisa et al., 2002; Bruser and Sanders, 2003)

Tab. 4.3 Cytoplasmic and periplasmic folding modulators

Classification	Name	Substrates/binding sites
Cytoplasmic folding modulators		
disaggregase	Hsp100 (ClpB)	segments enriched in aromatic and basic residues
folding/secretory chaperone	Hsp90 (HtpG)	unknown
folding chaperone	Hsp70 (DnaK) cofactors: DnaJ and GrpE	segments of 4–5 hydrophobic amino acids, enriches in leucine and flanked by basic residues
folding chaperone	Hsp60 (GroEL) cofactor: GroES	α/β folds enriched in hydrophobic and basic residues
holding chaperone	small heat shock proteins IbpA/B	unknown
PPIases	TF	8-amino-acid motif enriched in aromatic and basic residues
secretory chaperone	SecB	9-amino-acid residue enriched in aromatic and basic residues
Periplasmic folding modulators		
generic chaperones	Skp (OmpH)	outer membrane proteins and misfolded periplasmic proteins
	FkpA	broad substrate range
specialized chaperones	SurA	outer membrane proteins
	LolA	outer membrane lipoproteins
PPIases	SurA	outer membrane β-barrel proteins
	FkpA	broad substrate range
	PpiA, PpiD	outer membrane β-barrel proteins
proteins involved in disulfide bond formation	DsbA	reduced cell envelope proteins
	DsbB	reduced DsbA
	DsbC	proteins with nonnative disulfides
	DsbG	proteins with nonnative disulfides

Modified from Baneyx and Mujacic (2004).

exclusively deals with folded or partially folded proteins, where as both Sec- and SRP-dependent pathways handle preprotein that have not yet reached a native conformation.

4.5.4
Periplasmic Folding Modulators

The periplasm contains a single *bona fide* chaperone termed Skp that captures unfolded proteins as they emerge from the Sec translocation apparatus, and whose primary function is to assist the folding and membrane insertion of outer membrane proteins (Chen and Henning, 1996; Schafer et al., 1999). Other periplasmic folding modulators include the peptidyl–prolyl isomerases (PPIases) SurA, FkpA, PPiA and PPiD. Among these, FkpA has the most generic folding activity, and it combines PPIase and chaperone activity (Missiakas et al., 1996; Arie et al., 2001). SurA supports maturation of trimeric outer membrane proteins (Behrens et al., 2001). One of the features that distinguish the periplasm from the cytoplasm is its oxidizing environment. Indeed, in *E. coli*, stably disulfide-bonded proteins are only formed in the cell envelope where disulfide formation and isomerization is catalyzed by a set of thiol-disulfide oxidoreductases known as Dsb proteins (Hiniker and Bardwell, 2003; Kadokura et al., 2003). The various folding modulators in the periplasm and their substrate specificities are summarized in Tab. 4.3.

4.5.5
Cellular Stress Response

E. coli is subjected to a variety of stress conditions during its growth. The bacterium is able to survive and adjust to such stressful conditions by partitioning of the transcriptional space within the cell. A complex network of global regulatory systems with a multitude of molecular components ensures a coordinated and effective means to survive/adapt to stressful conditions. Such regulatory components include DNA, mRNAs, sRNAs, proteins such as DNA- and RNA-binding proteins, alternative σ factors and two-component systems, as well as small-molecular-weight molecules (e.g. ppGpp). These regulatory systems govern the expression of a plethora of further effectors that aim at maintaining stability of the cellular equilibrium under the various conditions.

There are six important stress response networks observed in *E. coli*. The heat shock response (controlled by the σ factor σ^{32}) (Arsene et al., 2000) and the envelope stress response (controlled by the σ factor σ^E and the Cpx two-component system) (Raivio and Silhavy, 1997; Alba and Gross, 2004) both result in an increased expression of chaperones and proteases in response to misfolded proteins. The cold shock response governs expression of RNA chaperones and ribosomal factors, ensuring accurate translation at low temperatures (Yamanaka, 1999). The general stress response depends on the σ factor σ^S, which controls the expression of more than 50 genes conferring resistance to many different stresses (Schweder et al., 2002). The (p) ppGpp-dependent stringent response reduces the cellular protein synthesis capacity and controls further global responses upon nutritional downshift (Cashel et al., 1996). The recA–lexA-mediated SOS response, which is activated in response to DNA damage, ensures the higher synthesis of the DNA repair enzymes (Houten, 1990; Sancar, 1996). There are other stress response networks such as oxidative stress, salinity stress and stress due to heavy metal accumulation (Farr and Kogoma, 1991; Imlay, 2003; Hantke, 2001). Among these various well-studied stress response networks, the heat-shock-like response and the stringent response are most commonly observed during recombinant protein overexpression. There are some reports regarding other stress response networks like the SOS response being activated during recombinant protein overexpression (Lee et al., 2002; Węgrzyn and Węgrzyn, 2002); however, not many details are known about the triggering events and mechanisms.

4.5.6
Stringent Response

The stringent response is induced when cells experience a sudden lack of aminoacylated tRNA – a situation which results in an immediate arrest of the synthesis of the stable RNA (Cashel et al., 1996). The stringent response was initially observed when the cells are starved of amino acids. It has also been observed to be associated with limitations in other nutrients, e.g. carbon, nitrogen and phosphorus, and a variety of other stresses. The stringent response is mediated by the accumulation of the alarmone (p) ppGpp, which is produced by the ribosome-bound RelA upon binding of uncharged tRNAs on ribosomes (Cashel et al., 1996). The response involves an excessive reprogramming of the gene expression pattern with the downregulation of a majority of genes encoding proteins of the transcription and translation machinery as well as half of those genes involved in amino acid biosynthesis (Cashel et al., 1996; Chang et al., 2002).

A shortage of amino acids can also occur during recombinant protein production if the composition of the product deviates considerably from the average *E. coli* protein. It has been hypothesized that increased protease activities following the induction of a protein with an unusual amino acid composition might be linked to a temporary depletion of the amino acid pool and the induction of a stringent-like response (Harcum and Bentley, 1993). Simulation results corroborate that the intracellular pool of certain amino acids might become depleted during recombinant protein overexpression (Harcum, 2002). Stress resulting from amino acid shortages can also be deleterious to the integrity of the recombinant protein by causing mis-incorporation of amino acids. Addition of the appropriate amino acid(s) at the time of induction can alleviate this stress and also reduce the degradation of the recombinant product (Ramirez and Bentley, 1993, 1999). This problem could also be overcome by growing the cells on complex media. In general, the complex media components were also found to enhance the stability of the recombinant protein.

4.5.7
Unfolded Protein Response

The heat shock response was first observed as a response to elevated temperature. However, this physiological response is observed not only at elevated temperatures, but is also due to several other adverse conditions such as exposure to certain chemicals (solvents, certain antibiotics), hyper-osmotic shock and as a response to overproduction of foreign proteins (Jenkins et al., 1988, 1991; Nystrom, 1994). Such challenges frequently lead to misfolded, unfolded or damaged proteins in the cell. These proteins may expose their surface hydrophobic patches which are normally buried inside the folded protein. These exposed hydrophobic residues can result in aggregation of proteins – an event that constitutes a serious threat to the organization and functioning of all components in the cell. To prevent this, heat shock response is triggered. Heat shock response is characterized by the increased expression of the chaperones and the ATP-dependent proteases. These proteins are involved in protein folding and degradation under both stressed and unstressed conditions. A general understanding of cellular protein folding and the folding modulators in the cell is necessary to study the effects of the perturbations caused by the overexpression of recombinant proteins on the normal functioning of the cell.

In *E. coli* two major regulons controlled by alternative σ factors σ^{32} and σ^E govern the heat shock response. These regulons respond to the protein misfolding in the cytoplasm and the extracytoplasmic space, respectively.

4.5.8
Cytoplasmic Response (Fig. 4.3)

The level of σ^{32} plays a vital role in the cytoplasmic heat shock response of the cell (Arsene et al., 2000). The σ^{32}-mediated stress response depends on the state of σ^{32} within the cell. σ^{32} exists either as a freely suspended protein, as a complex with RNA polymerase or as a complex with heat shock proteins DnaK, DnaJ and GrpE. Under nonstress and stressful conditions, the majority of the σ^{32} was found to be complexed with heat shock proteins and only a very small amount was free or in the form of holoenzyme. The σ^{32} complexed with the heat shock proteins (DnaK chaperone team) provides a potential reservoir of σ^{32} that can be readily tapped for gene expression without an increase in σ^{32} synthesis (Srivastava et al., 2001). In this way, the DnaK chaperone team can act in a buffering capacity so that the concentration of σ^{32} available for transcription of stress-related genes is not dependent only on the synthesis and degradation pathways. The increase in the σ^{32} level during temperature increase is due to the increased synthesis and stabilization of the normally unstable σ^{32}.

There are two mechanisms by which σ^{32} levels are increased when the temperature is raised. First, the translation of *rpoH* (the gene that codes for σ^{32}) mRNA increases immediately due to the change in the secondary structure of the mRNA (Morita et al., 1999), resulting in a fast 10-fold increase in the concentration of σ^{32}.

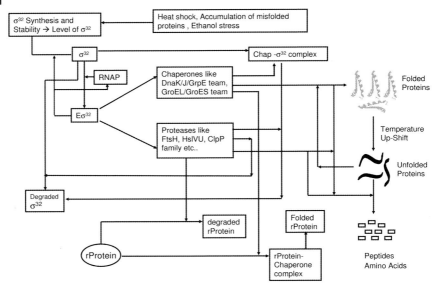

Fig. 4.3 Schematic of the cytoplasmic heat shock response.
The interaction of heat shock chaperones and proteases with
the recombinant protein is also shown.

Second, the σ^{32}, primarily in a sequestered, inactive form, complexed with the chaperones DnaK, DnaJ and GrpE becomes immediately available to mediate the stress response when subjected to a temperature up shift (Arsene et al., 2000).

Raising the temperature results in an increase in the cellular levels of unfolded proteins which titrate DnaK away from σ^{32}, resulting in more σ^{32} that is capable of binding to RNA polymerase and initiating the transcription of the heat shock genes. These proteins when folded improperly compete for the chaperones, thus titrating DnaK away from σ^{32} (Srivastava et al., 2000, 2001; Kanemori et al., 1994). The accumulation of high levels of heat shock proteins leads to the downregulation of the response. The DnaK also acts as a negative regulator of the heat shock response by presenting the sequestered σ^{32} to proteases such as FtsH.

Induction of recombinant protein synthesis leads to a rapid accumulation of the heterologous product. Due to high synthesis rates and the heterologous nature of the recombinant protein, it is normally recognized as a misfolded/abnormal protein. These misfolded/abnormal proteins compete for the chaperones and proteases which are otherwise required to regulate the level and activity of σ^{32}. The increase in abnormal proteins results in titration of chaperones and proteases away from σ^{32}, resulting in an increased level of free σ^{32}, which can bind to RNA polymerase. Thus, the activity of σ^{32} (characterized by the RNAP– σ^{32} complex) is elevated during the recombinant protein production (Kanemori et al., 1994).

Hoffmann and Rinas (2000) determined the level of DnaK during temperature-induced recombinant protein production and compared the levels with that of control cultivation. It was observed that the DnaK levels reached 2- to 3-fold higher

values during the temperature-induced recombinant protein production as compared to control cultivations (containing an empty expression vector). It was also
observed that there was a continuous heat shock response observed during temperature-induced recombinant protein expression, probably due to the stabilization of the otherwise unstable σ^{32}.

4.5.9
Extracytoplasmic Response

Regulation of the extracytoplasmic stress response (also called the envelope stress
response) differs from that of the cytoplasmic stress response in that the transcription factors directing the envelope response (σ^E) reside in the cytoplasm, whereas
the stress signals are generated in the envelope. Alternative   factor, σ^E, controls
genes that influence nearly every aspect of the cell envelope and is essential for
viability of the bacterium. Four key players in this signal transduction pathway
have been identified (RseA, RseB, DegS and YaeL) (Collinet et al., 2000;
Dartigalongue et al., 2001). RseA and RseB are the two negative regulators of σ^E
activity, and are encoded within the σ^E operon. They are intimately involved in
signal transduction through the inner membrane. RseA is a membrane-spanning
protein with a cytoplasmic N-terminal domain that functions as an anti- σ factor
of σ^E. RseB, a minor negative regulator of σ^E, is a soluble periplasmic protein that
binds to the periplasmic domain of RseA. Stress-induced activation of the σ^E
pathway requires the regulated proteolysis of RseA. DegS and YaeL are identified
as the principle proteases involved in the degradation of RseA. The other components of this stress response network include the periplasmic chaperones like
SurA, FkpA and SkpA, and proteases like DegP. The mechanism of activation of
the σ^E-mediated stress response is explained by the schematic in Fig. 4.4.

The unfolded protein response in the cytoplasm and periplasm are inter-related.
Under extreme conditions (like a shift in culture temperature to 43°C) the *rpoH*
gene, the gene which codes for σ^{32}, is actively transcribed by σ^E-bound RNA polymerase (Erickson and Gross, 1989; Wang and Kaguni, 1989). *RpoH* gene has four
promoters and the promoter recognized by σ^E is the only promoter responsible
for the transcription at extreme conditions (Erickson et al., 1987).

4.6
Strategies to Overcome Cellular Stress Response Induced due to
Recombinant Protein Overexpression

4.6.1
Coexpression of Folding Modulators

Coexpression of folding modulators has been widely used to enhance the yield of
soluble proteins in *E. coli* cytoplasm (various folding modulators are summarized
in Tab. 4.3). The beneficial effects of the increase in the intracellular concentration

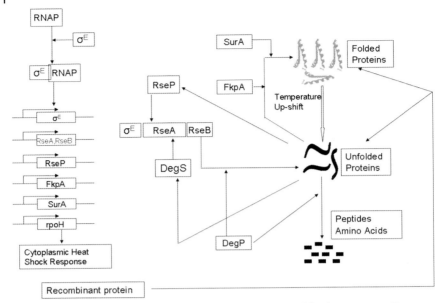

Fig. 4.4 Schematic of the extracytoplasmic stress response network in E. coli. Unfolded proteins are normally generated due to an increase in temperature. This causes a stress to the cell. These unfolded proteins activate the protease DegS by interacting with its PDZ domain. This activation in turn leads to series of proteolytic steps which finally liberates σ^E. The stress due to the accumulation of the unfolded proteins is also sensed by the interaction of the unfolded proteins with RseB, which renders RseA, the anti-σ factor, susceptible to proteolytic cleavage, finally resulting in the liberation of σ^E. The liberation of σ^E leads to the synthesis of a set of proteins required for the protein quality control in the periplasm. These proteins include the periplasmic folding modulators, i.e. the chaperones, PPIases and proteases. Based on their synthesis rates, structural properties and rate of export, secreted recombinant proteins could also be misfolded in the periplasmic space. These may activate one of the mechanisms that induce the extracytoplasmic stress response network, either through DegS or RseB. The induction of this response will finally determine the fate of the recombinant protein, especially if the protein is susceptible to periplasmic proteases.

of TF, DnaK–DnaJ and GroEL–GroES are well documented, and a number of plasmids compatible with the commonly used expression vectors are available (Baneyx and Palumbo, 2003). DnaK–DnaJ or TF overexpression is suitable to increase the solubility of proteins requiring the assistance of chaperones in the early stages of their folding pathway (Nishihara et al., 2000). For folding intermediates that rapidly transit through DnaK/TF or require help at the later folding stages, GroEL–GroES coexpression may be most beneficial. Technically, the GroEL–GroES encapsulation system should limit the usefulness of the system to proteins smaller than 60 kDa. Nevertheless, larger proteins may also benefit from GroEL–GroES coexpression, presumably because GroES-independent stabilization of partially folded domains by GroEL facilitates correct folding of the remainder of the chain (Ayling and Baneyx, 1996). If aggregation-prone intermediates are formed at both early and late stages of the folding pathway, coordinated coexpression of DnaK–DnaJ and GroEL–GroES may be required to maximize recovery of the target protein in a soluble form (Nishihara et al., 1998, 2000).

However, there are several studies in which coexpression of folding modulators fail to improve recombinant protein stability. The underlying mechanisms are unclear, but may be related to the need for timely interactions with specific folding modulators or to the substrate folding pathway itself. The way in which translation and folding are coupled varies significantly between the bacterial and eukaryotic systems (Netzer and Hartl, 1997). The functioning of the chaperone systems in eukaryotic systems is also significantly different from bacterial systems. Agashe et al. (2004) analyzed the effect of TF and DnaK system on the folding of the nascent polypeptides of multidomain proteins like firefly luciferase. However, the presence of these chaperones could not improve the yield of this protein. Binding of TF and DnaK to nascent firefly luciferase chains redirects the folding of this protein from an efficient eukaryotic cotranslational mode to a slower post-translational pathway that is accompanied by aggregation (Agashe et al., 2004).

4.6.2
Cell Conditioning

Preparing the cells for a stressful condition by transient exposure of the cells to a similar stress condition is termed "cell conditioning". In cell conditioning efforts, external stimuli such as a heat shock or ethanol shock is applied to the cells to alter the levels of various host proteins to improve the yield of recombinant protein. Dithiothreitol (DTT) and ethanol, which is capable of altering the cellular levels of chaperones and proteases, were used to evaluate the cell conditioning strategy (Gill et al., 2001). The addition of DTT 20 min before the induction of the recombinant protein was found to result in a 2-fold increase in the activity compared with the unconditioned controls. The analysis of the microarray and proteomic data during such cell conditioning strategies may be useful to identify key stress proteins for on-line monitoring and control of fermentation conditions.

4.7
Concluding Remarks

The analysis of the response of the bacterium *E. coli* to the overexpression of recombinant proteins shows a global adaptation and regulation. The metabolism, catabolism and the protein maintenance machinery in the cell were affected due to the overexpression of the recombinant protein. The response or the regulation of the global network in the cell depends on the nature of the protein expressed and the dynamics of the expression. Understanding the dynamics of the cellular response and designing appropriate strategies could help tackle such situations. The examples explained in this chapter illustrate the benefit of manipulating the global network in the cell to tackle stressful conditions. However, the available information on the changes in the transcript and proteome levels may not be sufficient to understand the cellular response. The time profile information on the transcript and proteome levels and the information on the dynamics of the signaling networks that connect the cause to the effect are important for understanding the response of whole complex system. Such an understanding would help in designing better expression systems and hosts to improve recombinant protein productivity in *E. coli*.

References

Agashe, V. R., Guha, S., Chang, H. C., Genevaux, P., Hayer-Hartl, M., Stemp, M., Georgopoulos, C., Hartl, F. U., Barral, J. M., *Cell* **2004**, *117*, 199–209.

Alba, B. M., Gross, C. A., *Mol Microbiol* **2004**, *52*, 613–619.

Arie, J. P., Sassoon, N., Betton, J. M., *Mol Microbiol* **2001**, *39*, 199–210.

Arsene, F., Tomoyasu, T., Bukau, B., *Int J Food Microbiol* **2000**, *55*, 3–9.

Ayling, A., Baneyx, F., *Protein Sci* **1996**, *5*, 478–487.

Baneyx, F., *Curr Opin Biotechnol* **1999**, *10*, 411–421.

Baneyx, F., Mujacic, M., *Nat Biotechnol* **2004**, *22*, 1399–1408.

Baneyx, F., Palumbo, J. L., *Methods Mol Biol* **2003**, *205*, 171–197.

Behrens, S., Maier, R., de Cock, H., Schmid, F. X., Gross, C. A., *EMBO J* **2001**, *20*, 285–294.

Bentley, W. E., Mirjalili, N., Anderson, D. C., Davis, R. H., Kompala, D. S, *Biotechnol Bioeng* **1990**, *35*, 668–681.

Bhattacharya, S. K., Dubey, A. K., *Biotechnol Lett* **1995**, *17*, 1155–1160.

Borth, N., Mitterbauer, R., Mattanovich, D., Kramer, W., Bayer, K., Katinger, H., *Cytometry* **1998**, *31*, 125–129.

Bruser, T., Sanders, C., *Microbiol Res* **2003**, *158*, 7–17.

Carrio, M. M., Villaverde, A., *FEBS Lett.* **2001**, *489*, 29–33.

Cashel, M., Gentry, D. R., Hernandez, V. J., Vinella, D., In: *Escherichia coli and Salmonella: Cellular and Molecular Biology*, Neidhardt, F. C., Curtiss III, R., Ingraham, J. L., Lin, E. C. C., Low, K. B., Magasanik, B., Reznikoff, W. S., Riley, M., Schaechter, M., Umbarger, H. E. (Eds.). ASM Press, Washington, DC, **1996**, p. 1458.

Chang, D. E., Smalley, D. J., Conway, T., *Mol Microbiol* **2002**, *25*, 289–306.

Chen, R., Henning, U., *Mol Microbiol* **1996**, *19*, 1287–1294.

Choi, J. H., Lee, S. Y., *Appl Microbiol Biotechnol* **2004**, *64*, 625–635.

Collinet, B., Yuzawa, H., Chen, T., Herrera, C., Missiakas, D., *J Biol Chem* **2000**, *275*, 33898–33904.

Dartigalongue, C., Missiakas, D., Raina, S., *J Biol Chem* **2001**, *276*, 20866–20875.

Delisa, M. P., Samuleson, P., Palmer, T., Georgiou, G., *J Biol Chem* **2002**, *277*, 29825–29831.

Dong, H., Nilsson, L., Kurland, C. G., *J Bacteriol* **1995**, *177*, 1497–1504.

Erickson, J. W., Gross, C. A., *Genes Dev* **1989**, *3*, 1462–1471.

Erickson, J. W., Vaughn, V., Walter, W. A., Neidhart, F. C., Gross, C. A., *Genes Dev* **1987**, *1*, 419–432.

Fahnert, B., Lilie, H., Neubauer, P., *Adv Biochem Eng Biotechnol* **2004**, *89*, 93–142.

Farr, S. B., Kogoma, T., *Microbiol Mol Biol Rev* **1991**, *55*, 561–585.

Flores, S., Ramon de Anda-Herrera, Gosset, G., Bolivar, F. G., *Biotechnol Bioeng* **2004**, *87*, 485–494.

Fraenkel, D. G., In: *Escherichia coli and Salmonella: Cellular and Molecular Biology*, Neidhardt, F. C., Curtiss III, R., Ingraham, J. L., Lin, E. C. C., Low, K. B., Magasanik, B., Reznikoff, W. S., Riley, M., Schaechter, M., Umbarger, H. E. (Eds.). ASM Press, Washington, DC, **1996**, p. 189.

Gellissen, G., Strasser, A. W. M., Suckow, M., In: *Production of Recombinant Proteins. Novel Microbial and Eukaryotic Expression Systems*, Gellissen, G. (Ed.). Wiley-VCH, Weinheim, **2005**, p. 1.

George, H. A., Powell, A. L., Dahlgren, M. E., Herber, W. K., Maigetter, R. Z., Burgess, B. W., Stirdivant, S. M., Greasham, R. L., *Biotechnol Bioeng* **1992**, *40*, 437–445.

Grengross, T. U., *Nat Biotechnol* **2004**, *22*, 1409–1414.

Gill, R. T., Delisa, M. P., Valdes, J. J., Bentley, W. E., *Biotechnol Bioeng* **2001**, *72*, 85–95.

Glick, B. R., *Biotechnol Adv* **1995**, *13*, 247–261.

Gottesman, S., *Annu Rev Genet* **1996**, *30*, 465–506.

Gottesman, S., *Curr Opin Microbiol* **1999**, *2*, 142–7.

Gottesman, S., *Annu Rev Cell Dev Biol* **2003**, *19*, 565–587.

Hantke, K., *Curr Opin Biotechnol* **2001**, *4*, 172–177.

Harcum, S. W., *J Biotechnol* **2002**, *93*, 189–202.

Harcum, S. W., Bentley, W. E., *Biotechnol Bioeng* **1993**, *42*, 675–685.

Harms, N., Luirink, J., Oudega, B., In: *Molecular Chaperones in the Cell*, Lund, P. (Ed.). Oxford University Press, New York, **2001**, p. 35.

Hardsock, C. E. III, Lewis, J. K., Leslie, I., Pope, J. A. Jr., Tsai, L. B., Sachdev, R., Meng, S. Y., *Biotechnol Lett* **1995**, *17*, 1025–1030.

Hellwig, S., Drossard, J., Twyman, R. M., Fischer, R., *Nat Biotechnol* **2004**, *22*, 1415–1422.

Hiniker, A., Bardwell, J. C. A., *Biochemistry* **2003**, *42*, 1179–1185.

Hoffmann, F., Rinas, U., *Biotechnol Prog* **2000**, *16*, 1000–1007.

Hoffmann, F., Rinas, U., *Biotechnol Bioeng* **2001**, *76*, 333–340.

Hoffmann, F., Rinas U., *Adv Biochem Eng Biotechnol* **2004**, *89*, 73–92.

Hoffmann, F., Weber, J., Rinas, U., *Biotechnol Bioeng* **2002**, *80*, 313–319.

Houten, B. V., *Microbiol Mol Biol Rev* **1990**, *54*, 18–51.

Imlay, J. A., *Annu Rev Biochem* **2003**, *57*, 395–418.

Jenkins, D. E., Schultz, J. E., Matin, A., *J Bacteriol* **1988**, *170*, 3910–3914.

Jenkins, D. E., Auger, E. A., Matin, A., *J Bacteriol* **1991**, *173*, 1992–1996.

Jürgen, B., Lin, H. Y., Riemschneider, S., Scharf, C., Neubauer, P., Schmid, R., Hecker, M., Schweder, T., *Biotechnol Bioeng* **2000**, *70*, 217–224.

Kadokura, H., Katzen, F., Beckwith, J., *Annu Rev Biochem* **2003**, *72*, 111–135.

Kanemori, M., Mori, H., Yura, T., *J Bacteriol* **1994**, *176*, 5648–5653.

Kim, J. Y. H., Cha, H. J., *Biotechnol Bioeng* **2003**, *83*, 841–853.

Lee, J., Kim, H.-C., Kim, S.-W., Kim S. W., Hong, S. I., Park, Y.-H., *Biotechnol Bioeng* **2002**, *80*, 84–92.

Lin, H. Y., Neubauer, P., *J Biotechnol* **2000**, *79*, 27–37.

Luli, G. W., Strohl, W. R., *Appl Environ Microbiol* **1990**, *56*, 1004–1011.

Ma, J. K. C., Drake, P. M. W., Christou, P., *Nat Rev Genet.* **2003**, *4*, 794–805.

Makrides, C. S., *Microbiol Rev* **1996**, *60*, 512–538.

Missiakas, D., Betton, J. M., Raina, S., *Mol Microbiol* **1996**, *21*, 871–884.

Morita, M., Kanemori, M., Yanagi, H., Yura, T., *J Bacteriol* **1999**, *181*, 401–410.

Netzer, W. J., Hartl, F. U., *Nature* **1997**, *388*, 343–349.

Nishihara, K., Kanemori, M., Kitgawa, M., Yanaga, H., Yura, T., *Appl Environ Microbiol* **1998**, *64*, 1694–1699.

Nishihara, K., Kanemori, M., Yanagi, H., Yura, T., *Appl Environ Microbiol* **2000**, *66*, 884–889.

Nystrom, T., *Mol Microbiol* **1994**, *12*, 833–843.

Oh, M. K., Liao, J. C., *Metab Eng* **2000**, *2*, 201–209.

Ow, D. S. W., Nissom, P. M., Philip, R., Oh, S. K. W., Yap, M. G. S., *Enzyme Microb Technol* **2006**, *39*, 391–398.

Rai, M., Padh, H., *Curr Sci* **2001**, *80*, 1121–1128.

Raivio, T. L., Silhavy, T. J., *J Bacteriol* **1997**, *179*, 7724–7733.

Ramirez, D. M., Bentley, W. E., *Biotechnol Bioeng* **1993**, *41*, 557–565.

Ramirez, D. M., Bentley, W. E., *J Biotechnol* **1999**, *71*, 39–58.

Rinas, U., *Biotechnol Prog.* **1996**, *12*, 196–200.

Sancar, A., *Annu Rev Biochem* **1996**, *65*, 43–81.

Sandén, A. M., Prytz, I., Tubulekas, I., Förberg, C., Le, H., Hektor, A., Neubauer, P., Pragai, Z., Harwood, C., Ward, A., Picon, A., Teixeira de Mattos, J., Postma, P., Farewell, A., Nyström, T., Reeh, S., Pedersen, S., Larsson, G., *Biotechnol Bioeng* **2003**, *81*, 158–166.

Schafer, U., Beck, K., Muller, M., *J Biol Chem* **1999**, *274*, 24567–24574.

Schmidt, M., Viaplana, E., Hoffmann, F., Marten, S., Villaverde, A., Rinas, U., *Biotechnol Bioeng* **1999**, *66*, 61–67.

Schweder, T., Lin, H. Y., Jürgen, B., Breitenstein, A., Riemschneider, S., Khalameyzer, V., Gupta, A., Büttner, K., Neubauer, P., *Appl Micorbiol Biotechnol* **2002**, *58*, 330–337.

Seeger, A., Schneppe, B., McCarthy, J. E. G., Deckwer, W. D., Rinas, U., *Enzyme Microb Technol* **1995**, *17*, 947–953

Sorenson, H. P., Mortensen, K. K., *J Biotechnol* **2005a**, *115*, 113–128.

Sorenson, H. P., Mortensen, K. K., *Microbiol Cell Fact* **2005b**, *4*, 1–8.

Soriano, E., Borth, N., Katinger, H., Mattanovich, D., *Metab Eng* **1999**, *1*, 270–274.

Srivastava, R., Cha, H. J., Peterson, M. S., Bentley, W. E., *Appl Environ Microbiol* **2000**, *66*, 4366–4371.

Srivastava, R., Peterson, M. S., Bentley, W. E., *Biotechnol Bioeng* **2001**, *75*, 120–129.

Vind, J., Sorensen, M. A., Rasmussen, M. D., Pedersen, S., *J Mol Biol* **1993**, *231*, 678–688.

Wang, Q. P., Kaguni, J. M., *J Bacteriol* **1989**, *171*, 4248–4253.

Weber, J., Hoffmann, F., Rinas, U., *Biotechnol Bioeng* **2002**, *80*, 320–330.

W grzyn, G., W grzyn, A., *Microb Cell Fact* **2002**, *1*, 2.

Wurm, F. M., *Nat Biotechnol* **2004**, *22*, 1393–1398.

Xu, Z., Knafels, J. D., Yoshino, K., *Nat Struct Biol* **2000**, *7*, 1172–1177.

Yamanaka, K., *J Mol Microbiol Biotechnol* **1999**, *1*, 193–202.

Part III   Preparative (Chromatographic) Methods

5

Ion-exchange Chromatography in Biopharmaceutical Manufacturing

Lothar Jacob and Christian Frech

5.1
Introduction

There is currently an increasing need for improved production processes for biological drugs. In the past, biotherapeutics like erythropoietin and growth hormones were generally only administered in small doses; the situation with therapeutic recombinant proteins like monoclonal antibodies is that they are required in far greater quantities. Among the biomolecules in use or in testing are nucleic acids and viral vectors for gene therapy and proteins as well as peptides. The yearly demand for such drugs can be as high as several hundred kilograms. This high demand has to be considered when establishing new purification schemes for commercial use of biologically derived biotherapeutics. However, producing high-quality, efficacious and safe proteins in sufficient amounts for the clinic is not trivial. Recovery and purification operations are the most time-consuming steps in manufacturing. Whether the products are derived from microbial or mammalian cells, the downstream purification scheme is expensive to develop. The development of a downstream purification process includes the responsibility of implementing appropriate and state-of-the-art chromatography media, techniques and equipment. All purification issues should consider technology transfer ability of the process to current Good Manufacturing Practice (cGMP) conditions. In any case, the aim of the downstream process is to use the shortest route from the biological source to the specified final product. The yield has to be maximized while at the same time the biological activity must be maintained.

The downstream process can be divided into different parts: the first step after the harvest is the capture of the target protein, followed by intermediate purification step(s) and a final polishing step before the formulation. The capture step, that is in many cases an ion-exchange column, has the objective to recover as much product as possible and to provide a product pool that is suitable for subsequent chromatography. The reduction of the volume (mainly by eliminating water) and the removal of other impurities has to be achieved during this step. The captured material is then purified by column chromatography using a process that removes

Bioseparation and Bioprocessing. Edited by G. Subramanian
Copyright © 2007 WILEY-VCH Verlag GmbH & Co. KGaA, Weinheim
ISBN: 978-3-527-31585-7

all host contaminants and product related impurities up to a certain level. The final polishing that is applied to separate degraded or modified forms of the product is often the most challenging step, because of the biochemical similarities that typically exist between the final product and the modified forms.

Finally, the end of the purification process is bulk formulation of the drug.

5.2
Ion-exchange Chromatography in the Downstream Processing of Proteins

Ion-exchange chromatography is the most widely used chromatographic method at the large scale and is extensively applied to protein separation.

Reasons for the popularity of ion-exchange chromatography include its (i) straightforward separation principle, (ii) ease of performance, (iii) high protein-binding capacity, (iv) versatility and (v) high resolving power.

5.2.1
Resins: Commonly Used Functional Groups and Recommended Buffers

Ion exchangers used in ion-exchange chromatography contain charged functional ion-exchange groups covalently attached to a solid matrix, either organic (e.g. cross-linked styrene–divinylbenzene or ethyleneglycol–methacrylate copolymers) or inorganic (most frequently silica gel support) (Tab. 5.1). The functional groups carry either a positive charge (anion exchangers) or a negative charge (cation exchangers) and retain ions with opposite charges by strong electrostatic interactions.

They in turn can be divided into those with weakly basic or acidic character or strongly basic or acidic character. With strongly basic or acidic ligands the

Tab. 5.1 Main packings commercially available for ion-exchange chromatography at large scale (in order to achieve higher flow rates and more profitability, synthetic polymers and highly cross-linked agaroses are advantageous)

Products for ion-exchange chromatography	Manufacturer	Base matrix
Fractogel® EMD media	Merck KGaA	methacrylate
Macroprep® media	Bio-Rad	methacrylate
Toyopearl® media	Tosoh Bioscience	methacrylate
Ceramic HyperD® Ion Exchange Sorbents	Pall (BioSepra)	ceramic bead/hydrogel "gel-in-a-shell"
XL Sepharose® media	GE Healthcare	agarose, dextran
Sepharose® FF media	GE Healthcare	agarose
Capto® media	GE Healthcare	agarose, highly cross-linked

Tab. 5.2 Commonly used functional groups and their properties

Functional group	Description	Operating range (pH)[a]	pK value/category
TMAE/Q	trimethylammoniumethyl	6–10	>13/strongly basic
DEAE	diethylaminoethyl	6–8.5	11/weakly basic
DMAE	dimethylaminoethyl	6–8.5	8–9/weakly basic
SO_3^-/SE/SP/S	sulfoisobutyl/sulfoethyl/ sulfopropyl/sulfonate	4–8	<1/strongly acidic
COO^-/CM	carboxy/carboxymethyl	5–8	4.5/weakly acidic

a Depends on the pH stability of the target protein.

functional groups are always present in ionized form, independent from the pH value in the specified operating range.

For example, quaternary amino groups (R_3N^+) are positively charged, while sulfonic acid groups (SO_3^-) are negatively charged. The pK values of quaternary amino groups are around 14, those of the sulfonate residues below 1. In addition, weakly basic types (pK=8–11) and weakly acidic types (pK=4–6) exist. The most common groups are summarized in Tab. 5.2 with their abbreviations and pK values.

Ion exchangers are more densely substituted than other adsorbents used in protein chromatography. The total ionic capacities are 100–500 $\mu mol\,mL^{-1}$ bed, which corresponds to a concentration of ion exchanging groups of 0.1–0.5 M.

The binding capacity for proteins (static protein capacity) is lower than the total ionic capacity, and depends strongly on the nature and size of the protein and the accessible surface area in the pores of the ion exchanger. Table 5.3 gives the total ionic capacity and the static protein-binding capacity for different ion exchangers used in manufacturing.

Under physiological conditions almost all proteins have a specific conformation which is responsible for the biological function of the molecule. This structure should remain completely intact during the biochromatographic separation. Therefore, most applications in ion-exchange chromatography take place in the presence of aqueous buffer systems to avoid protein denaturation. Any of the conventional buffers like Tris, Tricine, phosphate-buffered saline (PBS), glycyl-glycine, 2-(4-morpholino)ethanesulfonic acid (MES), 3-*N*-morpholino propansulfonic acid (MOPS), 4-(2-hydroxyethyl)-1-piperazineethanesulfonic acid (HEPES), bis-tris-propane–HCl or others can be used for ion-exchange chromatography. However, positively charged buffering ions should be used on anion exchangers to avoid an interaction or binding to the functional group. Therefore, Tris (pK_a 8.2) is preferred with Cl^- as counterion. For cation exchangers, the buffering ion should be negatively charged, e.g. carbonate, acetate or MES, and the counterion K^+ or Na^+. Phosphate buffers are generally used on both exchanger types. The buffer concentration is in the range of 10 up to 50 mM, usually. Additives like 10 mM 2-mercaptoethanol, 20% glycerol, 5 mM $MgCl_2$, 1 mM benzamidine–HCl, 1 mM dithiothreitol (DTT), 10 $\mu g\,mL^{-1}$ chymostatin, 2 mM ethylenediaminetetraacetic acid (EDTA), 0.5% Triton,

Tab. 5.3 Physicochemical properties of the different ion-exchange stationary phases for manufacturing as provided by the suppliers

Functional group	Ionic capacity (μeq mL^{-1} gel)	Protein binding capacity (mg potein mL^{-1} gel)	Mean particle diameter (μm)
Fractogel® EMD TMAE Hicap	60–150	160–200	65
Fractogel® EMD TMAE	30–50	80–120	65
Fractogel® EMD DEAE	60–90	80–120	65
Fractogel® EMD COO/SO$_3$	80–100	110–150	65
Fractogel® EMD SE Hicap	60–90	120–160	65
Q Sepharose® FF	180–250	120	90
S Sepharose® FF	180–250	70	90
DEAE Sepharose® FF	110–160	110	90
CM Sepharose® FF	90–130	30	90
Q Sepharose® XL	180–260	130	90
S Sepharose® XL	180–250	160	90
Q/SP Sepharose® Big Beads	180–250	70	20
Q/S Sepharose® HP	220	70	34
Capto® Q	160–220	100	90
Capto® S	110–140	120	90
MacroPrep® DEAE	100–250	35	50
MacroPrep® High Q	325–475	40	50
MacroPrep® High S	120–200	55	50
MacroPrep® CM	170–250	20	50
UNOsphere® Q	120	180	80
UNOsphere® S	250–310	60	80
Q F Ceramic Hyper D®	≥250	85	50
DEAE F Ceramic Hyper D	≥200	85	50
CM F Ceramic Hyper D®	250–400	60	50
S F Ceramic Hyper D®	≥150	75	50

0.25% sodium cholate or 0.25 M α-methyl-mannoside have been reported to support the chromatographic separation. The choice of potential buffers and additives is primarily dependent on the intended scale.

5.2.2
Principles of Ion-exchange Chromatography

Preparative ion-exchange chromatography is a technique for separating biomolecules based primarily on their electrostatic interactions with charged stationary-phase materials. It is simply an adsorption/desorption process utilizing the principle of ion exchange and is therefore reliant on the chemical properties of the solute molecules to be separated as well as on the stationary phases used. This technique is the most widely employed chromatographic separation technique in the biotechnology industry today. It is used for both capture and high-resolution separations.

Different generalizations of the simplest coulombic model describing the attraction of groups of opposite charge on the protein surface and the stationary phase, respectively, have been proposed.

They can be divided into stoichiometric and nonstoichiometric models. Stoichiometric models describe the multifaceted binding of the protein molecules to the stationary phase as a stoichiometric exchange of mobile-phase protein and bound counterions.

The stoichiometric displacement model (SDM) was first formulated by Boardman and Partridge [1], and has since then been used in several alternative formulations, all based on the stoichiometric concept [1–4]. The model is based on the assumption that ion exchange is the only mechanism of retention of the components studied and that the ion-exchange process can be modeled as a stoichiometric "reaction" described by the mass action principle. The binding process, in the absence of the specific salt binding effects, results in equilibrium of the eluents and the polyions of the sample between the mobile and the stationary phase.

The SDM [5–7] characterizes an ion-exchange process as:

$$P_0 + Z C_b \rightleftharpoons P_b + Z C_0, \tag{1}$$

where P_0 and P_b refer to protein concentration in the mobile and stationary phases, respectively, C_b and C_0 refer to the concentration of the bound and free salt ion, and Z is the number of bound ions displaced during the protein adsorption process. The protein retention (the retention factor k) under isocratic conditions can be related to the displacing ion concentration in mobile phase (C_{salt}) by:

$$\log k = \log A + Z \log(1/C_{salt}), \tag{2}$$

where $k = (t_R - t_{NR})/t_{NR}$, and t_R and t_{NR} are the retention times of the solute at retained and nonretained conditions, respectively. The term A is a protein-specific constant, which indicates the overall (specific and nonspecific interactions) affinity of the solute for the sorbent surface [8]. Equation (2) can be used to obtain the two

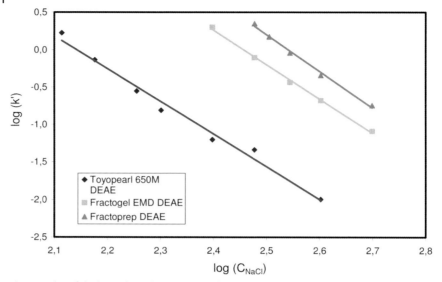

Fig. 5.1 Plot of the logarithm of the retention factor of β-lactoglobulin A obtained using different column packings versus the logarithm of the concentration of NaCl. The symbols are data from isocratic experiments, the lines are calculated results from the SDM.

protein-specific parameters, Z and A, by plotting k obtained from at a series of isocratic elution conditions (Fig. 5.1).

Z $(=z_P/z_S)$ is the ratio of the characteristic charge of the protein to the valence of the counterion and represents a statistical average of the electrostatic interactions of the protein with the stationary phase as it migrates through the column (Fig. 5.2).

The characteristic charge depends strongly on the three-dimensional (3-D) structure and charge distribution of the protein, which determine the surface amino acid residues which are in position to interact with the ion-exchange sorbent. The surface area containing the amino acid residues participating in the interaction is referred to as the chromatographic contact region or footprint of retention [9]. A single amino acid variation in the chromatographic contact region can have a marked effect on protein retention [10–12]. The residues in the contact region may interact directly with the sorbent or they may influence binding of other residues through electrostatic effects on the dissociation of neighboring residues and steric perturbation of hydrogen-bonded water molecules [2].

In addition, stationary-phase properties like the charged density of ion exchanger [13–15], base matrix materials and surface modification chemistries [16, 17] as well as the type of displacing ions of mobile phase [2, 18–21] can affect retention, elution, resolution, selectivity and recovery.

The model has been further modified to describe protein retention under linear gradient elution conditions (LGE model; see Section 5.3), as well as under

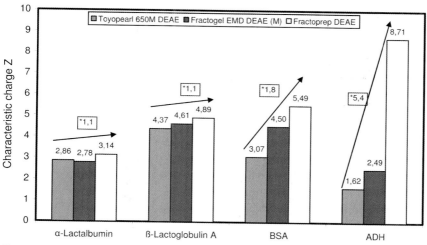

Fig. 5.2 Comparison of the characteristic charge *z* for four proteins on different stationary phases. For larger proteins the surface modification has a strong influence on the *z* value. BSA = bovine serum albumin, ADH = alcohol dehydrogenase.

nonlinear protein adsorption conditions [steric mass action (SMA) model] [22] for isocratic and gradient chromatography.

The SMA formalism [22, 23] has been used successfully to predict complex preparative chromatographic behavior of proteins in gradient and displacement ion-exchange systems [24–28]. By using this model, the adsorption of a protein in ion exchange can be described by its characteristic charge (Z) (i.e. the number of interaction sites with the resin), the equilibrium constant (K_{SMA}) for the exchange reaction between the protein and salt counterions, and the steric factor (δ), which is the number of sites sterically shielded by the adsorbed molecule (Fig. 5.3).

Although stoichiometric models are capable of accurately describing the behavior of ion-exchange chromatographic systems, they assume that the individual charges on the protein molecules interact with discrete charges on the ion-exchange surface.

Despite its widespread use, the SDM ignores certain phenomena generally thought to be important in determining the ion-exchange equilibrium of proteins, such as the diffuse nature of the electrical double layer adjacent to the adsorbent surface, the asymmetric charge distribution on the protein, structural changes of the protein upon binding and protein charge regulation, i.e. the change in protein charge caused by electrostatic effects during adsorption.

Although the SDM offers a simplified analysis of chromatographic behavior, the model is not strictly mechanistic, so the model parameters do not give a clear representation of the roles of properties of physical significance. In reality, retention in ion exchange is more complex and is primarily due to the interaction of the electrical fields of the protein molecules and the chromatographic surface.

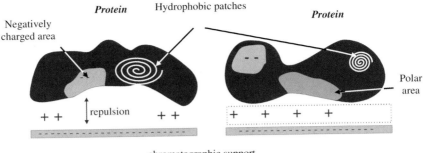

Sites shielded by protein are represented by the <u>steric factor</u>

<u>Characteristic charge:</u> the number of interaction sites of the protein with the charged ligands on the ion exchange surface

Fig. 5.3 Schematic representation of protein adsorption on an ion-exchange surface.

During the last decade, several nonstoichiometric models for describing protein retention as a function of the salt concentration in the eluent have been proposed. Two of the models are based on Manning's ion condensation theory [29, 30], originally formulated to estimate the properties of cylindrical polyelectrolytes, such as DNA, in a salt solution. More recently, theories used in colloid and surface chemistry to describe electrostatic and other interactions have been applied to describe retention properties of proteins in ion-exchange chromatography. In these theories, the electrostatic interaction is often calculated from solutions of the Poisson–Boltzmann equation for a system of given geometry [31–35].

Ståhlberg and coworkers [31–34] have developed the electrostatic interaction model, known as the slab model, which treats the retention process as a coulombic interaction between two charged flat plates. The model predicts that the logarithm of the retention factor varies linearly with the reciprocal square root of the ionic strength of the eluent. The slope of the line depends on the protein charge density, as well as on the interacting area between the protein and the stationary phase. A recent modification of the slab model includes the charges that are induced at the protein surface by the electrostatic field of the chromatographic surface and is therefore called the charge regulated slab model.

Roth and Lenhoff [35] solved the linearized Poisson–Boltzmann equation numerically for a sphere interacting with an oppositely charged surface by using a functional form suggested by the linear superposition approximation. The results were used to develop a mechanistic model within which the retention is related to protein and stationary-phase structural and functional parameters, as well as eluent composition [36]. The protein parameters are size and net charge, while incorporation of stationary-phase properties, i.e. the surface charge density and the short-range interaction energy, allows for a more mechanistic interpretation of SDM parameters.

Both the slab model and the mechanistic model make use of characteristic parameters of the protein, as well as of the stationary phase, to describe the reten-

tion process. The models should be a powerful tool to us in the selection of stationary phases and operating conditions.

The significance of charge heterogeneity on the protein molecule for protein
adsorption has been highlighted by the experimental observations of chromatographic separation of similarly charged proteins [37] and attractive interactions
between like-charged protein and adsorbent [38, 39] which cannot be easily
explained by a model reducing the protein charge distribution to a single net
charge or charge density. Protein modeling with the full 3-D structure of the proteins has therefore been performed, which provides a more detailed description
of the relevant molecular events in protein adsorption.

More accurate explanations have been proposed over the years, e.g. in relation
to charge asymmetry [5], calculated adsorption potential between the adsorbent
and full atomic models of the protein [40], and molecular electrostatics computa-
tions performed on protein structures to determine the average potential over the
molecular surface [11]. In the latter study the calculated binding equilibrium constant, found by averaging over the full 3-D configurational space, captures the
chromatographic differentiation of closely related cytochrome *c* variants. To obviate
the need for full sampling of protein configurations, calculations of interaction
free energies at short protein–adsorbent separation distances or of protein surface
potentials were found to yield reasonable semiquantitative descriptions of the
retention trends.

For proteins with more significant structural variations, as in the lysozyme
cytochrome *c* comparison, the accumulation of uncertainties due to other molecular forces can make even qualitative predictions of trends unreliable [12].

More detailed quantitative structure–property relationship (QSPR) models that
use linear free energy relationships have also been successfully used for the prediction of chromatographic retention [41–43]. A strong relation between the average
surface potential and retention times in cation- and anion-exchange chromatography was observed. However, it was shown, that by including more (and "less-
average") proteins in the data set of ion-exchange experiments one finds that the
relation does not hold generally and that the average surface potential may alone
by itself become a poor descriptor of the retention.

The QSPR approach with multiple protein descriptors addresses this type of
challenge and it was shown that it is possible to obtain a model that accounts
for the unusual retention of "less-average" proteins. Furthermore, due to the
ease of interpretability of the developed descriptor set it was possible to obtain
an explanation for this behavior. The QSPR approach has thus been successful
in both prediction and interpretability of the retention of structurally known
proteins [44].

An additional factor that emerges from the results is the synergistic role of the
protein structure and the stationary-phase structure and properties. The subtle role
of this effect, which might lead to multipoint attachment on some adsorbents, is
usually not adequately considered. The effect strongly depends on the distribution
of charges on the protein and the distribution of ligand spacing on the adsorbent.
It is very difficult to account for in any modeling effort even if the underlying

interactions are predominantly electrostatic and thus nominally amenable to calculation using the above mentioned methods.

Adsorption modeling with the full 3-D structure of the proteins to predict retention behavior can provide guidance for efficient process design. Current prediction models use proteins with well known 3-D structures and a simplified representation of the stationary-phase structure and surface.

When designing and optimizing an ion-exchange chromatography step the starting point is usually a complex sample mixture containing different unknown proteins (at least structurally unknown) and a selection of well known stationary phases (at least chemically known). QSPR models are therefore only of limiting applicability as long as only a few stationary phases are compared. The complex and diverse physicochemical characteristics of typical adsorbents have to be determined using different models of the ion-exchange process. It seems that simple SDMs are able to describe the differences in retention based on the most important adsorbent properties [45].

5.3
Development and Optimization Strategies

Bioseparations are typically difficult. All products of interest are labile and thus require mild processing conditions. They typically arrive at downstream processing contaminated with closely related species and the required product purity is usually very high. Safety is a major consideration. Perhaps the most severe problem is finding separating agents with both high capacity and selectivity for the product of interest. Regulatory considerations usually require that the basic separation scheme must be fixed very early in the overall development process. In addition to strict standards on the purity of a product, the US Food and Drug Administration requires validating the removal of various contaminants, i.e. host cell-related (DNA, endotoxin, protein, virus), product-related (oxidation, deamidation, acetylation, dimerization) and process-related (antibiotic, antiform, inducing agent) contaminants. Finally, the rising cost of health care is beginning to put severe economic constraints on processing cost. In summary, a safe and economic process must be found quickly somewhere in an extremely large parameter space which simply cannot be explored in depth.

To enable efficient process development rational strategies for screening and selection of chromatographic media and process conditions must be developed where one combines the experimental results with theoretical considerations and model calculations.

A chromatographic separation is developed through a number of steps including screening of different media and techniques to select appropriate candidates to investigate the retention behavior in dependence of the possible process variables that usually comprise particle size, pH, type and concentration of salt, solvents and additives, and temperature. When more material becomes available capacities must be estimated. Process optimization includes column size and column aspect ratio, flow rate, buffer composition, and gradient length. All this can of course be performed by the trial-and-error method, but ultimately model-assisted development should be beneficial.

Ion-exchange chromatography can be used as a capture, intermediate purification as well as a final polishing step. For product recovery (capture), the capacity and production rate are paramount. In addition, it is often necessary to process large volumes of relatively dilute feedstock, which often contain particles.

Step gradient is often employed for initial capture or recovery operations where the primary objective is to recover and to concentrate the bioproduct of interest. This operation is usually carried out in short, large-diameter columns packed with relatively large-particle diameter stationary-phase materials (above 90 μm). This mode of chromatography consists of first loading the column with a large volume of low-ionic-strength feed solution, followed by a wash step, and finally an elution step based on an elevated salt and/or modified pH eluent. This type of chromatography can be characterized as essentially an on/off operation. The most important stationary-phase characteristic in preparative step gradient chromatography is the dynamic capacity of the resin. Thus, these separations do not require very efficient columns and can be readily carried out in short "pancake"-type columns. These capture steps can be carried out by means of conventional packed-bed columns or with "expanded-bed"-type configurations.

Step gradient has many advantages, i.e. ease of operation, no need for complicated gradient pumping apparatus and concentration of the bioproduct. When the separation problem is more difficult, however, a higher resolution mode of operation is often required.

For high-resolution separations (intermediate purification and polishing), linear gradient chromatography is often employed. This operation is usually carried out in variable-diameter columns using a range of column lengths and particle diameters. It turns out that depending on the difficulty of the separation, it may be necessary to employ longer columns and/or materials of smaller particle diameter for these separations. This mode of chromatography is typically carried out by first loading the column with a relatively large volume of low-ionic-strength feed solution, followed by a continuous linear gradient of increasing ionic strength. This type of chromatography cannot be characterized as an on/off operation, and is thus dependent on the length of the column and the column efficiency. The most important stationary-phase attribute in preparative linear gradient chromatography is the selectivity of the resin for the given separation. In addition to the selectivity, the dynamic capacity of the stationary phase is also important for preparative linear gradient separations and must be considered when selecting a resin. The major advantage of linear gradient is its ability to resolve difficult separation problems.

As described earlier, it is important to screen a wide variety of ion-exchange resins to identify a material that has sufficient selectivity and capacity for a given separation as well as good rate factors. Selectivity and capacity are determined by the nature and concentration of the ionogenic groups, by the accessible surface area, and by the nature of the base matrix and coupling chemistry. Rate factors are primarily associated with mass transfer effects like external film transport and internal pore diffusion. As a result the particle size, the pore size and the pore network connectivity are critical. Knowledge of the mass transfer rates is needed to compare different stationary phases and to predict process performance [46].

The development of an ion-exchange chromatographic protein separation usually starts with small columns and at low loadings. During the development work the separation is optimized using linear gradient salt elutions. Binding capacity limits are determined and the influence of high loadings on the separation is investigated. Although LGE is best suited for very delicate and difficult separations, in many cases step elution (SE) is preferred for industrial-scale separations because of its simple operations. However, compared with LGE, it is not easy to determine proper operating conditions for process-scale SE [47–49].

Depending on the purpose of the separation (capture or purification) the protein load on the column is properly adjusted. Capture steps are often performed under nonlinear adsorption conditions i.e. high loadings, whereas intermediate and polishing steps are performed in the linear or quasi-linear range of the adsorption isotherm.

In linear chromatography (low loadings), the various solutes migrate down the column independent of each other. The retention is characterized by a linear isotherm and an absence of intersolute competition for the adsorption sites. Yamamoto and coworkers have employed a continuous-flow plate model to characterize and design linear gradient and step gradient operations [49] in linear chromatography.

Linear gradient elution experiments are performed at different GH values (gradient slope times the stationary-phase volume) at a fixed pH. The ionic strength at the elution time of the peak (I_R) is determined as a function of GH. The GH–I_R curves thus constructed with the use of a relatively small column do not depend on the flow velocity, the column dimension, the sample loading at nonoverloading conditions or the starting ionic strength I_0, provided that the sample is initially strongly bound to the column [49, 50].

The experimental GH–I_R data can usually be expressed by [49–52]:

$$GH = \frac{I_R^{(B+1)}}{[A(B+1)]},$$
(3)

Curves of GH versus I_R (Fig. 5.4) indicate the ionic retention strength at different gradient elution conditions and provide information about the nature of separation. The data is fitted with power law regressions curves, and the A and B parameters are determined.

The following relationship can be derived from the SDM:

$$A = K_e \times \Lambda^B,$$
(4)

where B is the number of sites (charges) involved in protein adsorption, which is basically the same as the Z number [5] or the characteristic charge [4] in the

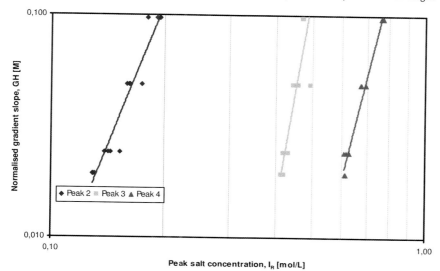

Fig. 5.4 $GH-I_R$ plots from LGE experimental data. The LGE experiments were carried out with various slopes g of the gradient. The ionic strength at the peak position I_R was then measured. Peaks 2–4 are three different egg white proteins, which were separated on the strong cation exchanger Fractogel SO_3^-.

literature, K_e is the equilibrium association constant and Λ is the total ion-exchange capacity. The following equation was derived from the ion-exchange equilibrium model [49–52] and Eq. (4):

$$K = K' + K_e\Lambda^B I^{-B}, \tag{5}$$

where K is the protein distribution coefficient, K' is the distribution coefficient of salt, and I is the ionic strength (salt concentration). Equation (5) provides the $K-I$ relationship, from which the SE condition can be determined [49] (Fig. 5.5). The details of the model and the model equations have been reported elsewhere [53].

The peak retention in the LGE can be predicted using Eq. (3) with A and B as experimental values. However, if the $GH-I_R$ curves are constructed as a function of mobile-phase pH, and determine the $pH-I_R$, $pH-B$ and $pH-K_e$ relationships, quite important information can be obtained on the retention (or biorecognition) and resolution of proteins as a function of the mobile-phase pH [54, 55]. In addition effective diffusivities can be extracted from experimental data and correlated with transport mechanism [46].

In LGE of protein, the same resolution can be obtained with various combinations of gradient slope, column length and flow rate based on the dimensionless variable O [50, 57] (Fig. 5.6):

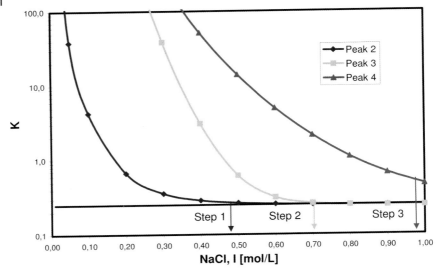

Fig. 5.5 Relation between the distribution coefficient K and the ionic strength I. The $K–I$ curves were calculated from the $GH–I_R$ curves in Fig. 5.4. The distribution coefficient of the salt is set to 0.25. The ionic strengths for stepwise elution of the proteins 2–4 are defined by the intercept points where $K = K'$.

$$O = \left(\frac{L \times I_a}{G \times (\mathrm{HETP}_{\mathrm{LGE}})} \right) \propto R_S^2, \tag{6}$$

where L is the column length, I_a is a dimensional constant equal to 1, G is the gradient slope normalized with respect to column void volume, and $(\mathrm{HETP})_{\mathrm{LEG}}$ is the plate height in the LGE and calculated with the elution curves (retention time and peak width) from the LGE experiments.

The resolution of a separation can be improved by choosing conditions that give a higher value of O. Small column LGE separations are possible for screening of optimization parameters. Attention must be paid during scale-up and scale-down procedures. This is because the extra-column broadening in a small-scale separation affects the shape of the gradient of the elution buffer [56]. Consequently, both the peak position and the peak width are likely to change.

After an O value that gives the desired resolution is identified, alternative operating conditions giving the same O can be calculated. In this way, O is used to specify various column geometries and flow rates that provide the required purity.

Although the same resolution can be obtained with various combinations of operating/column variables, there may be some optimum conditions because retention time t_R and retention volume V_R are a function of G and the mobile-phase velocity. "Iso-resolution curves" [57] can be used to determine optimized production scale operating conditions [54].

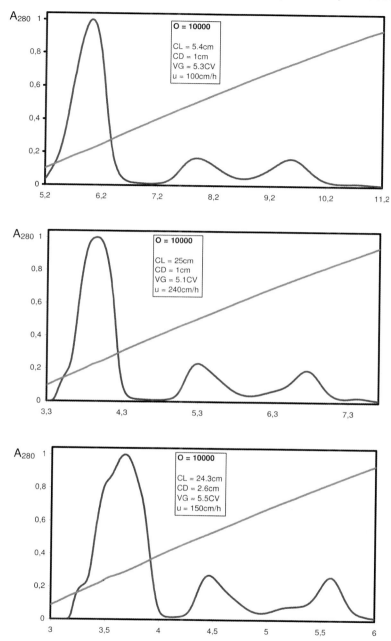

Column volumes

Fig. 5.6 Elution curves (chromatograms) of egg white proteins on a strong cation-exchange column (Fractogel SO_3^-). For all LGE experiments the O value was set to 10000. CL=column length, CD=column diameter, VG=gradient volume, u=linear velocity.

The LGE model has been applied to the scale-up of β-galactosidase purification in a 113-L column based on linear gradient data obtained with a 23-mL column [58].

Thus, for scale-up:

$$O = \left(\frac{L}{G \times (\mathrm{HETP}_{\mathrm{LGE}})} \right)_{\text{small column}} = \left(\frac{L}{G \times (\mathrm{HETP}_{\mathrm{LGE}})} \right)_{\text{large column}}. \tag{7}$$

While these approaches have proved useful for the scale-up of linear chromatography, the production rate in these ion-exchange systems can be dramatically enhanced by operating these columns under nonlinear conditions. However, increases in sample volume or protein concentration can change both peak retention and peak shape [59] if the protein is in a nonlinear portion of its isotherm. An alternative approach must be employed to model various modes of operation under nonlinear (high loading) conditions.

To study the effects of nonlinear chromatography, one needs a suitable isotherm that accurately describes nonlinear protein adsorption. Two commonly used expressions are the Langmuir isotherm and the SMA isotherm [22].

While the Langmuir isotherm has many advantages, it is inadequate for describing protein adsorption in ion-exchange systems. Protein adsorption is multipointed in nature and is affected by the presence of mobile-phase modifiers. The Langmuir isotherm is unable to accurately describe these features.

The multivalent ion-exchange model [60, 61] incorporates both the multipointedness of the adsorption process and the effect of the mobile-phase modifier on the process. It is derived from the SDM in conjunction with the law of mass action. Thus, the multivalent ion-exchange formalism predicts an isotherm of the form:

$$C = \frac{Q}{K_e} \times \left(\frac{C_{\mathrm{salt}}}{\Lambda - zQ} \right)^z, \tag{8}$$

where Q and C are the solute concentrations on the stationary and mobile phases, respectively, C_{salt} is the mobile-phase salt concentration, z is the characteristic charge of the adsorbing solute, K_e is the equilibrium constant for the ion-exchange reaction, and Λ is the total ionic capacity of the stationary phase.

To accurately describe competitive adsorption of a multicomponent sample the blocking of exchanger sites has to be taken into account. The shielding or steric factor (δ) accounts for this effect and is the basis for the SMA isotherm:

$$C = \frac{Q}{K_e} \times \left(\frac{C_{\mathrm{salt}}}{\Lambda - (z + \delta)Q} \right)^z. \tag{9}$$

The characteristic charge and the equilibrium constant define the linear part of the adsorption isotherm, and the steric factor accounts for all nonlinear adsorptive effects [22].

The SMA model was used to predict the behavior of mixtures in preparative chromatography, with a step gradient [48] or a linear gradient [4]. In addition, the model can predict sample displacement effects, which can have a profound effect on linear gradient separations under high loading conditions. Furthermore, the SMA formalism has been used along with iterative optimization techniques to develop optimal step and linear gradient separations.

No simple design equations are available and prediction of retention will require the use of numerical techniques. The effect of gradient elution on the production rate and the yield is addressed by systematically investigating various possible objective functions (e.g. production rates, recovery yields, solvent consumption).

5.4
Scale-up of Ion-exchange Chromatography

During early drug research, a high throughput of many samples is required with respect to their potency. However, after a specific protein has been identified to become a potential drug, more protein material is needed for further clinical testing. In contrast to the first material that is made in bioreactors containing a few hundreds of liters, the fermentation will be increased to several thousands of liters and huge volumes have to be processed [62, 63].

Scale-up is not a simple increase of the size and volume of the laboratory equipment. Usually, production-scale equipment, larger piping dimensions, larger column diameters, larger dead volumes and different types of pumps cannot be compared directly with bench-top workstations even if an optimal system configuration has been chosen. In particular, columns containing several hundreds of liters of resin are often customized, and column packing and qualification becomes a challenging part of the work. Due to decreased wall support, large columns behave completely different compared to pilot- or laboratory-scale columns.

The transfer of a method starts from the milligram level and ends up at the kilogram scale. The easiest way to scale-up the procedure is to change the column dimensions by keeping the height of the gel bed constant and increasing the column diameter. In many production suites large ion-exchange columns containing several hundreds of liters of media are installed. Column diameters usually are in the range between 60 and 160 cm or even more. The scale-up is typically done in two steps: the first step from laboratory to pilot plant is on the order of 50- to 100-fold and the final scale-up from the pilot plant to full-scale manufacturing is 10- to 50-fold [64]. A typical 60-fold scale-up on an anion exchanger is shown in Fig. 5.7. To demonstrate the similarity of the chromatogram obtained from a 20-cm inner diameter column the scale-up of a therapeutic protein was used as an example. During scale-up, the total load of protein per milliliter of resin should be identical and the linear flow rate has to be maintained.

The chosen matrix is the key to the success of chromatography and to process robustness. In contrast to the protein-binding capacities given by the suppliers of media, in the case of monoclonal antibodies some types of ion exchangers provide unexpected high binding capacities.

(A)

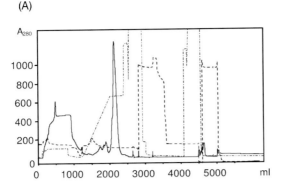

(B)

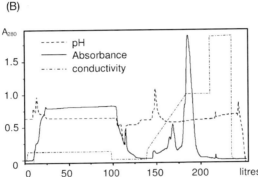

Fig. 5.7 To demonstrate the similarity of the chromatograms from a scale-up experiment a separation of a therapeutic protein is shown using a laboratory-scale column (A) and a pilot-scale column (B). (A) Fractogel EMD DEAE (M) (Merck KGaA), packed in a 2.6-cm inner diameter column, was loaded with a total protein amount of 1.3 g (sample volume: 815 mL) at a flow rate of 13.3 mL min^{-1}. (B) About 155 g of protein (sample volume: 96 L) injected onto a 20-cm inner diameter column containing the same type of resin gave nearly the same elution profile. The volumetric flow rate was adjusted to 47.1 L h^{-1}. The linear flow rate was 150 cm h^{-1}; bed height of both columns was about 26 cm (Buffer A: 50 mM NaP, pH 6.3; Buffer B: 50 mM NaP, 1 M NaCl, pH 6.3).

When choosing the media, long-term availability of the required amounts of media, lot-to-lot consistency, resin lifetime and supporting documentation have to be considered. The most relevant requirements on chromatographic media for production purposes are summarized in Tab. 5.4.

5.5
Application Areas

Ion-exchange chromatography is easy to scale-up to production scale and the results are very reproducible. The robustness of ion-exchange chromatography

Tab. 5.4 Main selection criteria and additional requirements on biochromatographic media (different prerequisites exist for manufacturing and laboratory scale)

Feature	Impact on manufacturing scale	Impact on laboratory scale
High dynamic capacity	+	–
Good recovery	+	–
Scalability	+	–
Cleanability	+	+
Good documentation (regulatory support file)	+	–
Lot-to-lot consistency	+	+
Long-term supply issues	+	–
Long shelf life	+	–
Large batch size	+	–
Certificates of analysis etc.	+	+
Possibility to do vendor audits (initially and periodically)	+	–
Longevity	+	–
Mechanical and chemical stability	+	–
Sufficient selectivity	+	+
Appropriate pore size	+	+
Reasonable costs	+	–

combined with the high protein-binding capacity makes this method very useful for most of the purification strategies, either for capture or intermediate steps, or even polishing. Many protein purification protocols contain one or more ion-exchange chromatography steps. When using small beaded ion-exchange chromatography media, a relatively good resolution can be achieved and, thus, occasionally ion-exchange chromatography is performed late in the procedure for polishing. The main task of this particular step is to remove possible impurities, such as structurally very similar or closely related forms of the product like aggregates, deamidated or oxidized product, or isoforms (Tab. 5.5). However, the most powerful application of ion-exchange chromatography is the capture step.

Principal applications of ion-exchange chromatography are purifications of cytosolic proteins [65–69], plant proteins [70, 71], plasma-derived or recombinant blood coagulation factors [72–77] and other proteins from different sources including recombinant proteins [78–86]. Membrane proteins and lipoproteins can also be isolated with this technique [87–90]. In addition, plasmid purification by ion-exchange chromatography at the production scale is described in the literature [91–94]. Recently, papers on ion-exchange chromatography of viruses and vaccines were published showing a crucial role of the weak anion exchanger Fractogel EMD DEAE [95–99].

Tab. 5.5 Main impurities that have to be removed during downstream processing

Process related	Product related[a]
Host cell proteins	Aggregates
DNA	Modified forms
Antibiotics	Degradation products
Endotoxins	Glycosylation variants
Solvents, additives, water	
Viruses	
Lysozyme (inclusion bodies)	

a Characterization is necessary for some product-related contaminants that cannot be removed.

Ion-exchange chromatography is the method of choice for the intermediate purification of monoclonal and polyclonal antibodies, but can also be used for capture. In contrast to purification strategies for plasma-derived antibodies, where weak ion exchangers are widely used, strong ion exchangers are preferred if monoclonal antibodies have to be manufactured. Antibodies have been reported to have isoelectric points between 5 and 8 and, therefore, can be purified at pH < pI on cation exchangers [100–102]. High purities and high recoveries can be obtained using weak cation exchangers carrying a carboxy group changing the degree of ionization depending on pH. However, the main ligand for cation-exchange chromatography is in many cases the sulfo group. Cation exchangers are able to remove unfolded forms of the monoclonal antibody as well as aggregates of the monoclonal antibody and small molecular weight impurities [103]. Both modes of ion-exchange chromatography are able to separate protein A leakage products and host cell proteins. Strong anion-exchange chromatography can efficiently remove DNA and, in addition, was described to contribute to the virus clearance.

References

1 Boardman, N. K., Partridge, S. M., *Biochem J* **1955**, *59*, 543.
2 Kopaciewicz, W., Rounds, M. A., *J Chromatogr* **1984**, *283*, 37.
3 Velayudhan, A., Horvath, Cs, *J Chromatogr* **1986**, *367*, 160.
4 Gallant, S. R., Vunnum, S., Cramer, S. M., *J Chromatogr A* **1996**, *725*, 295.
5 Kopaciewicz, W., Rounds, M. A., Fausnaugh, J., Regnier, F. E., *J Chromatogr* **1983**, *266*, 3.
6 Drager, R. R., Regnier, F. E., *J Chromatogr* **1986**, *359*, 147.
7 Drager, R. R., Regnier, F. E., *J Chromatogr* **1987**, *406*, 237.
8 Chicz, R. M., Regnier, F. E., *Anal Chem* **1989**, *61*, 2059.
9 Regnier, F. E., *Science* **1987**, *238*, 319.
10 Chicz, R. M., Regnier, F. E., *Anal Chem* **1989**, *61*, 2059.
11 Yao, Y., Lenhoff, A. M., *Anal Chem* **2004**, *76*, 6743.
12 Yao, Y., Lenhoff, A. M., *Anal Chem* **2005**, *77*, 2157.
13 Drager, R. R., Regnier, F. E., *J Chromatogr* **1987**, *406*, 237.

14 Wu, D., Waiters, R. R., *J Chromatogr* 1992, *589*, 7.

15 DePhillips, P., Lenhoff, A. M., *J Chromatogr* 2001, *933*, 57.

16 DePhillips, P., Lenhoff, A. M., *J Chromatogr* 2004, *1036*, 51.

17 DePhillips, P., Lagerlund, I., Färenmark, J., Lenhoff, A. M., *Anal Chem*, 2004, *76*, 5816.

18 Kopaciewicz, W., Rounds, M. A., Regnier, F. E., *J Chromatogr* 1985, *318*, 157.

19 Rounds, M. A., Regnier, F. E., *J Chromatogr* 1984, *283*, 37.

20 Brewer, S. J., Sassenfeld, H. M., *Trends Biotechnol* 1985, *3*, 119.

21 Ladiwala, A., Rege, K., Breneman, C. M., Cramer, S. M., *Langmuir* 2003, *19*, 8443.

22 Brooks, C. A., Cramer, S. M., *AIChE J* 1992, *38*, 1969.

23 Brooks, C. A., Cramer, S. M., *Chem Eng Sci* 1996, *51*, 3847.

24 Gallant, S. R., Vunnum, S., Cramer, S. M., *J Chromatogr A* 1996, *725*, 295.

25 Raje, P., Pinto, N. G., *J Chromatogr A* 1997, *760*, 89.

26 Natarajan, V., Cramer, S. M., *AIChE J* 1999, *45*, 27.

27 Ghose, S., Cramer, S. M., *J Chromatogr A* 2001, *928*, 13.

28 Gallant, S. R., *J Chromatogr A* 2004, *1028*, 189.

29 Melander, W. R., El Rassi, Z., Horváth, C., *J Chromatogr* 1989, *469*, 3.

30 Mazsaroff, I., Varady, L., Mouchawar, G. A., Regnier, F. E., *J Chromatogr* 1990, *499*, 63.

31 Stahlberg, J., Jönsson, B., Horvath, Cs, *Anal Chem* 1991, *63*, 1867.

32 Stahlberg, J., Jönsson, B., Horvath, Cs, *Anal Chem* 1992, *64*, 3118.

33 Stahlberg, J., Jönsson, B., *Anal Chem* 1996, *68*, 1536.

34 Jönsson, B., Stahlberg, J., *Colloids Surfaces B* 1999, *14*, 67.

35 Roth, C. M., Lenhoff, A. M., *Langmuir* 1993, *9*, 962.

36 Roth, C. M., Unger, K. K., Lenhoff, A. M., *J Chromatogr A* 1996, *726*, 45.

37 Chicz, R. M., Regnier, F. E., *Anal Chem* 1989, *61*, 2059.

38 Rounds, M. A., Regnier, F. E., *J Chromatogr* 1984, *283*, 37.

39 Lesins, V., Ruckenstein, E., *Colloid Polym Sci* 1988, *266*, 1187.

40 Noinville, V., Vidal-Madjar, C., Sebille, B., *J Phys Chem* 1995, *99*, 1516.

41 Mazza, C. B., Sukumar, N., Breneman, C. M., Cramer, S. M., *Anal Chem* 2001, *73*, 5457.

42 Song, M., Breneman, C. M., Jinbo, B., Sukumar, N., Bennet, K. P., Cramer, S., Tugcu, N., *J Chem Inf Comput Sci* 2002, *42*, 1347.

43 Ladiwala, A., Rege, K., Breneman, C. M., Cramer, S. M., *Langmuir* 2003, *19*, 8443.

44 Malmquist, G., Nilsson, U. Hjellström, Norrman, M., Skarp, U., Strömgren, M., Carredano, E., *J Chromatogr A* 2006, *1115*, 164.

45 Bruch, T., Graalfs, H., Jacob, L., Frech, C., submitted.

46 Carta, G., Uibera, A. R., Papst, T. M., *Chem Eng Technol* 2005, *28*, 1252.

47 Colby, C. B., O'Neill, B. K., Vaughan, F., Middelberg, A. P. J., *Biotechnol Prog* 1996, *12*, 662.

48 Gallant, S. R., Kundu, A., Cramer, S. M., *Biotechnol Bioeng* 1995, *47*, 355.

49 Yamamoto, S., Nakanishi, K., Matsuno, R., *Ion-Exchange Chromatography of Proteins*. Marcel Dekker, New York, 1988.

50 Yamamoto, S., *Biotechnol Bioeng* 1995, *48*, 444.

51 Yamamoto, S., Ishihara, T., *J Chromatogr A* 1999, *852*, 31.

52 Yamamoto, S., Ishihara, T., *Sep Sci Technol* 2000, *35*, 1707.

53 Yamamoto, S., Kita, A., *Trans IChemE* 2006, *84*, 72.

54 Ishihara, T., Yamamoto, S., *J Chromatogr A* 2005, *1069*, 99.

55 Yamamoto, S., *Chem Eng Technol* 2005, *28*, 1387.

56 Kaltenbrunner, O., Jungbauer, A., Yamamoto, S., *J Chromatogr A* 1997, *760*, 41.

57 Yamamoto, S., Kita, A., *J Chromatogr A* 2005, *1065*, 45.

58 Yamamoto, S., Nomura, M., Sano, Y., *J Chromatogr A* 1987, *396*, 355.

59 Yamamoto, S., Nakanishi, K., Matsuno, R., Kamibuko, T., *Biotechnol Bioeng* 1983, *25*, 1373.

60 Velayudhan, A., Horvath, C.,
J Chromatogr **1988**, *443*, 13.

61 Cysewski, P., Jaulmes, A., Lemque, R.,
Sebille, B., Vidal-Madjar, C., Jilge, G.,
J Chromatogr **1991**, *548*, 61.

62 Schenerman, M. A., Hope, J. N., Kletke,
C., Singh, J. K., Kimura, R., Tsao, E. I.,
Folena-Wasserman, G., *Biologicals* **1999**,
27, 203.

63 Cahill, M., Macniven, R., Hawkins, K.,
Gallo, C., Sernatinger, J., Myers, J.,
Notarnicola, S., Downstream GAb
Abstracts: "Reports from GAb 2000",
Barcelona, **2000**, Amersham
Biosciences 18-1150-47.

64 Rathore, A., Velayudhan, A., BioPharm
2003, Jan, 34.

65 Smith, C. R., Knowles, V. L., Plaxton,
W. C., *Eur J Biochem* **2000**, *267*,
4477.

66 Rozwadowski, K., Zhao, R., Jackman,
L., Huebert, T., Burkhart, W. E.,
Hemmingsen, S. M., Greenwood, J.,
Rothstein, S. J., *Plant Physiol* **1999**, *120*,
787.

67 Gregus, Z., Németi, B., *Toxicol Sci*
2002, *70*, 13.

68 Gross, W., Lenze, D., Nowitzki, U.,
Weiske, J., Schnarrenberger, C., *Gene*
1999, *230*, 7.

69 Shern, J. F., Sharer, J. D., Pallas, D. C.,
Bartolini, F., Cowan, N. J., Reed, M. S.,
Pohl, J., Kahn, R. A., *J Biol Chem* **2003**,
278, 40829.

70 Cacace, S., Schröder, G., Wehinger, E.,
Strack, D., Schmidt, J., Schröder, J.,
Phytochemistry **2003**, *62*, 127.

71 Zuurbier, K. W. M., Leser, J., Berger,
T., Hofte, A. J. P., Schröder, G.,
Verpoorte, R., Schröder, J.,
Phytochemistry **1998**, *49*, 1945.

72 Heidtmann, H.-H., Kontermann, R. E.,
Thrombosis Res **1998**, *92*, 33.

73 Fischer, B., Mitterer, A., Dorner, F.,
J Biotechnol **1995**, *38*, 129.

74 Radosevich, M., Zhou, F.-L., Huart, J.-J.,
Burnouf, T., *J Chromatogr B* **2003**, *790*,
199.

75 Burnouf, T., Goubran, H., Radosevich,
M., *J Chromatogr B* **1998**, *715*, 65.

76 Burnouf, T., *J Chromatogr B* **1995**, *664*,
3.

77 Fischer, B. E., Kramer, G., Mitterer, A.,
Grillberger, L., Reiter, M., Mundt, W.,
Dorner, F., Eibl, J., *Thromb Res* **1996**, *84*,
55.

78 Anspach, F. B., Spille, H., Rinas, U.,
J Chromatogr A **1995**, *711*, 129.

79 Panelius, J., Lahdenne, P., Heikkilä, T.,
Peltomaa, M., Oksi, J., Seppälä, I., *J Med
Microbiol* **2002**, *51*, 731.

80 Cruz, H. J., Conradt, H. S., Dunker, R.,
Peixoto, C. M., Cunha, A. E., Thomaz,
M., Burger, C., Dias, E. M., Clemente, J.,
Moreira, J. L., Rieke, E., Carrondo,
M. J. T., *J Biotechnol* **2002**, *96*, 169.

81 Sankala, M., Brännström, A., Schulthess,
T., Bergmann, U., Morgunova, E., Engel,
J., Tryggvason, K., Pikkarainen, T., *J Biol
Chem* **2002**, *277*, 33378.

82 Sprenger, G. A., Schörken, U., Sprenger,
G., Sahm, H., *Eur J Biochem* **1995**, *230*,
525.

83 Thuioudellet, C., Oster, T., Wellman, M.,
Siest, G., *Eur J Biochem* **1994**, *222*, 1009.

84 Sprenger, G. A., Schörken, U., Sprenger,
G., Sahm, H., *J Bacteriol* **1995**, *177*(20),
5930.

85 Arolas, J. L., Lorenzo, J., Rovira, A.,
Castellà, J., Aviles, F. X., Sommerhoff,
C. P., *J Biol Chem* **2004**, *280*, 12113.

86 Pagano, A., Cinque, G., Bassi, R., *J Biol
Chem* **1998**, *273*(27), 17154.

87 Puglielli, L., Mandon, E. C., Rancour,
D. M., Menon, A. K., Hirschberg, C. B.,
J Biol Chem **1999**, *274*(7), 4474.

88 Fricke, B., Buchmann, T., Friebe, S.,
J Chromatogr A **1995**, *715*, 247.

89 Grüber, G., Godovac-Zimmermann, J.,
Nawroth, T., *Biochim Biophys Acta* **1994**,
1186, 43.

90 Kashino, Y., *J Chromatogr B* **2003**, *797*,
191.

91 Eon-Duval, A., Burke, G., *J Chromatogr B*
2004, *804*, 327.

92 Diogo, M. M., Queiroz, J. A., Prazeres,
D. M. F., *J Chromatogr A* **2005**, *1069*, 3.

93 Urthaler, J., Schlegl, R., Podgornik, A.,
Strancar, A., Jungbauer, A., Necina, R.,
J Chromatogr A **2005**, *1065*, 93.

94 Urthaler, J., Buchinger, W., Necina, R.,
Acta Biochim Polon **2005**, *52*, 703.

95 Kamen, A., Henry, O., *J Gene Med* **2004**,
6, 184.

96 Arcand, N., Bernier, A., Transfiguracion, J., Jacob, D., Coelho, H., Kamen, A., BioProcess J **2003**, Jan/Feb, 72.

97 Burova, E., Ioffe, E., *Gene Ther* **2005**, *12*, 5.

98 Green, A. P., Huang, J. J., Scott, M. O., Kierstead, T. D., Beaupré, I., Gao, G.-P., Wilson, J. M., *Human Gene Ther* **2002**, *13*, 1921.

99 Ralston, R., Thudium, K., Berger, K., Kuo, C., Gervase, B., Hall, J., Selby, M., Kuo, G., Houghton, M., Choo, Q.-L., *J Virol* **1993**, *67*(11), 6753.

100 Denton, G., Murray, A., Price, M. R., Levison, P. R., *J Chromatogr A* **2001**, *908*, 223.

101 Bai, L., Burman, S., Gledhill, L., *J Pharm Biomed Anal* **2000**, *22*, 605.

102 Iyer, H., Henderson, F., Cunningham, E., Webb, J., Hanson, J., Bork, C., Conley, L., BioPharm **2002**, Jan, 14.

103 Necina, R., Amatschek, K., Jungbauer, A., *Biotechnol Bioeng* **1998**, *60*, 689.

6
Displacement Chromatography of Biomacromolecules

Ruth Freitag

6.1
Introduction

Sales of biotechnology products are expected to show double-digit annual growth rates over the next decades in several key sectors. The market for biopharmaceutical products, dominated by proteins (and peptides) for human therapeutics and diagnostics, looks especially promising. Increasingly, product isolation, the so-called "downstream process", is becoming the most costly part of such production processes. In the past it has often been stated that this is largely due to low productivity, e.g. in the case of mammalian cells, while most bacterial products were difficult to isolate because they were produced intracellularly, e.g. due to use of the popular *Escherichia coli* as production organism. However, in this regard we have seen much progress in recent years and even mammalian cells are now approaching or even surpassing the gram per liter production scale [1]. If anything, however, instead of diminishing, the relative contribution of the downstream process to the overall production cost and time is increasing. At first it may appear strange that the downstream process turns out to be a difficult part of bioproduction; however, the challenge facing biotechnology in that area is not so much one of general feasibility – almost any substance may be procured in pure form some way or the other – but that of operating within a certain financial and dimensional framework.

The difficulty of a given downstream process depends mainly on the complexity of the feed and the target molecule concentration therein together with the required final purity and composition. Feed product concentrations may vary between several grams per liter (antibiotics) and some milligrams per liter (recombinant blood factors). In order to gain approval for a pharmaceutical, for example, it must be guaranteed that a sufficient amount can be provided to satisfy the foreseeable medical need. Even if this may mean only a few kilograms per year, the sheer size of the product stream can pose a problem given the typical product concentrations. Moreover, substances which typically are produced at rather low product titers in the bioprocess tend to require the very highest levels of final purity, e.g. parentera-

Bioseparation and Bioprocessing. Edited by G. Subramanian
Copyright © 2007 WILEY-VCH Verlag GmbH & Co. KGaA, Weinheim
ISBN: 978-3-527-31585-7

lia intended for use in human beings. Concentration plus isolation are therefore the somewhat contradictory goals of many downstream processes in biotechnology. Intellectually, the downstream process is often divided into three distinct stages that differ somewhat in goal. These stages are capture, intermediate purification and final polishing. These three stages are typically preceded by a preparatory stage in which the product stream is rendered suitable for (standard) downstream processing. In particular, intermediate and final purification steps tend to depend heavily on chromatographic principles to achieve separation. This has largely to do with the high selectivity and "biocompatibility" of chromatographic operations. Concomitantly, however, chromatographic steps today contribute drastically to the difficulties and costs of the corresponding isolation process.

Preparative chromatography constitutes an established unit operation in the chemical industry for the separation of closely related substances. In the 1940s and 1950s, chromatography was, for instance, used by industry at fairly large scale to separate mixtures of rare earth oxides or hydrocarbons from crude oil [2]. Soon, however, the expensive chromatography was replaced in most areas by unit operations such as distillation, extraction or fractionation. A similar development is unlikely in the case of biopolymers such as proteins, which are often incompatible with these separation procedures. However, it should be noted that already truly large-scale processes and facilities dedicated to the production of antibiotics or to the fractionation of complex materials such as blood and milk, for example, rely on unit operations like extraction, filtration and precipitation rather than chromatography for purification, even at the price of forsaking certain valuable substances that are not recoverable by these procedures. While it is therefore improbable that chromatography as a separation principle will be replaced as easily in bioseparation as it has been in the petrochemical industry, the need to improve the productivity and reduce the costs of biochromatography is obvious given the growing needs of the biotech industry.

Most scientists and engineers in the life sciences and related areas are familiar with chromatography, albeit usually in its analytical form. This has certainly contributed to the predominance of the (overloaded) elution mode on the preparative scale. Using this approach, chromatographic units capable of producing tons of material per year have been build. A further increase in scale within a reasonable financial and technical framework seems at present unlikely. Instead, it may be unavoidable to revisit the complex field of theory of nonlinear chromatography to design more efficient preparative chromatographic separations. The recent success of simulated-moving-bed separations shows that this is possible. Moreover, modern computers together with numerical solutions for mathematical equations for which no analytical solution can be found, today make the use of theoretical calculations of nonlinear chromatographic systems much more accessible than even a decade ago. The corresponding computer programs already exist and can be put to use. In the meantime, the question of what constitutes the most versatile approach to preparative bioseparation awaits its final answer. In the intermediate scale, especially, displacement chromatography, a method where the components are resolved into consecutive zones of the pure and highly concentrated substances rather than into peaks, may become a serious competitor to overloaded elution chromatography for high-resolution, efficient biopolymer isolation schemes.

Examples for the successful exploitation of the displacement mode in prepara-tive chromatography date back to the beginning of industrial chromatography and biomolecules, such as amino acids, were among the first applications (for an excel-lent review of the earlier applications of displacement chromatography, see Cramer and Subramanian [3]). Before the 1980s, however, displacement chromatography was mainly used for the separation of rare earth oxides and isotope separation due to the limited efficiency of the available chromatographic systems and packings at that time [4, 5].

It was only when highly efficient high-performance liquid chromatography (HPLC) columns and systems became available in the 1980s, together with modern computers and the mathematical tools for dealing with the problems of nonlinear chromatography, that the displacement mode was rediscovered and applied to biomolecules, largely thanks to the dedicated work of Csaba Horvath and his group at Yale University.

6.2
Background and Basic Principle of Displacement Chromatography

The process of chromatography can be mathematically described by a mass balance together with appropriate initial and boundary conditions. The mass balance, the initial condition and the exit boundary condition apply to chromatography in general. For the typical chromatographic batch column we can write:

$$\text{Mass balance: } \partial c_i/\partial t + \phi \partial q_i/\partial t + u_o \partial c_i/\partial z = D_{\text{eff}} \partial^2 c_i/\partial z^2 \quad i = 1, 2, \ldots, n \quad (1)$$

where c_i is the mobile phase concentration, q_i is the stationary phase concentration, ϕ is the phase ratio, u_o is the linear flow velocity, D_{eff} is the dispersion coefficient, t is the time and z is the dimensionless column length.

$$\text{Initial condition: } c_i(0, z) = 0, 0 \leq z \leq L \quad i = 1, 2, \ldots, n \quad (2a)$$

$$\text{Exit boundary condition: } (\partial c_i/\partial z)_{z=L} = 0, \quad (2b)$$

where L is the column length.

As already suggested by Tiselius, three modes of chromatography, i.e. elution, frontal and displacement chromatography, can then be distinguished by the cor-responding three-inlet boundary conditions [6]. In particular, the sample mixture may enter the column as a Dirac or rectangular pulse (elution chromatography) or as a step function (frontal chromatography). The transition between these two modes is gradual. Highly overloaded elution chromatography, for example, shows certain aspects of frontal chromatography. In the third mode of chromatography, the displacement mode, the rectangular sample pulse is followed by a step function of another substance, dubbed "displacer", which adsorbs strongly to the stationary face surface.

In this particular case, the inlet boundary condition is:

$$c_i(t, 0) = c_{o,i} \quad 0 < t \leq \tau \quad i = 1, 2, \ldots, n-1 \text{ (sample)} \tag{3a}$$

$$c_D(t, 0) = c_{o,D} \quad H(t - \tau) \text{ (displacer)}, \tag{3b}$$

where $H(t)$ is the step function and τ is the duration of feed introduction.

The following Gedankenexperiment may be useful to understand the basic principle of displacement chromatography. Imagine a substance A distributed in a two-phase system consisting of a solid adsorbent and a second fluid phase. Under these conditions, a dynamic equilibrium establishes itself between the relative amount of A bound to the adsorbent and the relative amount of A dissolved in the liquid phase. This equilibrium over a concentration range is commonly described by the corresponding equilibrium isotherm. The interaction of many biologicals with chromatographic stationary phases is characterized by favorable (convex upward) isotherms and the well-known Langmuir adsorption model or one of the many extensions thereof often presents a reasonable first approximation of the experimentally observed interaction. For a Langmuir isotherm (Eq. 4a) the adsorbed amount q_A is directly proportional to the liquid concentration c_A as long as this concentration is fairly low (linear part of the adsorption isotherm). If the liquid-phase concentration increases, the increase in q_A decreases with increasing c_A, up to a point where it becomes independent of the fluid phase concentration (nonlinear part of the adsorption isotherm). If a second substance B is introduced to the system, the situation becomes potentially more complex. As long as the adsorption of both species can be described by linear adsorption isotherms (conditions typical for analytical chromatography), the interaction of the two substances with the solid phase can be considered independently. This renders the theory of linear (analytical) chromatography fairly accessible. However, this is no longer the case under the (strongly) nonlinear conditions prevailing in preparative chromatography. Under these circumstances, a direct competition for the binding sites will ensue and the amount of bound A will also depend on the liquid phase concentration of B (see Eq. 4b), the equilibrium isotherm of substance A has been "suppressed" by substance B under nonlinear conditions.

$$q_A = \frac{ac_A}{1 + bc_A} \tag{4a}$$

$$q_A = \frac{ac_A}{1 + b_A c_A + b_B c_B}, \tag{4b}$$

with $a = q_{max}b$ and $b = k_a/k_d$, where q_{max} is the maximum stationary-phase concentration, k_a is the rate constant of the adsorption reaction and k_d is the rate constant of the desorption reaction.

In displacement chromatography, this competition for the binding sites is used to drive the separation. The first step of a displacement separation (Fig. 6.1) is the adsorption of the substance mixture on the column under conditions favorable to binding. During loading some degree of separation is already achieved due to a frontal chromatographic effect. A considerable portion of the stationary-phase capacity may be exploited during that phase.

After the relevant feed components have been adsorbed, a solution containing the displacer is pumped through the column. A displacer should bind more strongly to the stationary phase under chromatographic conditions than the sample components. It is therefore able to compete successfully for the binding sites with any of them. As the displacer front advances, the number of binding sites available to the sample compounds decreases, and the more strongly bound substances will begin to displace and push ahead the less strongly bound ones. Finally, in a system governed by favorable isotherms, all sample components are focused into consecutive zones of the pure substances lined up according to stationary-phase affinity. The so-called "displacement train" or "scaled isotachic state" has been formed. Under these (highly idealized conditions) the borders between adjacent zones are self-sharpening. If for some reason a molecule travels into the zone ahead of its own bulk zone, it will be amongst molecules with a lesser stationary-phase affinity than itself, i.e. it will be strongly adsorbed and retained until overtaken by its own zone. The opposite is true for a molecule staying behind its zone for some reason. It will be immediately dislocated and moved ahead until rejoined to the proper zone. Following the breakthrough of the displacer front, the column has to be regenerated and re-conditioned for further use.

The velocity of the displacement train and thus the speed of the separation is dictated by that of the displacer front, u_D. According to the material balance

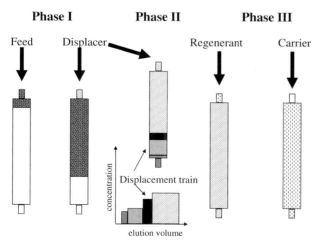

Fig. 6.1 Schematic presentation of the steps involved in a typical separation by displacement chromatography.

argument of de Vault [7], the velocity of the latter is in the case of ideal chromatographic conditions given by:

$$u_D = \frac{u_o}{1 + \phi(\Delta q_D/\Delta c_D)}, \qquad (5)$$

with Δq_D and Δc_D corresponding to the differences in the displacer concentrations – adsorbed and dissolved – before and in the displacer zone. As the concentrations in front of the displacer zone are zero, $\Delta q_D/\Delta c_D$ for a given displacer concentration c_D corresponds to the ratio q_D/c_D defined by the single-component displacer isotherm.

As a result of the enforced isotachicity of the various substance zones, the individual ratios $\Delta q_i/\Delta c_i$ (or q_i/c_i) for the various substances must equal q_D/c_D. An "operating line" with a slope given by $\Delta q_D/\Delta c_D$ determines the entire system and in particular the concentration in each sample zone, c_i, depends solely on the relation between the respective component isotherm and the displacer isotherm (Fig. 6.2). By changing the displacer concentration, one changes both the speed of the

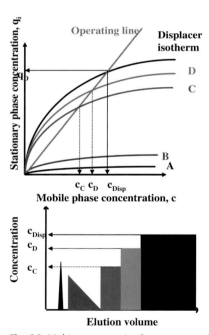

Fig. 6.2 Multicomponent isotherm system of the feed components and the displacer. The corresponding fully developed displacement train is shown below. The concentration in the substance zones of the displacement train depends on the intersection of the operating line with the respective substance's multicomponent isotherm. The steepness of the operating line is defined by the isotherm and the carrier concentration of the displacer.

separation $[u_D = f(c_D)]$ and the concentration of all substance zones. Note that substances whose adsorption isotherms are only touched or not intersected at all by the operating line will not be displaced. Instead, such substances elute ahead of the displacement train.

The length of each substance zone in the displacement train depends on the original amount of the substance in the feed. The respective feed concentration, on the other hand, is of little consequence to the development of the displacement train. According to the ideal mode of displacement chromatography (see Section 6.3), this is also the case for trace components and in the case of highly diluted feeds. In practice, dispersive effects will prevent trace components from reaching their theoretically predicted concentration maximum; however, the application of displacement chromatography to isolate and enrich trace components prior to detailed analysis has been demonstrated in a couple of real-life applications (see Section 6.6.5). Moreover, since column loading takes place under conditions favorable to adsorption, even highly diluted feeds can be given an excellent prognosis in terms of recovery, concentration factor and purity; as long as suitable conditions are chosen, the final product concentration can be considerably higher than in the feed. Just as important, however, in terms of preparative protein chromatography is the fact that the concentration in the substance zone may easily be kept below some critical (aggregation, denaturation, viscosity, e.g. Refs. [8, 9]) level, since a concentration plateau rather than a "peak" is produced in displacement chromatography. It should also be noted that due to the efficient use of the stationary phase, the displacement approach constitutes an efficient way to quickly purify a few milligrams of protein on an analytical instrument. This option has recently been demonstrated for the processing of some chemically synthesized dinucleotide (5′,5′) polyphosphates by reversed-phase displacement chromatography [10]. Another major advantage of displacement chromatography with regard to preparative applications is the fact that feed, displacer and regenerant are introduced as simple step functions. Displacement chromatography is therefore much easier adapted to continuous separation of multicomponent mixtures than (gradient) elution chromatography, e.g. by continuous annular chromatography [11–13]. Contrary to (gradient) elution chromatography, the displacer stays behind the substance zones rather than intermingling with the target molecule.

In spite of these advantages, only a few applications of displacement chromatography for industrial protein isolation are known (see Section 6.7). This comparative lack certainly has to do with the "problem areas" of displacement chromatography, i.e. the monitoring of the displacement train, the necessity of a suitable displacer, and the perceived difficulties in modeling and optimizing this exclusively nonlinear chromatographic mode. Monitoring the displacement train, i.e. differentiating between the consecutive, highly concentrated substance zones, has in the past been difficult and time consuming. Typically fractions were collected and analyzed afterwards. This has negative consequences for the process time, while the potential for automation is low in such a system. The use of a two-dimensional chromatographic system, where an analytical HPLC automatically provides an on-line analysis of the effluent of the preparative column in short intervals, is a possible alternative to the off-line analysis [14]. Modern analytical HPLC systems allow us to carry out the required analysis within seconds, while state-of-art process control software can use these data to control the automatic collection of the desired substances at a given quality, e.g. according to a pre-set purity and/or concentration threshold. The particular challenges of modeling biopolymer displacement chromatography and suitable biopolymer displacers will be discussed below.

6.3
Modeling and Simulation of Displacement Chromatography

A chromatographic separation is governed not only by the (thermodynamic) interaction equilibria, but also by the kinetics of this interaction, by mass transfer phenomena, various extra column effects (e.g. contributing to dispersion), and putative secondary interactions including protein–protein interactions, aggregation and denaturation. The modeling of chromatographic separations is more difficult for nonlinear than for linear conditions, due to the fact that the behavior of the sample components can no longer be considered independently. The successful interaction of one molecule lowers the chances of another to bind and *vice versa*. Moreover, this is not only the case for other sample components, but also (often completely ignored) for all other solutes, e.g. also the mobile-phase modifiers added to enforce desorption in (overloaded) gradient elution chromatography. Enrichment effects, but also zone distortion as a result of such competition for interaction has been observed and analyzed (e.g. by the groups of Horvath and Velayudhan [15, 16]). Related considerations have led to the development of the steric mass action (SMA) model for nonlinear chromatography (see Section 6.3.5).

The kinetics of mass transfer and the surface reaction can rarely be neglected in biopolymer chromatography, and must be incorporated into the model, e.g. via the mass balance equation [17, 18]. The resulting numerical algorithms have promoted the understanding and even the design of nonlinear chromatographic separations, e.g. with regard to the effects of chromatographic parameters such as the column dimensions, the particle diameter and porosity, the mobile phase flow rate, and the composition and concentration of the feed (and in our case also of the concentration and heterogeneity of the displacer) [19, 20]. For an excellent introduction into the numerical modeling of displacement chromatography, see Guiochon and coworkers [21].

6.3.1
The Ideal Model of Displacement Chromatography

The first detailed analysis of nonlinear multicomponent systems was given by Glückauf [22], who provided an analysis of the displacement phenomenon and discussed effects of solute and displacer concentrations as early as 1935. In the late 1960s, Helfferich and coworkers presented the first algorithm for a mathe-matical description of displacement chromatography based on the theory of inter-ference originally developed for stoichiometric ion-exchange systems [23]. Rhee and coworkers later developed a similar theory based on the theory of systems of quasi-linear partial equations and the method of characteristics [24]. These earlier approaches as well as many recent ones assume ideal chromatography. Ideal chromatographic conditions prevail when it is legitimate to assume a separation exclusively controlled by the sorption equilibrium thermodynamics, plug flow of the mobile phase, i.e. the total absence of dispersion effects, and infinite column efficiency, the absence of mass transfer or kinetic effects, and a system that is thermodynamically consistently described by multicomponent Langmuir iso-therms. Under such conditions the mass balance simplifies to:

$$\partial c_i/\partial t + \phi \partial q_i/\partial t + u_\circ \partial c_i/\partial z = 0 \quad i = 1, 2, \ldots, n. \tag{1a}$$

In the case of a nonlinear multicomponent system, a series of mass balance equations is written for the various compounds and coupled via the multi-component Langmuir isotherms. The system is then transformed into a set of simple – and at that time already solvable – algebraic functions using the so-called h-transformation (Helfferich and coworkers) or ω-transformation (Rhee and coworkers). Although the characteristic parameters cannot be determined explic-itly for systems of more than two substances, their calculation by simple numerical methods is possible. In ideal displacement chromatography the isotachic state is always reached at some point of time and distance. The development of the displacement train is usually shown in a normalized distance–time diagram. The column length necessary for the development of the displacement train can be taken directly from this diagram.

The only experimental parameters needed for the treatment of displacement chromatography within the hermeneutics of the ideal model are the equilibrium isotherms of the relevant substances and the isotherm of the displacer. In 1985 the model was used by Frenz and Horvath to simulate the separation of proteins by high-performance displacement chromatography [25]. The model was found to correctly describe the most important aspects of displacement chromatography and the influence of parameters such as the sample and displacer concentration on a given separation. This is especially the case under highly nonlinear conditions and for small molecules. The model can, however, be of only limited worth in a practical situation. This is not a problem of displacement chromatography *per se,* but a general calamity of nonlinear chromatography especially in the case of (bio)polymers and their sometimes erratic adsorption behavior.

6.3.2
Kinetic Models for Displacement Chromatography

One reason for the success of the ideal model of displacement chromatography in simulating experimental results is due to the fact that conditions were employed in the corresponding experiments that approach ideal conditions to a high degree, e.g. high-efficiency columns (several thousand plates per meter), strongly nonlinear conditions (high sample concentration, large amounts) and separations of small molecules (fast mass transfer and reaction kinetics). However, even in such cases the abrupt discontinuous concentration changes ("shocks") predicted by the ideal model will be smoothed out into sharp jet continuous changes in concentration ("shock layers"), since dispersive effects (molecular and eddy diffusion, mass transfer and reaction kinetics, etc.) counteract the equilibrium thermodynamics. The implementation of dispersive effects into the simulations requires a model that somehow takes the limited column efficiency into account. Moreover, in a more realistic model of nonlinear chromatography, two aspects have to be considered. One is the physical transport of the compounds through a fixed bed of porous particles, the other is the surface reaction between the compounds and the stationary phase. Phenomena like axial dispersion as well as mass transfer and kinetic resistance can be incorporated into the mass balance equation, e.g. by the introduction of dispersion coefficients, pore/overall column efficiency parameters and even radial velocity profiles [17, 18]. Numerous kinetic models are available to account for the band profile development in chromatography [17, 26, 27]. Combined with isotherm data these should, in principle, allow the simulation and concomitantly the *in silico* development and optimization of the separation.

In nonlinear chromatography peak shape and spreading are thus complex functions of the equilibrium isotherms, the mass transfer parameters (film mass transfer coefficients and intraparticle diffusivity), as well as certain design (particle size and column length) and operating parameters (pulse size, feed concentration and flow rate) [28–32]. For the purpose of modeling, the process is usually divided into three discrete steps, i.e. mass transfer from bulk liquid to the outer surface of the stationary phase particles (film diffusion resistance, external mass transfer resistance), movement by diffusion into the pores of the adsorbent (pore diffusion, internal mass transfer resistance) and binding to the adsorptive surface (surface reaction resistance). The kinetic models differ in their degree of flexibility in taking these effects into account.

6.3.2.1 The Equilibrium-dispersive (ED) or Transport-dispersive (TD) Model in Displacement Chromatography

When the mass transfer resistances are small and have a minor influence on the profiles, models like the ED or TD model can be used [17, 33–35]. Such models lump several effects together. They are easy to apply, require only a few simply determined experimental parameters and, as a consequence, enjoy much popularity. However, the predictive value of such models is limited, since the involved lumped mass transfer and dispersion parameters are physically meaningless. In some cases these parameters have, for example, shown an inexplicable dependency on the sample concentration [36].

The ED or semi-ideal model assumes that the dispersive effects have simply a modifying influence on the final band profile. For this purpose, a constant bulk axial dispersion coefficient, D_{bulk}, is introduced into the ideal mass balance equation:

$$\partial c_i/\partial t + \phi \partial q_i/\partial t + u_o \partial c_i/\partial z = D_{bulk}\partial^2 c_i/\partial z^2 \quad i = 1, 2, \ldots, n. \tag{1b}$$

The dispersion coefficient, which is the only experimental parameter besides the component isotherms required by the model, is for plate numbers above 100 plates per meter easily calculated from the plate height, H (plate number, N) of the column:

$$D_{bulk} = Hu_o/2 = Lu_o/2N, \tag{6}$$

with $N = L/H$.

It is assumed that the plate height of a given column is identical under linear and nonlinear conditions.

There are no closed-form analytical solutions to the ED model. However, the comparative simplicity of the model facilitates the calculation of numerical solutions, using computation methods such as finite differences, finite elements or collocation. Using numerical calculations, the band profiles can be calculated for any specified parameters and initial conditions. Compared to a similar calculation by the ideal model, the basic features of a displacement train stay the same, as the dispersive effects do not introduce new phenomena. In particular, features such as the length and height (concentration) of the zones stay the same, only the shocks between zones are transformed into shock layers. A shock layer has similar properties (velocity, etc.) as the shock; its thickness is determined by the dispersive effects, i.e. the plate number of the column.

6.3.2.2 Complex Kinetic Models for Displacement Chromatography

As long as the various dispersive effects are dominated by a single one, e.g. axial dispersion, the ED (TD) model can be expected to give good results. However, especially in biopolymer chromatography, this is rarely the case. Instead, several phenomena must be considered more or less individually and, in cases like this, the ED (TD) model is not capable of providing helpful simulations. Instead, a more complex kinetic model should be used. In such models the mass balance equation is combined with a kinetic equation relating the rate of variation of the concentration of each component in the stationary phase to its concentration in both phases and to the equilibrium concentration in the stationary phase. An important decision concerns the question of which degree of complexity is necessary in a given situation. Various models have been proposed, which vary mainly in the choice of the kinetic rate expressions.

The general rate (GR) model is by far the most comprehensive one [17, 26]. In this model, axial dispersion and mass transfer resistances are calculated individually. In many ways it is a very attractive model, as it allows the development of equations based on assumptions concerning the actual physical behavior of the proteins under chromatographic conditions. The GR model does, however, require knowledge of a large number of experimental values that are difficult to determine.

The lumped pore diffusion (POR) model is intermediate in complexity between the GR and the ED/TD models, and is in fact a simplification of the GR model [21, 37]. The POR model needs the same set of parameters as the GR model, but can be solved much faster. As Kaczmarski and coworkers were recently able to demonstrate [37], the POR model can replace the GR model without loss in predictive power when:

$$\text{Pe} > 100 \quad \text{and} \quad \text{St}/\text{Bi} > 5,$$

where Pe is the Peclet number ($uL/DL\varepsilon_e$), St is the Stanton number ($k_{ext}a_pL\varepsilon_e/u$), Bi is the Biot number ($k_{ext}d_p/2D_{eff}$), u is the mobile phase velocity, d_p is the average particle diameter, L is the column length, a_p is the external surface area of the adsorbent particles, ε_e is the external porosity, k_{ext} is the external mass transfer coefficient and D_{eff} is the effective diffusion coefficient.

Most recently we have witnessed the development of hybrid models combining both physical and empirical aspects. The generalized run-to-run control (GR2R) strategy suggested by Cramer and coworkers [38] for displacement and overloaded elution chromatography belongs into this category. The GR2R approach may be used to control chromatographic processes in the presence of sporadic and auto-correlated disturbances. In this approach the physical model is used to determine the initial parameters of the nonlinear empirical model, which is updated after each run using a nonlinear recursive parameter estimation method. The updated empirical model is then used in the control algorithm (predictive control) to estimate operating conditions for the next batch. According to the authors the GR2R strategy outperforms process operation under fixed optimal conditions in the presence of various disturbances.

6.3.3
The Shock Layer Theory

A shock layer of a certain thickness is unavoidable in experimental displacement chromatography. The resolution, but also the productivity of a given separation in turn depends strongly on the shock layer thickness. According to Guiochon [21], the shock layer thickness, $\Delta\eta_t$, between two successive zones in the isotachic train is given by:

$$\Delta\eta_t = \left(\frac{(1+K_D)^2 D_{ax}}{K_D u_o^2} + \frac{1}{k_f} \right) \frac{1+\alpha}{1-\alpha} \ln \left| \frac{1-\theta}{\theta} \right|, \tag{7}$$

where $K_D = k_D'/(1+b_D c_D)$, k_D' is the capacity factor of the displacer, k_f is the film mass transfer coefficient, α is the separation factor (k_i'/k_j'), D_{ax} is the axial dispersion coefficient and θ is the characteristic parameter. It is usually assumed that the dispersion coefficients and the rate constants are equal for all compounds. Otherwise the shock layer thickness lays between the values calculated for the two compounds.

According to Eq. (7), $\Delta\eta_t$ depends directly on the axial dispersion and the mass transfer coefficient of the two components, their separation factor, as well as on the concentration and retention factor of the displacer [39]. The shock layer thickness between the zones does not depend on the compounds' retention factors or their concentration in the feed/sample. The separation factor, α, is defined as the quotient between the retention factors, k', of any two compounds. Its influence on a displacement would be difficult to account for by the ideal or even the ED model. By the shock layer theory, it can be shown that the column length required to achieve full displacement increases rapidly as the separation factor approaches unity, as the shock layer thickens in proportion to $(\alpha + 1)/(\alpha - 1)$ and increases dramatically as α decreases toward unity. Displacement chromatography is therefore not superior to elution chromatography in resolving binary mixtures with a very small value of α. In such a case, the shock layers would encompass the entire displacement train, while the time required to achieve isotachic conditions would result in a very low throughput.

The influence of parameters such as the column's length and plate height, the particle diameter, the carrier flow rate or the sample size and concentration on the separation can be predicted from the shock layer theory. These predictions only apply to the established displacement train. Column efficiency is just as important in displacement chromatography as in elution chromatography, since the shock layer thickness decreases with increasing plate number. Parameters influencing the plate height (flow rate, particle diameter) have a direct influence of the shock layer thickness and thus of the chromatographic result. For a given column diameter, the column length required to achieve the displacement equilibrium increases with absolute amount of the components. Note that the loading factor for which the isotachic train is formed remains constant.

Differentiation of Eq. (7) shows that the shock layer thickness becomes smallest for $K_D = 1$, i.e. for:

$$k'_D = 1 + b_D c_D,$$
(8a)

or:

$$c_D = (k'_D - 1)/b_D.$$
(8b)

In other words, the shock layer is wide at low displacer retention and/or at high displacer concentration. For $k'_d \gg 1 + b_d c_d$ the shock layer thickness increases linearly with increasing k'_d. As a consequence, displacer retention factor and displacer concentration can only be optimized together. The optimal mobile phase flow velocity is given by [40]:

$$u^D_{opt} = \sqrt{\frac{D_{ax}(1 + K_d)^2 k_f}{K_d}}.$$
(9)

Contrary to the situation in elution chromatography, in displacement chromatography the optimum mobile-phase velocity depends not only on the axial dispersion and mass transfer resistance, but also on displacer properties such as the displacer's retention factor and the concentration [39]. Differentiation of Eq. (9) shows that an optimal u_{opt} is again obtained in the case of $K_d = 1$. In practical terms, the means that the optimum flow rate will usually be lower in displacement than in elution chromatography.

The shock layer concept also has consequences for the experimentally observed enrichment of trace compounds. According to the theory, a trace component will be considerably enriched in displacement chromatography, often much more so than in elution chromatography. However, when the zone width reaches the same order of magnitude as the shock layer thickness, no further enrichment takes place. Instead, the zone width stabilizes. This also applies to impurities of the displacer. Any displacer impurity acts as an additional (trace) compound in the displacement train. Impurities with an equilibrium isotherm below that of a certain compound of the separation mixture will contaminate the displacement train [41]. If the separation factors between the bulk displacer and the impurities are close to unity, formation of the fully developed displacement train will be difficult. Homogeneous displacers are therefore preferable.

Finally, the shock layer concept can help to predict the so-called "over-displacement effect". According to the ideal model of displacement chromatography, the only possible effect of an increase in the displacer concentration is an increase in the concentration of the substance zones with a concomitant shortening. Under experimental conditions, however, the relative amount of target molecules found in the shock layer increases together with the shortening of the substance zone. The term "over-displacement" describes a situation where bands are so narrow that no concentration plateau of the pure substance is observed while the two shock layers in front and at the end of the zone touch.

6.3.4
Isotherm Models

It has be stated before that for a simulation of multicomponent separations in chromatography it will be necessary to couple the various mass balance equations via the multicomponent adsorption isotherms describing the equilibrium distribution of the components between the mobile and the stationary phase. At higher surface concentrations, when the competition of the various substances for the adsorption sites can no longer be ignored, these interactions become exceedingly difficult to model over a sufficiently wide concentration range. This is especially the case for biopolymers, where adsorption is a complex process, which may, for example, involve multipoint interaction, conformational changes, interaction of the components with each other (adsorbed and in solution) or multiple retention mechanisms. The slow diffusion rate of such large molecules makes the attainment of true adsorption equilibrium within the practical timescale of chromatographic separations highly unlikely.

6.3.4.1 The Langmuir Algorithm

In ion-exchange chromatography, in particular, stoichiometric models such as the stoichiometric displacement model (SDM) [42] or the SMA model [43] have been proposed to take some of these effects into account. In most cases, however, the Langmuir model is used to describe the adsorption behavior [17]. This model has been derived originally from a simple kinetic consideration of the adsorption of small molecules from a gaseous phase to a surface. Later a thermodynamic derivation has been proposed. The basic hypotheses of the model are the following:

- The molecules are adsorbed on a fixed number of defined sites.
- Each site accepts one (and only one) molecule.
- Adsorption occurs in the form of a monolayer.
- All sites are energetically equal.
- There is no interaction between the adsorbed molecules.

The adsorption reaction rate can then be described as $k_a c(1 - q)$ and the desorption reaction rate as $k_d q$. Under the assumption of equilibrium, these two equations can be used to derive the well-known Langmuir equation (Eq. 4a). This simple approach often describes the experimentally observed behavior well hence its continued popularity. Variants of the Langmuir approach, such as the bi-Langmuir, the Langmuir–Freundlich or the extended Langmuir (multilayer) model, can be used to allow the consideration of binding sites of different strength. It has also been proposed that experimentally observed deviations from the proposed shape of the Langmuir isotherm, such as an initial inflection, may simply be taken into account via the numerical value of the b_1 parameter [44]. Most theoretical treatments to date rely therefore on the multicomponent Langmuir formalism (Eq. 4b) for the description of the nonlinear multicomponent system, which can easily be constructed for each component in question, once the individual Langmuir constants, b_i and a_i, have been derived from the respective single component isotherms.

However, the Langmuir multicomponent approach has been shown to violate the Gibbs–Duhem law of thermodynamics [45]. The correlation with the experimental results is also often not satisfactory [17, 46]. The separation factors will hardly ever be constant over the entire concentration range, especially in biopolymer chromatography. The conformational changes and the interaction with already adsorbed molecules will lead to a change in interaction energy and thus to an "S"-shaped adsorption isotherm. Obviously this cannot be accounted for by the Langmuir-isotherm model, which assumes constant separation factors. Moreover, since the adsorption energy tends to increase with molecular mass, while the saturation capacity decreases in the same direction, a larger molecule will often be characterized by an isotherm with a steeper initial slope, but at a lower saturation plateau than a similar but smaller molecule. The corresponding single-component isotherms would cross and a separation gap develop in the system. If the operating line crosses the isotherms within this gap, "displacement azeotropes" are formed and no separation takes place [46, 47]. If it intersects on the right- or the left-hand side of the gap, a separation of the two compounds by displacement is possible. However, the order of the substances in the displacement train is opposite in the two cases. The occurrence of such azeotropes has been observed experimentally, e.g. by Carta and Dinerman for the separation of α-aminobutyric acid and isoleucine on Dowex 50W-X8 resin [48] and by Kim and Cramer for protein separations in the immobilized metal-affinity chromatography (IMAC) mode [49]. A similar reason was proposed by Kasper and coworkers for their inability to resolve a mixture of antithrombin III (AT III) and bovine serum albumin (BSA) on hydroxyapatite columns under certain experimental conditions [50].

Several more suitable approaches to protein isotherm formulation such as the LeVan and Vermeulen isotherm derived from the theory of the ideal adsorbed solution (IAS) [51] can be found in the literature. The direct determination of multicomponent isotherms has also been proposed, but is experimentally rather involved, if possible at all.

6.3.5
The SMA Model

Over the last decade, a model (the SMA model) has been developed by Brooks and Cramer [43] for the particular needs of the predictive simulation of displacement separations. While initially developed for ion-exchange chromatography, the model has since been extended to the description of IMAC [52] as well as dye–ligand (Cibacron Blue) affinity displacement chromatography [53, 54] and should in theory be applicable to any type of adsorption chromatography governed by a single interaction mechanism. The model describes nonlinear ion-exchange chromatography of large molecules (proteins) simply by the law of mass action assuming ideal chromatographic conditions. In such it is kin to the multivalent ion-exchange formalism proposed by Velayudhan and Horvath [55] and the stoichiometric displacement model used by Regnier and coworkers [56]. Innovatively, however, it takes into account that a large molecules will not only (actively) interact with certain adsorptive sites on the stationary phase surface, but will also (passively) cover other of these interaction sites simply due to its bulk. Most importantly, the model takes into account the fact that a salt gradient is induced in front of the displacer, creating a salt microenvironment for each component zone [57]. This gradient causes changes in the displacement train, which are otherwise difficult to explain.

The SMA model is only able to simulate the fully developed displacement train; simulations of the developing train or the effect of the dispersive effects are beyond its scope. Within its limitations, however, the reported agreement with the experimental results is good. Nonideal effects such as aggregation or changes in the tertiary structure of the protein, but also other potential deviations from the ideal stoichiometric case such as van der Waals and electrostatic interactions between adsorbed proteins and the salt ions or secondary (e.g. hydrophobic) interactions between the proteins and the adsorbate also escape consideration. In most cases these can be ignored. If not, the combination of the SMA approach with the non-ideal surface solution (NISS) model proposed by Li and Pinto [58] may be used.

The SMA formalism is specifically designed for representing multicomponent protein-salt equilibria in ion-exchange chromatography based on the following assumptions:

- The solution and adsorbed phases are thermodynamically ideal allowing the use of concentrations instead of activities.
- The multipoint nature of protein binding can be represent by an experimentally determined characteristic charge.
- Competitive binding can be represented by the law of mass action where the electroneutrality on the stationary phase is maintained.
- The binding of large molecules causes a steric hindrance of salt counterions bound to the adsorptive phase, which become unavailable for exchange with other solutes.
- The effect of the co-ion can be neglected in the ion-exchange process.

Assuming that electrostatic interaction is the only mechanism involved during adsorption, the stoichiometric exchange of a polyelectrolyte (protein) and the small counterions can be represented by:

$$c_P + v_P\overline{q_S} \Leftrightarrow q_P + v_P c_S,$$
(10)

where v is the characteristic charge, and subscripts P and S refer to protein and salt respectively. The overbar denotes salt ions available for free exchange with other solutes.

The equilibrium constan is defined as:

$$K_P = \left(\frac{q_P}{c_P}\right) * \left(\frac{c_S}{\overline{q_S}}\right)^{v_P}.$$
(11)

Electroneutrality on the stationary phase requires that:

$$\Lambda \equiv \overline{q_S} + (v_P + \sigma_P) * q_P.$$
(12)

where Λ is the ion capacity of the column and σ is the steric factor of the protein.

The following implicit isotherm can then be written for the protein by combining Eqs. (11) and (12):

$$c_P = \left(\frac{q_P}{K_P}\right)\left(\frac{c_S}{\Lambda - (v_P - \sigma_P)q_P}\right)^{v_P}.$$
(13)

Physically, the characteristic charge represents the number of ion-exchange groups on the stationary phase surface involved in the ion-exchange reaction with a given protein molecule, the steric factor represents the number of counterions on the adsorbent surface, which are unavailable for exchange with other molecules in solution due to the shielding by the adsorbed protein, while the equilibrium constant is a measure of the affinity of the macromolecule to the stationary phase.

Experimentally, the characteristic charge, v, of the proteins and the displacer can be calculated from isocratic elution data using the following formula:

$$\log k' = \log(\phi K_P \Lambda^{v_P}) - v_P \log c_S. \tag{14}$$

Equation (13) can be rewritten in the following manner to calculate the steric factor, σ, from points in the nonlinear range of the respective adsorption isotherms:

$$\sigma = \left(\frac{\left(\dfrac{c_S}{\left(K_P \dfrac{c_P}{q_P} \right)^{1/v_P}} - \Lambda \right)}{-q_P} \right) - v_P. \tag{13a}$$

Alternatively the following formula can be used to calculate the steric factor of strongly adsorbing substances:

$$\sigma = (\Lambda / q_{max}) - v. \tag{13b}$$

While K and v can thus be calculated from linear chromatography experiments, the steric factor must be determined under nonlinear conditions.

Based on the SMA parameters, two relations, i.e. the operating regime plot and the dynamic affinity plot, can be established to help process development (Fig. 6.3). The dynamic affinity plot is a graphical method to described solute stationary phase affinities in ion-exchange displacement chromatography [59]. In particular, a dynamic affinity parameter λ is derived from the linear SMA parameters of the solutes and the characteristic velocity of the displacer front, respectively the slope of the operating line defined by q_D/c_D (Δ):

$$\lambda = \sqrt[v_i]{\frac{K_i}{\Delta}} \quad or \quad \log K_i = \log \Delta + \log \lambda_i v_i. \tag{15}$$

A plot of $\log K$ versus v for all components in the system thus yields a series of lines with individual slopes $\log \lambda_i$ and $\log \Delta$ as the common intercept on the y-axis (Fig. 6.3a).

(a)

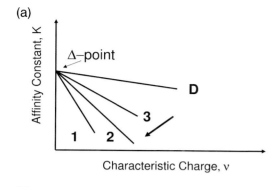

(b)

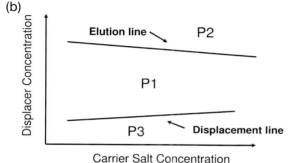

Carrier Salt Concentration

Fig. 6.3 (a) Presentation of the dynamic affinity plot according to the SMA model of a displacer and a three-substance mixture to be separated. The arrow points in the direction of decreasing stationary-phase affinity. This plot predicts a displacement train in the order of 1–2–3–D for the corresponding experiment. (b) Presentation of the operating regime plot according to the SMA model. The plot is created for a protein of interest P. If conditions (salt/displacer concentration) corresponding to P1 are chosen, the protein will be displaced; if conditions corresponding to P2 are chosen, the protein will elute ahead of the displacement train; if conditions corresponding to P3 are chosen, the protein will desorb somewhere inside the displacer zone.

For each compound the corresponding affinity plot defines two regions – a region below the dynamic affinity line, where solutes have a lower dynamic affinity than the molecule of interest, and a region above the affinity line, where solutes have a higher dynamic affinity and can therefore displace the molecule of interest. The dynamic affinity plot can be used to illustrate and predict the elution order in a displacement separation, but also makes predictions concerning the ability of a putative displacer to displace a given target molecule possible.

The operating regime plot defines boundaries between chromatographic modes [60], in particular the boundary between displacement and desorption (displacement line) and the boundary between displacement and elution (elution line) (Fig. 6.3b). The first line of the operating regime plot is calculated by selecting values for c_D and substituting them into Eq. (8) in order to obtain the corresponding critical salt concentration c_S':

$$c'_S = (K_D/\Delta)^{1/v_D}\{\Lambda - [(v_D + \sigma_D)c_D\Delta_D]\}. \tag{16}$$

The elution line is calculated by modifying values for the displacer's partition coefficient Δ (q_D/c_D) based on Eq. (17a and b). The critical displacer concentration at which elution of the protein in the induced gradient occurs, is given by Eq. (17a), while the corresponding critical salt concentration c''_S is given by Eq. (17b).

$$c_D = \{\Lambda - [1 - (K_D/\Delta)^{1/v_D}(\Delta/K_P)^{1/v_P}]\}/\{(\Delta/K_P)^{1/v_P}[\Lambda - ((v_D + \sigma_D)\Delta)]\} \tag{17a}$$

$$c''_S = (K_D/\Delta)^{1/v_D}\{\Lambda - [(v_D + \sigma_D)c_D\Delta]\}. \tag{17b}$$

If conditions c_D and c_S are chosen that for a given target molecule fall between the displacement and elution lines, displacement is possible. If conditions for a given molecule are above the elution line, this substance will elute ahead of the displacement train. If conditions are below the displacement line, this substance will either be retained on the column or desorb within the displacer zone (Fig. 6.4). Especially in the case of small displacer molecules, displacement, elution and desorption region show a pronounced dependency on the displacer and salt concentrations. Selective displacement may become possible in such cases, where mainly the target molecule is displaced, while most impurities either elute ahead of the displacement train or desorb in the displacer zone.

More recently, the SMA model has been incorporated into a number of kinetic models to predict nonlinear chromatography. In Ref. [61], the POR model is used; in Refs. [62, 63], the SMA model is used together with a solid film linear driving force model to describe the effects of feed load, flow rate, particle diameter, initial

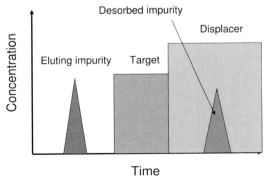

Fig. 6.4 Schematic presentation of the principle of selective displacement chromatography (conditions defined by the operating regime plot according to the SMA model, see also Fig. 6.3(b). Only the compound of interest is displaced, impurities elute either ahead of the displacement train or a redesorbed only in the displacer zone.

salt concentration and the displacer partition ratio on the separation. The latter approach was also the basis for a comparison of linear gradient and displacement separations in ion-exchange systems [64]. It was shown that for "easy" separations both modes are equally effective, but that displacement has advantages as the separations become more challenging (low separation factors). For such mixtures production rates could be significantly higher in displacement chromatography. The SMA formalism can be used as the basis of an iterative optimization scheme for the identification of optimum operation conditions, e.g. for various resin materials, but also to predict the significance, of diverging, converging or parallel affinity lines on the performance [65].

6.4
Technical Aspects and Process Development

Practical biopolymer displacement chromatography often takes place under conditions that were explicitly excluded in the theoretical treatment of the displacement phenomenon, e.g. in the presence of secondary equilibria (denaturation, aggregation, reactions), in non-Langmuirian systems characterized by complex, sometimes crossing or unfavorable isotherms, under conditions were mass transfer and reaction kinetics are dominant and cannot be treated simply as additions to the axial dispersion, but also under conditions were the relevant parameters are noticeable concentration dependent. The growing number of successful applications of displacement chromatography in bioseparation shows, however, that displacement chromatography is less limited by these circumstances than one would assume from the theoretical considerations (see also Section 6.7).

The development of a separation by displacement chromatography will usually involve the following steps: (i) choosing the stationary phase/chromatographic mode, (ii) choosing the displacer, (iii) choosing/optimizing the carrier composition, (iv) adjusting the column length/sample size, (v) adjusting the flow rate and perhaps (vi) adjusting the temperature, as an increase in temperature may result in beneficial effects such as an decrease in viscosity or improved mass transfer and reaction kinetics. The optimization of a displacement separation remains a challenge and simulations are still only of limited help. Some of the factors governing practical method development in displacement chromatography are discussed below. Due to its overwhelming importance, the discussion of suitable displacers for biopolymer displacement chromatography will be put in a separate section of this chapter (Section 6.5).

6.4.1
The Stationary Phase/Interaction Mode

The vast majority of the published protein displacement separations have been done in the ion-exchange mode, while most peptides and other smaller biomolecules have been separated in the reversed-phase mode (see also Section 6.7 for examples). Occasionally other stationary phases have been used, including IMAC, antibody exchange, hydroxyapatite or hydrophobic-interaction chromatography phases. When choosing a stationary phase, some attention should be paid to the availability of the bulk material. As the column length is an important parameter in displacement chromatography (see Section 6.4.3), the length of the available prepacked columns may not be optimal.

The particle diameter of the stationary phase is another important parameter, since mass transfer effects have a substantial effect in biopolymer displacement chromatography. So far discussions of the effect of particle diameters in displacement chromatography are inconclusive. Smaller particles stand for higher efficiencies, i.e. lower theoretical plate heights, but higher backpressures. While Felinger and Guiochon [66] argue for an optimum ratio of the squared particle diameter over column length for each sample composition, Subramanian and coworkers [67] found their separation efficiencies unchanged up to particle diameters of 90 μm. Theoretically derived arguments also predict a minimum size below which the displacement kinetics become rate limiting.

Since mass transfer can be even more limiting in displacement than in elution chromatography, stationary phases for biopolymer chromatography that improve this particular feature may become interesting for displacement chromatography. Examples include the "perfusion" chromatography beads from PerSeptive Biosystems and the continuous bed column (UNO™ column) from Bio-Rad. The stationary phases used in perfusion chromatography contain uncommonly large pores, so-called throughpores, through which a convective fluid flow is possible. Access of the inner particle surface becomes less dependent of the slow diffusion process and usually significantly less mass transfer resistance is observed. However, the mobile-phase flow velocity needs to be quite high, much higher than usual in traditional displacement chromatography, for the perfusion phenomenon to occur. Perfusion displacement chromatography has been used to separate the genetic variants of β-lactoglobulin at a flow rate of 4 mL min^{-1} [68]. Thus, 18 mg was prepared within 90 s using heparin as displacer. The authors claimed a resolution similar if not better to that achievable in conventional displacement chromatography performed at much lower flow rates.

A continuous bed or monolithic column may be envisioned as a porous polymer rod through which the mobile phase flows. Many mass transfer limitations are no longer observed in such columns. Their van Deemter curve usually shows an unusual low optimum, since the nodule size seems to determine the A term. Moreover, the column efficiencies remain more or less constant even at elevated flow rates. The preparative application of this concept is somewhat handicapped by the fact that such columns cannot be scaled up beyond the few milliliter scale, due to difficulties in handling the polymerization at larger scales. Here the superior exploitation of the column capacity by displacement compared to elution chromatography may be most relevant. An anion-exchange UNO column (column volume 1 mL) has, for example, been successfully used for the separation of whey proteins by displacement chromatography [69]. Compared to similar experiments on traditional beaded supports, the shock layer between the two protein zones was narrow in spite of the shorter column length (3 cm for the UNO column versus 5 cm for the traditional column). The possibility of using elevated flow rates with this column was investigated in a subsequent paper by the same group [70] where it was shown that the flow rate could indeed be elevated by one order of magnitude for such columns without any deterioration of the separation quality.

In a systematic comparison of particulate and monolithic columns (both Bio-Rad) using the SMA model as a means to aid and accelerate method development, Freitag and Vogt found better agreement between the predictions and the experimental results in the case of the monolithic column [71]. This was ascribed to the reduced intraparticular mass transfer resistance. When the SMA model was subsequently used to develop a separation of cationic proteins (ribonuclease, chymotrypsinogen) using poly(diallyl-dimethylammonium chloride) (polyDADMAC) as displacer on both monolithic and particle-based strong cation-exchanger columns [72], the SMA parameters and, especially, the operating regime plot was inconclusive in the case of the monolithic column. Conditions suggested by the plot did not translate into successful displacement separations. That such a separation was possible, albeit using different and empirically derived conditions, was subsequently shown by the same authors. An attempt to model breakthrough curves of such monolithic columns in comparison to the particle-based ones using a combination of the SMA approach with the lumped pore diffusion model is described in Ref. [61]. The authors found the approach equally suited to both stationary phase morphologies, presumably due to a better incorporation of the mass transfer phenomena in the model. For both column morphologies the characteristic charge needed some fitting as a function of the mobile phase composition. Finally, in 2001, Ghose and Cramer proposed a reaction-dispersive SMA formalism, which takes the enhanced mass transfer properties of monolithic materials into account and which could be utilized successfully for simulating displacement separations on such column materials [73]. Kinetics and consequently mobile phase salt concentrations are very important variables in such separations, and, contrary to conventional wisdom, elevated salt concentrations may improve the separations in the case of monolithic columns.

Apart from their superior mass transfer properties, monolithic columns also should be chosen in displacement chromatography whenever molecules of vastly different size have to be separated. In such cases the competition for the binding sites may be overlaid by differences in the accessibility of the interactive surface. Such an effect has been described in the separation of plasmid DNA (4.7 kb) from *E. coli* endotoxins and proteins [9]. While this separation was not possible on the particle based strong anion-exchanger column, it was no problem when the UNO Q column was used. Poly(acrylic acid) (PAA) served as displacer in both cases.

6.4.2
Composition of the Mobile Phase/Flow Rate

In displacement chromatography it is often advised to use a mobile phase that supports strong binding of the substances and the displacer to the stationary phase. However, it should be kept in mind that the separation factor rather than the individual retention factors determine the separation of any two components and that even the displacer retention factor cannot be considered individually, but must be optimized in relation to the displacer concentration (see Section 6.3.3). While a higher displacer concentration means a faster separation and high product concentration factors, it also diminishes the length of the zones and thus increases the amount of substance found in the shock layers. In practical terms, however, the solubility of the displacer will usually be the limiting factor in that regard. Low retention factors (normalized retention times) for all involved components including the displacer may actually improve displacement separations. The optimum values may well rest between 1.2 and 2.0 [66]. Given the tendency of biopolymers to show all-or-nothing binding, the deciding factors in choosing the mobile phase in biodisplacement chromatography are often less chromatographic and more biological in nature, i.e. prevention of denaturation, interaction/aggregation or the solubility of the target molecule.

Some discrepancy can also be found in the discussion of the optimal flow rate. Typically flow rates used in practical displacement chromatography, especially when porous particle-based stationary phases are considered, are 2–10 times lower than in elution chromatography. Zhu and Guiochon convincingly argue that the optimum mobile-phase linear velocity for minimum shock layer thickness and maximum recovery yield does not only depend on the axial dispersion and mass transfer resistance as in elution chromatography, but also very much so on the retention factor and the concentration of the displacer [39]. According to this treatment, the optimal flow rate in displacement chromatography will be lower than that of the corresponding elution chromatography and in most cases of biopolymer displacement chromatography, this optimum flow rate will be much too low to be of practical relevance. Most authors use flow rates between 0.1 and 0.5 mL min^{-1} in biopolymer displacement chromatography in the semi-preparative scale (typical column volumes: few milliliters). At higher flow rates, the quality of the separation tends to decrease rapidly, although flow rates of up to 1 mL min^{-1} have occasionally been used with success even with analytical-scale columns [47]. However, these lower flow rates in displacement chromatography do not necessarily translate into lower throughputs or productivities. In the detailed investigation by Gerstner [74] it was shown for the separation of oligonucleotides by displacement and elution chromatography that although the flow rate could be twice as high in the elution mode and the feed per run similar (due to the fact that at 15 L the elution column was 3 times as large as the displacement column), only 311 runs were required to produce 5 kg of product in the displacement versus 467 runs in case of the elution mode. The number of production days was 26 in the displacement compared to 39 days in the elution mode. Due to the higher productivity, and the lower consumption of solvent, chemicals and labor as well as to the much better recovery, the production costs for this amount of product would amount to $3 657 000 in the elution case compared to only $2 677 000 in the displacement case.

6.4.3
Column Length/Sample Size

When the occurrence of an adsorption azeotrope can be ruled out, there should be an optimum column length for a given feed load/sample size. The optimization of column length and sample size should thus be interactive. When the column length is increased beyond that required for the development of the displacement train, the "quality" of the separation in terms of the sharpness of the displacer front or the border between two neighboring zones is not further improved. Often it is easier to adjust the sample size to the dimensions of a given column, rather than optimize the column length for a given sample. Displacement chromatography of proteins has been shown on columns as short as 3 cm. On the other hand, the absolute amount of protein separated, as well as the separation time and the column's backpressure for a given flow rate increase with column length. Finally, it should also be considered whether the full development of the displacement train is always necessary. While this does maximize the recovery yield, it does not necessarily correspond to the highest productivity. Throughputs may, for example, be higher when the column is not long enough to allow the full development of the displacement train and the mixed zones are recycled. The decision will depend on the position of the target molecule in the displacement train and the stability of the molecule.

6.5
Displacers for Biopolymer Displacement Chromatography

The displacer is a unique feature of displacement chromatography. The importance of the displacer for a successful displacement separation can hardly be overestimated. In general, the ideal displacer should have the following features. It should be nontoxic and stable. It should combine high solubility in the carrier with a high binding tendency towards the stationary phase. Column regeneration must nevertheless be possible. In addition, the displacer should be highly uniform, since displacer impurities/heterogeneity may make column regeneration difficult or pollute the substance zones depending on their relative affinity. Detectability, costs and the possibility to recycle the displacer are other considerations. Ideally the displacer should also be easy to remove from the recovered substance fractions. For pharmaceutical applications it might even be necessary to sterilize the material.

The type of molecule that may be considered as a displacer strongly depends on the chromatographic interaction. Displacement chromatography of comparatively small biologicals such as amino acids, peptides and small proteins (antibiotics, insulin) is usually carried out in the reversed-phase mode. Hydrophobic substances such as 2-(2-butoxyethoxy)ethanol (BEE), decyltrimethylammonium bromide, cetyltrimethylammonium bromide (cetramide), benzyldimethyldodecylammonium bromide, dodecyloctyldimethylammonium chloride and palmitic acid are possible displacers in these cases [3]. Nucleotide and nucleoside mixtures have been separated using a similar combination of reversed-phase stationary phases and hydrophobic displacers. The separation of oligonucleotides is also possible in the anion-exchange mode using dextran sulfate as displacer [74].

6.5.1
Protein Displacers

The situation is more difficult in the case of biopolymer and, in particular, protein displacement chromatography. In this case it is often assumed that a protein displacer needs to be a large molecule itself, all the more so since little to nothing was known until recently about the physicochemical basis of a "good" biopolymer displacer [75] and elution order under linear chromatographic conditions was often confused with ability to displace. Cases where the authors claimed to have used a comparatively small substance as displacer, e.g. Ref. [76], were countered by the argument that extremely heterogeneous polymer preparations were used, which contained high-molecular-weight substances acting as the "real" displacers.

Protein displacement chromatography is currently almost exclusively done in the ion-exchange mode. Reversed-phase separations are less suited to preparative protein purification in general, since many proteins denature under reversed-phase conditions. Other stationary phases have been used only occasionally. Hydrophobic-interaction chromatography, a very popular option in preparative elution chromatography, is almost unknown in displacement chromatography, presumably largely due to a lack of suitable displacers, i.e. mildly hydrophobic, water-soluble polymers. A few cases of hydrophobic-interaction displacement chromatography have been reported, using mainly proteins as displacer. For example, the separation of a mixture of ribonuclease and lysozyme on a hydrophobic-interaction chromatography column (TSK-Butyl NPR) has been achieved using BSA as displacer [77]. Shukla and coworkers have proposed the use of small charged (and hence water-soluble) molecules containing several short alkyl and/or aryl groups, such as benzyltributyl ammonium chloride or benzethonium chloride as protein displacers in hydrophobic-interaction chromatography [78]. The method has since been applied to the purification of an industrial protein mixture [79], see also below. Ruaan and coworkers proposed a triblock copolymer [$A_{12}M_4B_{12}$, with A = methacrylic acid, M = methyl methacrylate, B = 2-(dimethylamino)ethyl methacrylate] to displace trypsinogen and α-chymotrypsinogen from butyl-Sepharose phases [80].

In the case of the much more common ion-exchange displacement chromatography, a number of (semi-)synthetic polyelectrolytes such as chondroitin sulfate, dextran sulfate, carboxymethyl starch, alginate, Eudragit, Nacolyte 7105 and polyethyleneimine (PEI) have been used as protein displacers [81]. Since 1978, Torres and Peterson have promoted the use of (modified) carboxymethyldextrans (CMDs) for that purpose [82]. (Semi-)synthetic polyelectrolytes have the advantages of being cheap, available in large variety and often suited to sterilization (e.g. by filtration). However, they are generally only available in heterogeneous mixtures varying considerably in molecular mass and structure. Displacer heterogeneities and impurities pose grave problems, since the stationary-phase affinity of the different molecules varies and the less well-retained displacer molecule fractions contaminate the product zones. In addition, the high-molecular-mass fractions of the displacer preparation have very high stationary phase affinities and are therefore almost impossible to remove from the column. Solubility tends to be restricted for large polyelectrolytes, while many synthetic polymer (displacer) preparations contain toxic impurities (e.g. residual monomer molecules). The high viscosities of concentrated polymer solutions pose another problem.

In almost any case a protein that is known to bind exceptionally well to a given stationary-phase material may then be used as a displacer of less-well-bound proteins. Protamine was, for example, suggested as a protein displacer in cation-exchange displacement chromatography [83], while heparin may act as a nontoxic displacer in anion-exchange methods [84]. While proteins as protein displacers yield important data, they are also expensive, fragile and generally not suited for large-scale pharmaceutical applications. However, they do represent a homogeneous molecule population and are susceptible to detection by ultraviolet (UV)/visible analysis. Their removal from the product is also comparatively well understood.

In recent years certain low-molecular-weight substances have also been discussed as putative protein displacers in the ion-exchange mode. In fact, it was suggested as early as 1990 that low-molecular-mass molecules may act as protein displacers, and that such small displacers have advantages in terms of column regeneration and fine-tuning of the separations [14, 76, 85–87]. In general, small molecules will be more sensitive to the actual chromatographic conditions and, in the case of ion-exchange chromatography, most specifically to the salt content of the mobile phase than large displacer molecules. They have the advantage of a precise molecular mass and hence a homogenous stationary-phase affinity. As small molecules they are characterized by superior mass transfer properties and even at high concentration they do not increase the viscosity of the mobile phase to the same extent as a polymer would. Low-molecular-weight displacers can easily be removed from the protein product by filtration or dialysis and are often cheaper than the corresponding larger ones. Today it is generally accepted that small molecules in the range of several hundred to several thousand grams per mole can be effective displacing agents for much larger molecules. In fact when alginate was used as displacer in anion-exchange chromatography [88] it was the shorter polysaccharide chains obtained by acid hydrolysis that gave the best results. Using such displacers it was possible to separate the two variants of β-lactoglobulin that are identical in size and vary only by 0.1 units in their isoelectric points.

Examples of low-molecular-weight protein displacers include modified amino acids [N-α-benzoyl-L-arginine ethyl ester (BAEE)] [89], low-molecular-weight dextran sulfates [87] and pentosan polysulfate (molecular weight 3000 g mol^{-1}), which were used for the separation of the genetic variants of β-lactogobulin [85].

Polyvinylsulfonic acid (molecular weight 2000 g mol^{-1}) [88] and pentaerythritol based dendritic polymers (molecular weight 480–5100 g mol^{-1}) were used as displacers of basic proteins such as α-chymotrypsinogen and cytochrome *c* on cation-exchanger materials [86]. Antibiotics such as neomycin B and streptomycin A have been suggested as displacers especially of very strongly retained proteins such as lysozyme in cation-exchange displacement chromatography [90]. Low-molecular-weight displacers for anion-exchange displacement chromatography such as 1,2-benzene disulfonic acid, *p*-toluene sulfonic acid, 1,5-naphthalene disulfonic acid, pentane sulfonic acid, methane sulfonic acid and phytic acid as well as their performance under various conditions are discussed in [91, 92]. Modified ethylene glycols (molecular weight 1000–10 000 g mol^{-1}) and small chelating agents such as ethylene glycol-bis(β-aminoethylether)-*N,N,N',N'*-tetraacetic acid (EGTA; molecular weight 380.4 gmol^{-1}) and imminodiacetic acid (IDA; molecular weight 133.4 g mol^{-1}) were suggested as displacers of recombinant proteins and whey proteins from hydroxyapatite columns [14, 50, 69], while imidazole, but also *N*-protected histidines and to a lesser extent tryptophan, have been suggested for IMADC [93]. The SMA model, particularly the operating regime plot, has been especially useful to demonstrate the potential but also the limits of low-molecular-weight displacers in protein displacement chromatography [60], while the affinity plot can yield important information on effects of the displacer concentration.

A comparison of the results obtained with small displacers to those obtained for similar, but larger, molecules shows that it is charge density and hence adsorption energy rather than size and absolute number of interaction points that determines the quality of a displacer. It has also been observed, however, that the behavior of small displacers depends to a much higher degree on the chromatographic conditions, e.g. the salt content of the mobile phase. As a consequence the switch from displacer to elution promoter is more likely in the case of these small substances. This sensitivity to the process conditions may also present an advantage, as was shown by the development of selective displacement chromatography (see Section 6.6.3).

6.5.2
Theoretical Considerations in Displacer Behavior and Design

The systematic design of a protein displacer requires a well-developed understanding of the parameters that influence the efficacy of such molecules. The database for ion-exchange displacement chromatography is comparatively well developed and some general guidelines can be proposed, which should in principle also apply to other types of chromatography.

Size is one of the best-investigated displacer properties in ion-exchange displacement chromatography. Since affinity increases with the number of interaction points, large polyelectrolytes have high, sometimes too high, stationary-phase affinities. By comparison, small molecules tend to have lower affinity constants, lower characteristic charges (number of charges per molecule), lower steric factors (number of shielded sites on the stationary phase surface) and, hence, higher normalized characteristic charges (characteristic charge per repeating unit). The last point has consequences for the induced salt gradient, since for a given number of repeating units, the small molecules displace relatively more salt ions with the predictable effects on the protein isotherms and the displacement train. It has also been observed that under otherwise identical conditions, low-molecular-mass displacers perform better, if carriers with medium rather than low salt concentrations are used. According to some authors [43, 85], this can be explained by the reduction of the target protein's stationary-phase affinity by the induced salt gradients. High-molecular-mass ion-exchange displacers (polyelectrolytes), on the other hand, generally work equally well at low salt concentration, since the isotherms of such molecules are almost independent of the local salt environment. As displacers, this renders them less dependent on the process parameters, but also makes column regeneration more difficult. If this is indeed the case, then by manipulating the characteristic charge relative to the mass of the displacer one could fine-tune the dynamic affinity of the displacer and the salt gradient induced by it.

Steric factor and characteristic charge are not fixed molecular properties of a given displacer. They depend also on the surface chemistry (charge density and distribution) of the stationary phase and hence can be optimized from that side. A comparison of the performance of various low-molecular-mass displacers on three typical types of ion-exchangers, i.e. a poly(methacrylate)-based material (PMA), a hydrophilized poly(styrene-divinylbenzene)-based material (PS-DVB) and an agarose-based material, gave the following indications [94, 95]. The affinity of the displacer increases with the number of charges. While specific electrostatic interactions with the stationary phase are most important for the binding of polyelectrolytes to ion-exchange stationary phases, secondary interactions (e.g. hydrophobic, π–π) are also possible. The extent and type of such secondary interactions depends on the stationary phase chemistry. In the case of the above-mentioned three types of ion exchanger it was found, for instance, that the introduction of a hydrophobic region increased the displacer affinity in case of the PMA and the PS-DVB materials. Aromatic groups were only effective in the case of the PMA-based column. In the case of agarose, no affinity increased was observed in either case, which is not surprising given the highly hydrophilic nature of this material. Flexible linear molecules may have advantages over branched or cyclic structures with equal charge densities, while the spreading of charges may actually improve affinity. In a more recent discussion of homologous series of triazine- or phloroglucinal-based displacers in combination with styrene-divinylbenzene- and methylmethacrylate-based stationary phases, the importance of aromaticity/hydrophobicity, but also of the positioning of the aromatic group within the low-molecular-weight displacer was once more demonstrated [96]. In the same paper the authors demonstrate how molecular descriptors such as the $\log P$ descriptors can be used to characterize putative displacers.

6.5.3
The Rational Design of Protein Displacers

In spite of a much improved understanding of the factors governing displacer performance, few if any of the substances used as protein displacers have been synthesized with that explicit goal in mind. Torres and Peterson were the first to try to synthesize a dedicated polymer-type displacer. Starting in 1978 [82], they chemically modified high-molecular-mass CM-Ds to obtain a series of (heterogeneous) molecules with graded affinity to anion-exchange materials. It was possible to remove these displacer molecules from the protein fractions by hydrophobic-interaction chromatography if necessary [97]. CM-Ds are relatively inexpensive materials. They are produced under controlled conditions and cover a wide range of affinity. In the early 1980s, fractions with fairly narrow ranges of affinity became available and it was tried to commercialize these substances [98, 99]. Heterogeneous mixtures of CM-D were successfully used to separate such complex samples as guinea pig serum, mouse liver cytosol and alkaline phosphatase from *E. coli* periplasm [100].

Following these early efforts, the year 1995 can be considered as the starting point of rational displacer design, since at least three different types of synthetic displacer were introduced that year. Quintero and coworkers [101] proposed a method for the dedicated design of a series of homologous generic displacers for chiral stationary phases (Pirkle-type, Cyclobond II). These displacers contain an anchoring phenyl group, which fits in the cavity of α-cyclodextrin. A second section of the molecule carries carboxyl and carbonyl groups that form multiple hydrogen bonds with the secondary hydroxyl groups of α-cyclodextrin. Finally, a solubility adjusting tail section (alkanoate group with variable chain length) is present, which regulates the displacer's solubility. The retention and adsorption properties of these displacers were found to be a function of the size of the alkanoic group. Longer side-chains resulted in higher retention factors and hence increased retention. In the same year, Freitag and Breier introduced a series of poly(ethylene glycol) (PEG)-based linear and dendritic polymers (molecular weight $1000–50\,000\,\mathrm{g\,mol^{-1}}$) that were modified to carry chelating groups at the end [14]. These molecules were used as protein displacers in anion-exchange and hydroxyapatite chromatography. A few years later, Vogt and Freitag introduced their building block method for target-orientated displacer synthesis by copolymerization [102]. This approach allows the production of a displacer for a given stationary phase and/or separation problem. In a first application, this method was used to produce a thermo-precipitable displacer, which is soluble in water at temperatures below a certain critical temperature, but rapidly and quantitatively precipitates if this temperature is surpassed by even the fraction of a degree. The exact value of the critical solution temperature can be adjusted from between 10 and 90 °C, while the displacing character of the molecules stays roughly the same. The precipitation/resolution is a very fast reaction. The polymers can run through the cycle several hundred times, the tendency for unspecific coprecipitation, for example, of the proteins can be kept low. The use of thermo-responsive polymers either as improved protein displacers or for column shielding was also discussed in a contribution by Kumar and coworkers [103]. The same group has also proposed thermo-responsive polymers of *N*-vinylamidazole as protein displacers in IMAC and dye–ligand chromatography (Blue Sepharose) [104]. In the latter case, however, PEI was an even better displacer.

A somewhat similar approach was been taken by Patrickios and coworkers [105] in collaboration with Cramer's group. Group transfer polymerization was used to produce di- and tri-block polymethacrylates consisting of a sequence of positively charged groups at one end, a sequence of negatively charged ones at the other and a neutral, hydrophobic middle block. Due to their polyampholytic nature, such polymers show isoelectric points much like proteins. The authors claimed that their polymers were suitable protein displacers for anion, but also for cation-exchange displacement chromatography, depending on the relative ratio of positively and negatively charge groups in the molecule and the pH of the carrier, while the ampholytic nature of the molecules facilitated recovery, e.g. by precipitation at the displacer's isoelectric point. Finally, Cramer's group have proposed pentaerythritol-based dendritic polyelectrolytes and protected amino acids as displacers for cation-exchange protein displacement chromatography [86, 106].

Freitag's group has promoted the use of polyDADMAC as a well-characterized protein displacer in cation-exchange displacement chromatography [107, 108]. Using specific synthetic conditions, this polycation can be produced in a very homogeneous form (linear, polydispersity below 1.5, defined mass). This allowed in particular an investigation of the influence of the displacer's mass on the stationary phase affinity and displacer efficiency [109]. Contrary to expectations, but in accordance with polyelectrolyte theory, it was the smallest polyDADMAC of the series that proved to be the most efficient displacer.

Most recently, a technique for high-throughput screening and quantitative structure–efficacy relationship models has been proposed by Cramer's group [110, 111]. The method requires only batch ion-exchange experiments using standard proteins. In 2005, it was extended to a multidimensional high-throughput screening protocol that allows investigating the effect of various parameters such as the displacer chemistry and concentration, the resin chemistry and the mobile-phase salt counterion concentration on the efficacy and selectivity of displacement separations [112]. The predictive value of the method was demonstrated, and the approach can be expected to facilitate and accelerate future development of selective and productive displacement separations.

6.6
Special Variants of Displacement Chromatography

Most applications of displacement chromatography are more or less straightforward variants of the experimental setup as outlined above. However, apart form the standard scheme and application area, a number of derivations and highly specialized forms of displacement chromatography exist, which may also be useful to the biotechnologist and applicants in related fields.

6.6.1
Spacer Displacement Chromatography

In a typical displacement chromatography the mixture components are focused into consecutive individual zones by means of a displacer. Torres and Peterson's group, in particular, developed two interesting variants of this approach. In so-called spacer displacement chromatography, substances of varied affinity for the stationary phase are added to the mobile phase [100]. The components of the spacer/displacer mixture vary in their adsorption energy within the range of the feed components. As a result, the more strongly bound molecules of the mixture act as displacer in a manner similar to that of any ordinary displacer, while the less strongly bound ones act as spacers between the target molecule zones. Since the spacers are usually chosen to be non-UV-active, they facilitate the monitoring of the protein zones and allow in theory at least to cut out a certain component of the complex feed [100]. It can be debated, however, whether this is really an advantage or whether the spacer/displacer impurities often found in the substance zones are more of a disadvantage.

6.6.2
Complex Displacement Chromatography

The second variant developed by Torres and Peterson's group is so-called complex displacement chromatography. In this mode, instead of directly competing for the binding sites, the "displacer" attaches itself to the bound target molecules and thereby lowers their stationary phase affinity [113]. When a sufficient amount of displacer has bound to the protein, the complex is released from the stationary phase. Complex displacement chromatography is related to ordinary displacement chromatography less through its mechanism and more through the chemicals used. A typical application of complex displacement chromatography was the isolation of an antibody (cationic protein) using a cation-exchange column and substances such as the CM-Ds ordinarily used as displacers in anion-exchange displacement chromatography as complex displacers by Torres and Peterson's group [113].

6.6.3
Selective Displacement Chromatography

The efficacy of low-molecular-weight displacers depends much more than that of high-molecular-weight displacers on both the mobile phase salt and the displacer concentration. This effect is exploited in selective displacement chromatography as proposed by Cramer's group [114]. This method enhances the resolving power of the displacement approach by establishing conditions under which mainly the product is displaced, while impurities of lower binding strength are eluted ahead of the displacement train in the induced salt gradient and impurities of higher binding strength are either retained on the colum or desorbed in the displacement zone, see also Fig. 6.4 (Section 6.3.5). In order to define suitable operation conditions, the SMA approach is used. The method has been demonstrated for both cation and anion displacement chromatography, taking typical model mixtures such as lysozyme/cytochrome c (cation-exchange chromatography) and α-lactalbumin/β-lactoglobulin A and B (anion-exchange chromatography) as examples. Crucial to the application of selective displacement chromatography is the operating regime plot (see Section 6.3.5). This plot defines for each protein the conditions under which it is eluted/displaced (elution line) or displaced/desorbed (displacement line). It is now necessary to find conditions (mobile-phase salt/ displacer concentration) for which the protein of interest is displaced, while most (crucial) impurities are either eluted or desorbed. For this approach both the linear and the nonlinear SMA parameters have to be determined for the displacer. For all proteins only the linear parameters are required.

6.6.4
Thin-layer Displacement Chromatography (TLDC)

TLDC and forced-flow TLCD (FF-TLDC) are especially connected with Kalasz and coworkers [115, 116]. In this approach, the sample mixture is applied as a spot to the standard planar stationary phases. The use of spacers, i.e. the spacer displacement chromatography approach, is necessary to keep the substance zones apart and available for visualizing. Sudan Black dye has successfully been used to that purpose. TLDC has, for example, been used to isolate the metabolites of radiolabeled Deprenyl [117] and its metabolites in rats' urine, and for the screening of ecdysteroids (a class of important steroid hormones) from plants [118].

6.6.5
Analytical Aspects of Displacement Chromatography

Displacement chromatography is correctly considered a predominantly preparative technique. As such it may, however, become a useful tool in analytical (bio-)chemistry. Many analytical procedures in biochemistry, molecular biology and related fields involve the separation of a complex mixture, e.g. a peptide digest, prior to a closer analysis of the individual components of the mixture or at least the resulting less-complex subsets of the mixture. The displacement process, which focuses even trace components into highly concentrated zones while enriching all mixture components to a high extent, is a prime choice for such sample pretreatment steps. At the same time no previous knowledge of the exact physical nature of the "impurity" is necessary. Trace components, which may be difficult to isolate by conventional chromatographic methods, can be obtained in sufficient amounts to allow chemical characterization by displacement chromalography.

Ramsey and coworkers [119] demonstrated this most elegantly for β-naphthyl-amine containing an impurity at the parts-per-million level. Diethylphthalate was used as displacer. In the case of a biosynthetic human growth hormone produced at BioWest Research, trace components polluting the product at levels of as low as 0.1% were made accessible to an characterization be mass spectrometry [120]. Displacement chromatography also played a key role in the discovery of two previously unknown amino acids, amarine and feline [120].

Displacement chromatography has also repeatedly been used instead of conventional elution chromatography for concentration/separation in hyphenated techniques. Mhatre and coworkers used a low-angle laser light scattering photometer for the on-line determination of the molecular weight of proteins separated by displacement chromatography [121]. Frenz and coworkers used such a hyphenated system [displacement chromatography in connection with continuous-flow fast atom bombardment- and electrospray ionization (ESI)-mass spectrometry (MS)] to analyze the components of a tryptic digest of a recombinant growth hormone [122, 123]. Cetramide served as displacer for the peptide mix on a C_{18} reversed-phase column. The exploitable capacity of the column could be increased by a factor of 50–100 compared to elution chromatography used in a similar setup. The resolution was equal to that of the previously used chromatographic method. Several of the collected fractions showed a completely different spectrum from those seen in the preceding or following fractions of the major peptides of the digest. Presumably, these fractions contained peptides fragments of incorrectly expressed growth hormone molecules, whose detection would otherwise have been difficult. A microsystem compatible to flow rates in the microliter per minute range has been suggested for direct displacement chromatography-MS coupling [124]. The selective enrichment of trace components in a tryptic digest of recombinant human growth hormone has also been demonstrated by Horvath's group [125, 126] using elution-modified displacement chromatography prior to ESI-MS and ESI-MS/MS. Up to 400-fold enhancement was demonstrated for some low abundance traces (femtomole per microliter level) in such complex peptide mixtures covering a wide dynamic concentration range.

The displacement approach could also be a much more suitable "first-dimension" proteome analysis scheme than either electrophoresis or elution chromatography, due to the fact that the concentration of a given component in the original sample is of little consequence. Substances varying by several orders of magnitude in concentration can be co-enriched by displacement chromatography without interference. No matter how long the zone of a major component, it will always be followed or preceded by the individual zone of a minor or trace component. Tag-along effects and hidden peaks are much mess likely.

Finally, displacement chromatography has been used to study interaction mechanisms. Jozwiak and coworkers used the displacement approach (displacer mecamylamine) to verify that a number of noncompetitive inhibitors of the $\alpha_3\beta_4$ nicotinic acetylcholine receptor did indeed bind to the same site on the immobilized receptor [127].

6.6.6
Miscellaneous

The concept of displacement chromatography has been adapted to centrifugal partitioning chromatography [128] and continuous annular chromatography [12] (see also Chapter 9). In case of centrifugal partition chromatography, an ionic liquid, i.e. benzalkonium chloride, was used as strong anion exchanger together with iodine as displacer. Multigram quantities of three hydroxycinnamic acid regioisomers were obtained. Sharp shock layers were observed and a numerical separation model was proposed as a tool for preliminary process validation and further optimization. In case of the continuous annular chromatography, a displacement separation of three standard proteins (two whey proteins and soybean trypsin inhibitor) previously developed as batch process for a 2-mL column [129] was transferred to the continuous annular chromatography system (0.5-L column) using the loading factor concept [12]. Separations of similar quality in terms of final product purity and recovery yield were obtained using the continuous system.

6.7
Applications of Displacement Chromatography for Separations in Biotechnology

Although displacement chromatography is far from being an accepted standard operation in industrial bioseparation, results accumulated over the last decade and a half have demonstrated that the technique can be a powerful tool for the purification of antibiotics, peptides and proteins. Certain features of displacement chromatography make the method even ideally suited to application in proteomics or the efficient isolation of a few grams of pure protein using analytical equipment in the biopharmaceutical industry. While the number of applications of displacement chromatography for bioseparation is still small compared to the elution mode, it is, however, already much too large to discuss each case in detail. Instead we have tried to compile some of the more relevant published examples of displacement separations of crude biomolecule mixtures with practical relevance in Tabs. 6.1–6.4. Applications using standard mixtures to investigate the fundamental parameters and/or to demonstrate the validity of simulations are not included. The most pertinent examples for each type of application are in addition briefly discussed in the text. The listed conditions apply to the actual displacement steps. In some cases different flow rates or temperatures were used during sample loading.

6.7.1
Separation and Isolation of Amino Acids, Peptides and Antibiotics

Modern peptide displacement chromatography started early in the 1980s and is closely connected to Horvath and coworkers. Reversed-phase chromatography

Tab. 6.1 Amino acids, peptides and antibiotics

Target molecule	Conditions	Comments	Reference
Synthetic peptides	method: RP-DC column: analytical size, various C$_{18}$ materials displacer: BEE, decyltrimethylammonium bromide, decyltrimethylammonium bromide, cetyltrimethylammonium bromide	peptides were purified and concentrated, amino acids eluted ahead of the displacement train, up to 5 g recovered from a 500-mL feed in a single run	130
Peptide, Merrifield synthesis-type product	method: multidimensional, RP-DC and IE-DC column, RP-DC: LiChrosorb RP-18, 250 × 4 mm, LiChroprep RP-18, 20, 40, 80 mm column, IE-DC: Mono-Q, 50 × 5 mm, 10 µm, Q Sepharose, 350 × 10 mm flow rate: 0.5 mL min^{-1} (two-column system) displacer, RP-DC: benzyltributylammonium chloride displacer, IE-DC: ammonium citrate	scale up from 100 mg to 35 g of product (final purity >90%), no sample pre-treatment necessary	131
Synthetic peptide (malaria vaccine)	method: RP-DC, 23 °C column: Aquapore RP-18 250 × 4.6 mm, 7 µm, 30 nm flow rate: 0.1 mL min^{-1} displacer: benzyldimethyl-dodecylammonium bromide	up to 50 mg purified in a single experiment (purity >95%)	131
Synthetic peptides intended for Plasmodium vivax malaria sero-epidemiology	method: RP-DC after lyophilization, gel filtration and IE-elution chromatography column: Vytac 218TP5 C$_{18}$, 250 × 4 mm, 5 µm, 30 nm flow rate: 0.1 mL min^{-1} displacer: BEE	advantages over elution chromatography demonstrated; up to 107 mg of crude mixture processed (final purity: 85%)	163
Mellitin (honey bee venom peptide)/ synthetic variants	method: RP-DC, 40 °C column: Hy-Tach C-18, 105 × 4.6 mm, 2 µm, non-porous flow rate: 0.2 mL min^{-1} displacer: benzyldimethyl-hexydecylammonium chloride	5 mg mellitin isolated in 20 min from a 10-mg mixture	133

Tab. 6.1 (Continued)

Target molecule	Conditions	Comments	Reference
Crude α-/β-melanocyte-stimulating hormone mixtures	method: RP-DC, 22 °C column: C_{18} based on Hypersil silica, 250 × 4.6 mm, 5 μm flow rate: 0.1 mL min^{-1} displacer: benzyldimethyl-dodecylammonium bromide	30 mg separated in a single run	164
Insulin (bovine, porcine)	method: RP-DC column: Nucleosil C-8, 5 μm flow rate: 0.1 and 0.2 mL min^{-1} displacer: cetramide	semi-preparative protocol; up to 500 mg of raw insulin could be purified (proinsulin level <100 ppm)	134
Synthetic luteinizing hormone-releasing hormone	method: RP-DC column: RP1 DisKit column, 250 × 4.6 mm, 5 μm flow rate: 0.5 mL min^{-1} displacer: DisKit displacer	displacement kit containing column, carrier, displacer and regenerant was employed (BioWest research); peptide concentration was kept *below* a critical value preventing aggregation and precipitation	132
N^a-9-fluorenoxy-carbonyl-*S*-trityl-cysteine derivative (Fmoc-Cys-Trt)-OH)	method: normal-phase-DC column: LiChrocart RP-18, 250 × 4 mm, 10 μm (mg scale) compressed 20, 40, 80 mm columns, LiChroprep Si 60 silica, 25–40 μm (g scale) flow rate: 0.1 mL min^{-1} (mg scale)/2 mL min^{-1} (g scale) displacer: benzyltributylammonium chloride	scale up from 100 mg to 38 g	165
Commercial polymyxin B sulfate	method: RP-DC column: LiChrosorb RP-8 C_8, 250 × 4.6 mm, 5 μm flow rate: 0.1 mL min^{-1} displacer: dodecyloctylammonium chloride	product concentrations between 10 and 20 mg mL^{-1} were reached	136
Oligomyxins A, B and C	method: RP-DC column: LiChrosorb RP-18, 250 × 4.6 mm, 5 μm flow rate: 0.1 mL min^{-1} displacer: palmitic acid	processing of highly hydrophobic substances (carrier 75% methanol in water)	137
Cephalosporin C from fermentation broth	method: RP-DC, 35 °C column: Zorbax BP C-18, 350 × 4.6 mm, 5 μm flow rate: 0.1 mL min^{-1} displacer: BEE	5 mL of culture supernatant processed in 20 min	138

Chromatography abbreviations: DC, displacement chromatography; IE, ion exchange; RP reverse phase.

Tab. 6.2 Protein separations

Target molecule	Conditions	Comments	Ref.
Crude β-galactosidase (industrial enzyme) from *A. oryzae*	method: weak anion IE-DC column: two TSK DEAE-5PW columns in series, 75 × 7.5 mm flow rate: 0.2 mL min^{-1} displacer: chondroitin sulfate	displacement mode superior to elution in throughput and waste production	140
Human serum proteins	method: anion IE-DC column: DEAE Bio-Gel A, DEAE Selectagel, 140 × 5.5 mm displacer: gradient of CM-D with increasingly higher content of carboxyl groups	general method for fractionation of complex protein mixtures (spacer displacement chromatography); fractions are low in salt and can be directly analyzed by electrophoresis; methods for displacer modification are discussed	82
Human Gc-2 globulin	method: two-anion-IE-DC at different pH followed by elution chromatography on hydroxyapatite column: DEAE Sephacel, 7 mL flow rate: 5 mL h^{-1} spacer/displacer: different CM-D	protein isolation from blood; 6 mL of serum (400 mg protein) gave 0.5 mg pure substance; alternatively, 4 mg of the protein had previously been recovered from 34 L of serum using 13 elution chromatography and electrophoresis steps	166
Alkaline phosphatase from *E. coli* periplasm	column: DEAE-5PW displacer: CM-D		100
Monoclonal antibodies from ascites	method: complex displacement chromatography on anion exchanger complexer/displacer: CM-D	scale up from 1 to 450 mL, final purity at largest scale: 79%	113
Guinea pig serum	method: spacer displacement chromatography column: Fractogel DEAE 650S flow rate: 10 mL h^{-1} spacer/displacer: heterogeneous mixture of CM-D	displacement on medium resolution adsorbent	100

Tab. 6.2 (Continued)

Target molecule	Conditions	Comments	Ref.
Mouse liver cytosol proteins	method: spacer displacement chromatography column: DEAE cellulose flow rate: 5 mL h^{-1} spacer/displacer: mixture of CM-D	displacement on low-resolution microgranular cellulose	100
Lactate dehydrogenase from beef heart	*method I*: weak anion IE-DC column: Tris Acyl DEAE, 91 × 10 mm and 250 × 10 mm displacer: carboxymethyl starch *method II*: weak anion IE-DC in connection to affinity chromatography (Cibacron Blue) column: Tris Acyl DEAE, 250 × 10 mm flow rate: 0.5 mL min^{-1} displacers: chondroitin sulfate C, alginate, and Eudragit L and S	investigation of scale-up parameters (column dimensions, protein load); comparison to elution chromatography, comparison of displacers	143, 144
Industrial recombinant human growth hormone	method: anion IE-DC	displacement as final polishing step	120
Thrombolytic protein from cell culture supernatant	*method I*: strong cation IE-DC column: 88 × 5.0 mm, 8 µm flow rate: 0.1 mL min^{-1} displacer: DEAE dextran *method II*: antibody-exchange-DC column: 100 × 4.6 and 50 × 4.6 mm, 5µm flow rate: 0.1 mL/min displacer: DEAE dextran (10 KDa)	capacity is better in case of the of the antibody-exchange phase, resolution better in case of the cation exchanger	145, 146
rh-AT III from Chinese hamster ovary cell culture supernatant	method: hydroxyapatite-DC column: 250 × 4 mm, 2 µm, 100 nm flow rate: 0.1 mL min^{-1} displacer: EGTA	crossing isotherms; final purity: 90%, recovery: 84%; by comparison, a quantitative separation was not possible in the elution mode	14, 50
Technical dairy whey	method: strong anion IE-DC column: BioScale Q 2, 52 × 7 mm, 10 µm flow rate: 0.1 mL min^{-1} displacer: PAA (molecular weight 6000 g mol^{-1})	dairy whey was processed directly; yield: α-lactalbumin 78%, β-lactoglobulins 92%; concentration factor: 3	69

Tab. 6.2 (*Continued*)

Target molecule	Conditions	Comments	Ref.
Antigene vaccine protein from an industrial process stream	method: DEAE-IE-DC column: fast-flow Sepharose, 10 × 290 mm flow rate: 1 mL min⁻¹, 25 °C carrier: 20 mM Tris, pH 7.0 displacer: various	selective displacement chromatography	148
rh-BDHF from six closely related (1 amino acid difference) variants and *E. coli* proteins	method: RP-DC column: various C₄ flow rate: varied carrier: varied displacer: varied, to be preferred: tetrahexylammonium chloride	development analytical scale (column 250 × 4.6 mm) scale up to process scale (column 250 × 50 mm) removal of *E coli* proteins possible	91, 147
Monoclonal IgA from hybridoma cell culture supernatant	method: strong anion IE-DC column: Bio-Rad Q2, 52 × 7 mm, 10 μm flow rate: 0.1 mL min⁻¹ carrier: 20 mM Tris, pH 8.0 displacer: PAA		142
Industrial protein mixture containing three proteins differing at their C-terminus from their aggregate impurities yet maintaining their relative ratio	method: hydrophobic-interaction chromatography-DC column: octyl Sepharose 4 FF, hydrophobic-interaction chromatography resource ISO, Phenyl 650 M resin flow rate: varied carrier: varied displacer: varied, preferred PEG-3400, surfactant "Big Chap"	high-throughput screening used to identify suitable displacer; PEG and surfactants are possible protein displacers; productivity increased compared to established gradient elution protocol	79

Chromatography abbreviations: DC, displacement chromatography; IE, ion exchange; RP reverse phase.

Tab. 6.3 Isomers

Target molecule	Conditions	Comments	Reference
Isobufen DDATHF (5, 10-di-deazatetrahydro-folic acid)	method: normal-phase and RP-DC, 4 °C column: Cyclobond I, 250 × 2 mm, two in series flow rate: 0.2 mL min⁻¹ displacer: 4-tert-butyl-cyclohexanon flow rate: 0.3 mL min⁻¹ displacer: cetramide	quantitative separation even for α-values of 1.08; purities/yields were similar to elution for the less retained isomer, better for the more retained one	151, 152
Binary isomer mixtures of epirubicin and doxorubicin	method: RP-DC column: Kromasil KR 100-10 C₁₈, 250 × 4.6 mm, 10 μm flow rate: varied carrier: varied displacer: 30 mg mL⁻¹ benzethonium chloride	Separation of 30 mg on analytical column; final purity epirubicin >99%, recovery 60%	154
Enantiomers of 1,2-O-dihexadecyl-rac-glycerol-3-O-(3,5-dinitrophenyl)-carbamate	method: normal-phase-DC, 15 °C column: Pirkle-type naphthylalanine silica, 250 × 4 mm, 5 μm flow rate: 0.5 mL min⁻¹ displacer: 3, 5-dinitrobenzoyl ester of n-heptanol		167
DL-Methinonine β-naphtylamide	method: enantioselective DC column: poly-L-valyl groups as chiral selectors on modified silica, 250 × 4.6 mm flow rate: 0.2 mL min⁻¹ displacer: DL mandelic acid		168
Regioisomers of hydroxycinnamic acids	method: anion-exchange displacement centrifugal partition chromatography column: ionic liquid (benzalkonium chloride) carrier: 20 mM Tris, pH 8.0 displacer: iodine	use of centrifugal partition chromatography; numerical separation model for process development proposed	128

Chromatography abbreviations: DC, displacement chromatography; IE, ion exchange; RP reverse phase.

Tab. 6.4 Miscellaneous

Target molecule	Conditions	Comments	Reference
Nucleotides	method: RP-DC displacer: butanol		169

Nucleosides	method: RP-DC
	displacer:
	    benzyltributylammonium
	    chloride
			169

Methylesters of
    polyunsaturated fatty
    acids (EPA, DHA)
	method: RP-DC, 30 °C
	column: C_{18} silica
	flow rate: 0.5 mL min^{-1}
	displacer: oleic acid
	carrier 1: acetonitrile:water
	    80:20
	carrier 2: acetonitrile:water
	    90:10
		difficult carrier
		selection: low
		solubility of polar
		compounds
		desired during
		binding, high
		solubility for
		high displacer
		concentration
		during
		separation;
		solution: two
		carrier system;
		final purities
		>90%,
		concentration
		factors 4–13
			156

Phosphatidylcholine (PC)/
    phosphatidylethanolamine
    (PE) from soybean
    phospholipids
	method: RP-DC
	column: C_{18} silica
	flow rate: 0.2 mL min^{-1}
	displacer: ethanolamine
	carrier: dichloromethane:
	    methanol (9:1)
		yields of 100%
		pure (from two
		component
		mixture) PE:
		94.8%, PC:
		87.9%; Cycle
		time: 195 min;
		Purity/
		throughput/
		recovery from
		soybean, PE:
		80.2%/65.7 mg
		h^{-1}/70.9%, PC:
		90.5%/272.6 mg
		h^{-1}/88.3%;
		optimization of
		loading,
		concentration,
		flow rate by
		orthogonal test
		design and
		statistical method
			170, 171

Oligonucleotides	method: anion IE-DC column: 10 in × 10 cm flow rate: 2.1 L min^{-1} (250 cm h^{-1}) displacer: dextran sulfate	large-scale approach	[74]

Tab. 6.4 (Continued)

Target molecule	Conditions	Comments	Reference

Mixture of 20mer antisense oligonucleotides	method: anion IE-DC
	column: Poros HQ/M,
	    4.6 × 100 mm, 20 μm
	flow rate: 0.2 mL min⁻¹
	carrier: 20 mM NaOH +
	    500 mM NaCl
	displacer: varied
		several high-affinity, low-molecular-weight displacers studied; application of the SMA model; very high purities obtained; possible application for process scale purification	158

20mer phosphorothioate oligonucleotide from closely related impurities	method: anion IE-DC
	column: Source 15Q,
	    100 × 4.6 mm, 15 μm
	flow rate: 0.2 mL min⁻¹
	carrier: 20 mM NaOH +
	    500 mM NaCl
	displacer: amaranth
		investigation of column load; successful scale up to high column loadings combined with high purity (92%) and yields (86%); SMA model used to define conditions; both conventional and selective displacement possible)	159, 160

Dinucleoside (5′,5′) polyphosphates (mixture from chemical synthesis)	method: RP-DC column: C₁₈, 300 × 8 mm, 4 μm flow rate: 0.1 mL min⁻¹ carrier: 40 mM TEAA displacer: n-butanol (100 mM)	use of analytical equipment for semi-preparative work	10
Plasmid DNA (4.7 kb, E. coli), endotoxin, protein (holo transferin)	method: strong anion IE-DC column: UNO Q (monolith), 7 × 52 mm flow rate: 0.1 mL min⁻¹ carrier: 20 mM Tris, pH 8.0 displacer: PAA	separation not possible on porous particle-based column (protein peak overlays displaced DNA/endotoxin zone) presumably due to differences in accessibility of the interactive surface leading to a loss in competition for the binding sites	9

Chromatography abbreviations: DC, displacement chromatography; IE, ion exchange; RP, reverse phase.

dominates this particular area of displacement separations and the separation of the product of a (solid-phase) peptide synthesis from its closely related byproducts remains one of the typical applications. A pertinent example is the use of reversed-phase displacement chromatography for a preparative separation of the peptides synthesized by immobilized carboxypeptidase Y by Cramer and Horvath in 1988 [130]. Displacers of varied hydrophobicity were used, including BEE and decyltrimethylammonium. The peptides were both isolated and concentrated from the diluted aqueous mixtures; 5.2 g were procured from a 500-mL feed in a single chromatographic run.

In 1991, Viscomi and coworkers developed a large-scale multidimensional reversed-phase and ion-exchange displacement method to isolate a synthetic peptide fragment corresponding to the fragment 163–171 of human interleukin-β [131]. The efflux of the reversed-phase column was transferred directly to the ion-exchange column. From 100 mg to 35 g of the Merrifield synthesis-type product could be processed. In the reversed-phase mode benzyltributylammonium chloride was used as displacer; in the ion-exchange mode, ammonium citrate. Peptide purities were greater than 90%. The same group used reversed-phase displacement chromatography to isolate a synthetic peptide containing two epitopes of circumporozoite protein of *Plasmodium falciparum*; 50 mg of this promising malaria vaccine was brought to more than 95% purity using benzyldimethyldodecylammonium bromide as displacer [131]. Controlling rather than maximizing the product concentration was a problem in the peptide isolation described by Jacobson [132], since the conventional elution chromatographic protocol was troubled by aggregation and precipitation effects.

Mellitin, a 26-amino-acid residue peptide from honey bee venom, and its synthetic variants were isolated by Kalghatgi and coworkers [133] using benzyldimethylhexydecyl ammonium chloride as displacer. A nonporous C_{18} material (2 μm diameter) was used as stationary phase. Since intraparticular mass transfer is not possible in such materials, a very fast separation was possible. A semi-preparative protocol for the isolation of bovine and porcine insulin by reversed-phase displacement chromatography was proposed by Vigh and coworkers [134]. Up to 100 mg was purified using Cetramide as displacer. The level of proinsulin could be maintained below 100 ppm in the collected fractions.

Although less common, the separation of amino acids by displacement chromatography is possible. Zammournni and coworkers proposed the use of carbon dioxide dissolved in water as a displacer of amino acids from strong anion exchangers in a theoretical simulation for some two-amino-acid mixtures [135]. Particular emphasis is put in this contribution on the possible effect of dissociation equilibria in solution, but also the ion-exchange equilibria present in the system, which also influence the quality of the separation.

The separation of antibiotics is another common application of reversed-phase displacement chromatography. Kalasz and Horvath [136] separated more than 100 mg of commercial polymyxin B sulfate into its constituents using an aqueous solution of dodecyloctylammonium chloride as displacer. Product concentrations between 10 and 20 mg ml^{-1} were reached. Oligomyxcins A, B and C have been separated using palmitic acid as displacer [137]. Cephalosporin C has been isolated from a fermentation broth, using BEE as displacer [138]; 5 mL of the culture supernatant could be processed on a 4.6 × 350 mm column within 20 min.

Recently, the separation of closely related peptides by reversed-phase displacement chromatography has been reinvestigated by Velayudhan's group, especially taking into account the question of the role and effect of displacer impurities in the separation [139]. Other parameters such as displacer chemistry and concentration, mobile-phase composition, and flow rate were investigated with the aim of optimizing productivity. The limited solubility of the feed components in the carrier posed one potential problem in the separation that had to be considered via the displacement conditions. While the displacer impurities were of little consequence for the separation quality, a change in carrier composition during loading (high binding conditions) and displacement (lower binding conditions) proved beneficial.

6.7.2
Protein Separation

Among the first applications of displacement chromatography to protein isolation was the purification of crude β-galactosidase from *Aspergillus oryzae* on a weak anion exchanger with chondroitin sulfate as displacer [140]. A comparison with the elution mode demonstrated the superiority of displacement chromatography in terms of the utilization of the stationary- and mobile-phase capacity, throughput, and waste production.

Torres and Peterson's group have published repeatedly on applications of their CM-D displacers to practical protein separation. The separation of human serum proteins was among the earliest [86]. The low salt content of the fractions, unusual for ion-exchange protein chromatography, was noted as a major advantage, since it allowed their direct analysis by electrophoresis. In 1985, the isolation of a protein, later identified as Gc-2 globulin, was achieved [141]. The protein was isolated from the blood serum from psoriasis patients, where it had previously been located only as a spot in a two-dimensional electrophoresis. Two anion-exchange displacement chromatographic steps at different pH levels followed by elution chromatography on hydroxyapatite were used to process 6 mL of serum containing a total of around 400 mg protein and more than 100 different proteineous compounds. The separation resulted in the isolation of 0.5 mg of the pure protein. The solids-free serum sample could be applied directly to the first displacement column. Previously, 4 mg of that very Gc-2 protein had been isolated from 34 L of serum using 13 chromatographic and electrophoretic steps. The CM-D displacers were also used in the isolation of alkaline phosphatase from *E. coli* periplasm and that of monoclonal antibodies from ascites [100, 113]. Since the monoclonal antibodies carry a positive net charge under physiological conditions, complex displacement chromatography had to be used in the latter case. The method was scaled up from 1 to 450 mL; the recovery was 79% in the largest scale. Spacer displacement chromatography, on the other hand, was used to separate guinea pig serum and mouse liver cytosol proteins using medium- to low-affinity adsorbents [100]. In the case of a monoclonal immunoglobulin A, the efficient isolation of this fraction – containing the total of the monomers, dimmers and polymer – was possible by ion-exchange displacement chromatography using PAA as displacer [142]. Both purification and concentration of the IgA fraction was achieved.

Technical dairy whey containing all milk components, save the casein fraction, was separated by anion-exchange displacement chromatography using PAA as displacer [69]. The technical dairy whey contained 3.45 g L^{-1} α-lactalbumin and 12.65 g L^{-1} β-lactoglobulin together with some other UV-active components. The whey was directly applied to the anion-exchanger column without further preparation. Compared to the feed, both whey proteins were concentrated by a factor of 3 during the displacement separation. Yields were 78% for the α-lactalbumin and 92% for the β-lactoglobulin. A similar separation using hydroxyapatite displacement chromatography described in the same paper was unsuccessful due to the lack of a suitable displacer.

Ghose and Mattiasson have used displacement chromatography for the recovery of lactate dehydrogenase from beef heart [143]. Process scale up in terms of column dimensions and protein load was investigated. Carboxymethyl starch was used as displacer. In a second paper [144], chondroitin sulfate C, alginate and Eudragits were compared as alternative displacers, and the displacement used in connection to a subsequent affinity chromatographic step on a Cibacron Blue Sepharose CL 4B column. Eudragits L and S were demonstrated as readily available, cheap, nontoxin protein displacers.

Displacement chromatography has also been used repeatedly for the isolation of recombinant proteins from cell culture supernatants. More recently, Kim has described the use of cation-exchange and antibody-exchange displacement chromatography for the isolation of a proprietary thrombolytic protein from crude fermentation broth, containing among others albumin, insulin, transferrin, aprotinin, methatrexate and bovine serum [145, 146]. DEAE–dextran was used as displacer in both cases. The recovery of recombinant human AT III (rh-AT III), an anticoagulant of high pharmaceutical interest, produced by Chinese hamster ovary (CHO) cells in a serum-free culture medium was possible by displacement chromatography on hydroxyapatite using EGTA as displacer [14, 50]. The problem of crossing nonlinear single-component isotherms and their influence on the chromatographic result was encountered during the optimization of that separation. The recovered rh-AT III still contained 10% other proteins, mostly transferrin. BSA was present only in trace amounts. Of the AT III activity originally detected in the feed, 84% was recovered by the displacement separation. For comparison, the same column was also used in the elution mode. A quantitative separation of the rh-AT III was not possible under these circumstances. Anion-exchange high-performance displacement chromatography has also been the method of choice for the purification/polishing of an industrial recombinant human growth hormone [120].

The separation of recombinant brain-derived human neurotrophic factor (rh-BDNF) from closely related variants (six, differing by only a single amino acid) and *E. coli* proteins has been discussed by Cramer's group in a series of papers ([91, 147] and references within). The removal of such closely related impurities generally presents a problem in the downstream process of recombinant proteins and is usually achieved with shallow gradients at low column loading, i.e. reduced throughput/productivity. Reversed-phase (C_4) chromatography was used in combination with tetrahexylammonium chloride as displacer. Displacement chromatography was able to achieve high yield and purity at high column loadings. Compared to the established gradient elution protocol, productivities could be considerably increased (8-fold for 20 µm particles, 4.5-fold for 50 µm particles). The production rate of the single-step, pilot-scale process was 23 times higher than that of the current standard elution chromatography process based on a two-step ion-exchange chromatography/hydrophobic-interaction chromatography protocol.

One of the rare applications of hydrophobic-interaction chromatography in displacement chromatography was the purification of an industrial protein mixture using PEG and surfactants as displacers [79]. The goal was the separation of a mixture of three proteins that differed in the C-terminus from their aggregate impurities while maintaining the same relative ratio of the three proteins in the final product as in the original feed. Compared to the established gradient elution process, productivity was increased by switching to displacement chromatography.

In 1998, Cramer and coworkers [148] demonstrated the potential of the SMA model in the development of a displacement separation of an antigenic vaccine protein from an industrial process stream. Selective displacement was used, i.e. conditions were chosen such that mainly the product was purified by displacement, while some impurities eluted before the displacement train and others are desorbed after the breakthrough of the displacer front. The product was of equal purity to that obtained by a conventional elution chromatography protocol. The choice of the displacer and the operating conditions were much aided by the SMA model. In the same year, Frey's group [149] proposed a hybrid method that combined chromatofocusing using a self-sharpening pH front and displacement chromatography, albeit only for the separation of a two component model protein mixture (β-lactoglobulin A and B) by anion-exchange displacement chromatography.

6.7.3
Separation of Isomers

While the large-scale separation of isomer mixtures in the pharmaceutical industry today is a typical application of the SMB technology, displacement chromatography has also been recognized as a method for the efficient separation of optical and structural isomers at high throughputs and concentrations, perhaps at the intermediate rather than large scale. In particular, the quick separation of an isomer mixture at the laboratory scale may be one area of application for the displacement approach. Vigh and coworkers used cyclodextrin–silica columns for the separation of positional and geometrical isomers [150]. α- and β-Cyclodextrin materials have been used in the normal and reversed-phase mode to separate substances of pharmaceutical relevance such as isobufen [151], a nonsteroidal anti-inflammatory agent, and 5,10-dideazatetrahydrofolic acid (DDATHF) [152]. 4-Tert-butylcyclohexanon was used to displace the isobufens, Cetramide to displace the DDATHF. Quantitative separation was possible even for isomers with retention factor ratios (α values) as small as 1.08. For the less-retained isomer, purities and yields comparable to those achievable with overloaded elution chromatography on the same stationary phases were found, while the performance of the displacement mode was superior with regard to the more retained isomer. A series of homologous displacers of varied affinity for Cyclobond II columns (α-cyclodextrin–silica) was introduced by the same group in 1995 [153]. Qi and Huang investigated the separation of a binary isomer mixture, epirubicin and doxorubicin, by reversed-phase displacement chromatography using benzethoniium chloride as displacer [154]. Amounts of 30 mg could be processed on an analytical column. Epirubicin was obtained in a purity above 99% and a recovery of 60%. The possibility to use high feed loads, the low solvent costs and the high resolution of the method were given as particular advantages of the displacement approach in this context.

In a more theoretical approach, Rathore and Horvath have simulated the separation of two compounds (e.g. *cis/trans* isomers) interconverting during the separation by a reversible first-order reaction (e.g. isomerization reaction) taking a phenylalanine–proline dipeptide as an example [155]. The dependence of yield and production rate on the temperature, column length, flow velocity and displacer concentration was evaluated both by simulations assuming Langmurian adsorption behavior and experimentally. Optimal conditions for yield varied from those determined for optimal production rate. The temperature was the most important parameter in both cases.

6.7.4
Miscellaneous

Displacement chromatography has also been used for the isolation of many other substances. Huang and Jin [156] reported on the use of displacement chromatography to purify methyl esters of polyunsaturated fatty acids. Oleic acid was used as displacer. The dilemma of carrier selection, i.e. enhancement of the adsorptive forces due to a low solubility for polar compounds versus high solubility for a polar compound, i.e. the displacer, during displacement, was solved by the use of a two-stage displacement process. Carrier 1 was acetonitrile:water 80:20 and carrier 2 was acetonitrile:water 90:10.

As already mentioned, displacement chromatography has also been suggested as an economically sound way for the large-scale purification of oligonucleotides [74]. In 1983, Horvath and coworkers separated nucleotide and nucleoside mixtures on reversed-phase columns using butanol and benzyltributylammonium chloride, respectively, as displacers [157]. An application note to that method has been published by PerSeptive Biosystems. The use of high-affinity, low-molecular-weight displacers (all sulfonic acids-based, affinity enhanced by the presence of hydrophobic groups) was proposed by Shukla and coworkers for the preparation of antisense oligonucleotides (mixtures of 20mers) supplied by Isis Pharmaceuticals [158]. Very high purities could be obtained. In subsequent papers, amaranth [159] and saccharin [160] were suggested as displacers of antisense oligonucleotides from anion-exchange columns. Saccharin can also serve as protein displacer.

Reversible chemical reactions are another area of application for displacement chromatography. Such reactions require the removal of the product in order to reach high conversions. An interesting application in this context is the combination of displacement chromatography with packed-bed enzyme reactors for integrated reactor/separator schemes. In one pertinent example, a packed-bed enzyme reactor with immobilized carboxypeptidase Y was used by Cramer and coworkers [161] in tandem with displacement chromatography for the preparation of N-benzoyl-L-arginyl-L-methioninamide from N-benzoyl-L-arginin and L-methioninamide. The enzyme cartridge was operated in the recirculation mode and the reaction mixture was separated using BEE as displacer on a C_{18} column. The unreacted L-methioninamide was recycled to the reactor. The system was operated in parallel and 460 mg of the product (purity above 99%) was obtained within 24 h. A similar system was used by el Rassi and Horvath [162] for the separation of the reaction mixture in the ribonuclease T_1-catalyzed synthesis of GpU from cyclic GMP in the presence of a large excess of uridine. Nearly 100 mg of GpU with a purity of 99.7% could be produced within 2.5 h. n-Butanol was used as displacer.

The potential of displacement chromatography as a means of quick workup of a complex mixture in the research laboratory has been demonstrated by Jankowski and coworkers [10]. When for their work on dinucleoside (5′,5′) polyphosphates, a new group of hormones controlling biological processes, it was necessary to synthesize the commercially unavailable ones by chemical synthesis, displacement chromatography in the reversed-phase mode, flanked by preparative reversed-phase chromatography and gradient anion-exchange chromatography, allowed the rapid production of 0.1–0.2 g amounts of the desired substances. n-Butanol was used as displacer.

6.8
Conclusions

Displacement chromatography is an exclusively nonlinear form of chromatography. This makes it more difficult to apply and optimize. However, much progress has been made over the last two decades both in the development of computer tools and mathematical algorithms for the simulation of nonlinear chromatography. In addition, the question of what constitutes a suitable displacer for typical applications in protein or peptide separation and how to prepare them has been addressed on both a theoretical and practical basis. Applying the displacement approach today may still be challenging, but has become considerably easier as a result of these efforts. Concomitantly, the many advantages of displacement chromatography should be considered, most prominently the ability to use the stationary-phase capacity to an extreme extent, often allowing semi-preparative purifications on analytical columns, the ability to work isocratically with just a single displacer step even in the case of protein chromatography, but also the possibility to co-enrich and separate a large number of substances that differ considerably in concentration, in any valuation of this chromatographic mode.

References

1 Wurm, F., *Nat Biotechnol* **2004**, *22*, 1393–1398.

2 Guiochon, G., *J Chromatogr A* **2002**, *965*, 129–161.

3 Cramer, S. M., Subramanian, G., *Sep Purif Methods* **1990**, *19*, 31–91.

4 Spedding, F. H., Powell, J. E., Fulmer, E. I., Butler, T. A., *J Am Chem Soc* **1950**, *72*, 2354–2361.

5 Araki, H., Umeda, M., Enoleida, Y., Yamamoto, I., *Fusion Eng Des* **1998**, *39*, 1009–1013.

6 Antia, F. D., Horvath, C., *Ber Bunsenges Phys Chem.* **1989**, *93*, 961–968.

7 DeVault, D., *J Am Chem Soc* **1943**, *65*, 532–540.

8 Jakobson, J., *ACS Symp Ser* **1993**, *529*, 77–84.

9 Freitag, R., Vogt, S., *Cytotechnology* **1999**, *30*, 159–167.

10 Jankowski, J., Potthoff, W., Zidek, W., Schlüter, H., *J Chromatogr B* **1998**, *719*, 63–70.

11 de Carli, J. P., Carta, G., Byers, Ch. H., *AIChE J* **1990**, *36*, 1220–1228.

12 Giovannini, R., Freitag, R., *Biotechnol Prog.* **2000**, *18*, 1324–1331.

13 Hilbrig, F., Freitag, R., In: *Bioseparation and Bioprocessing*, G. Subramanian (Ed.). Wiley-VCH, Weinheim, **2007**, pp. 257–288. (Chapter 9 of this volume.)

14 Freitag, R., Breier, J., *J Chromatogr* **1995**, *691*, 101–112.

15 Velayudhan, A., Horváth, C., *Ind Eng Chem Res* **1995**, *34*, 2789–2795.

16 Velayudhan, A., Ladisch, M. R., *Ind Eng Chem Res* **1995**, *34*, 2805–2810.

17 Bellot, J. C., Condoret, J. S., *J Chromatogr* **1993**, *657*, 305–326.

18 Phillips, M. W., Subramanian, G., Cramer, S. M., *J Chromatogr* **1988**, *454*, 1–21.

19 Gu, T., Tsai, G.-J., Tsao, G. T., *Adv Biochem Eng* **1993**, *49*, 45–71.

20 Katti, A. M., Guiochon, G., *J Chromatogr* **1988**, *449*, 25–40.

21 Guiochon, G., Golshan-Shirazi, S., Katti, A., *Fundamentals of Preparative and Nonlinear Chromatography.* Academic Press, New York, **1994**.

22 Glückauf, E., *Proc R Soc London Ser A* **1946**, *186*, 35.

23 Helfferich, F., James, D. B., *J Chromatogr* **1970**, *46*, 1–28.

24 Rhee, H.-K., Aris, R., Amundson, N. R., *First Order Partial Differential Equations II: Theory and Applications of Hyperbolic Systems of Quasilinear Equations.* Prentice Hall, Englewood Clifs, **1989**.

25 Frenz, J., Horvath, C., *AIChE J* **1985**, *31*, 400–409.

26 Piatkowski, W., Gritti, F., Kaczmarski, K., Guiochon, G., *J Chromatogr A* **2003**, *989* 207–219.

27 Weaver, L. E., Carta, G., *Biotechnol Prog* **1996**, *12*, 342–355.

28 Garke, G., Hartmann, R., Papamichael, N., Deckwer, W. D., Anspach, F. B., *Sep Sci Technol* **1999**, *34*, 2521–2538.

29 Duri, B. A., McKay, G., *J Chem Tech Biotechnol* **1992**, *55*, 245.

30 Chen, W.-D., Sun, Y., *J Chem Ind Eng (China)* **2002**, *53*, 88.

31 Zhang, S., Sun, Y, *AIChE J* **2002**, *48*, 178–186.

32 Horstmann, B. J., Chase, H. A., *Chem Eng Res Des* **1989**, *67*, 243–254.

33 Sajonz, P., Guan-Sajonz, H., Zhong, G., Guiochon, G., *Biotechnol Prog* **1997**, *12*, 170–178.

34 Miyabe, K., Guiochon, G., *J Chromatogr A* **2000**, *866*, 147–171.

35 Seidel-Morgernstern, A., Jacobson, S. C., Guiochon, G., *J Chromatogr* **1993**, *637*, 19–28.

36 Kaczmarski, K., Antos, D., Sajonz, H., Sajonz, P., Guiochon, G., *J Chromatogr A* **2001**, *925*, 1–17.

37 Kaczmarski, K., Antos, D. J., *J Chromatogr A* **1996**, *756*, 73–87.

38 Nagrath, D., Bequette, B. W., Cramer, S. M., *AIChE J* **2003**, *49*, 82–95.

39 Zhu, J, Guiochon, G., *J Chromatogr* **1994**, *659*, 15–25.

40 Zhu, J., Ma, Z., Giochon, G., *Biotechnol Prog* **1993**, *9*, 421–428.

41 Zhu, J., Katti, A., Guiochon, G., *Anal Chem* **1991**, *63*, 2183–2188.

42 Dose, E., Jakobson, S., Guiochon, G., *Anal Chem* **1991**, *63*, 833–839.

43 Brooks, C. A., Cramer, S. M., *AIChE J* **1992**, *38*, 1969–1978.

44 Huang, J.-X., Guiochon, G., *J Colloid Interface Sci* **1989**, *128*, 577–591.

45 Ruthven, D. M., *Principles of Adsorption and Adsorption Processes*. Wiley, New York, **1984**.

46 Antia, F. D., Horvath, C., *J Chromatogr* **1991**, *556*, 119–143.

47 Subramanian, G., Cramer, S. M., *Biotechnol Prog* **1989**, *5*, 92–97.

48 Carta, G., Dinerman, A. A., *AIChE J* **1994**, *40*, 1618–1628.

49 Kim, Y. J., Cramer, S. M., *J Chromatogr* **1991**, *549*, 89–99.

50 Kasper, C., Breier, J., Vogt, S., Freitag, R., *Bioseparation* **1996**, *6*, 247–262.

51 Frey, D. D., *J Chromatogr* **1987**, *409*, 1–13.

52 Kim, Y. J., *Bioseparation* **1995**, *5*, 295.

53 Zhang, S., Sun, Y., *J Chromatogr* **2002**, *957*, 89–97.

54 Zhang, S., Sun, Y., *Ind Eng Chem Res* **2003**, *42*, 1235–1242.

55 Velayudhan, A., Horvath, C., *J Chromatogr* **1988**, *443*, 13–29.

56 Kopaciewicz, W., Rounds, M. A., Fausnaugh, J., Regnier, F. E., *J Chromatogr* **1983**, *266*, 3.

57 Brooks, C. A., Cramer, S. M., *J Chromatogr* **1995**, *693*, 187–196.

58 Li, Y., Pinto, N. G., *J Chromatogr A* **1994**, *658*, 445–457.

59 Brooks, C. A., Cramer, S. M., *Chem Eng Sci* **1996**, *51*, 3847–3860.

60 Gallant, S. R., Cramer, S. M., *J Chromatogr A* **1997**, *771*, 9–22.

61 Jozwik, M., Kaczmarski, K., Freitag, R., *J Chromatogr* **2005**, *1073*, 111–121.

62 Natarajan, V., Cramer, S. M., *AIChE J* **1999**, *45*, 27–37.

63 Natarajan, V., Bequette, B. W., Cramer, S. M., *J Chromatogr* **2000**, *876*, 51–62.

64 Natarajan, V., Ghose, S., Cramer, S. M., *Biotechnol Bioeng* **2002**, *78*, 365–375.

65 Natarajan, V., Cramer, S. M., *J Chromatogr* **2000**, *876*, 63–73.

66 Felinger, A., Guiochon, G., *J Chromatogr* **1992**, *609*, 35–47.

67 Subramanian, G., Phillips, M. W., Jayaraman, G., Cramer, S. M., *J Chromatogr* **1989**, *484*, 225–236.

68 Gerstner, J. A., Morris, J., Hunt, T., Hamilton, R., Afeyan, N. B., *J Chromatogr* **1995**, *695*, 195–204.

69 Vogt, S., Freitag, R., *J Chromatogr* **1997**, *760*, 125–137.

70 Vogt, S., Freitag, R., *Biotechnol Prog* **1998**, *14*, 742–748.

71 Freitag, R., Vogt, S., *J Biotechnol* **2000**, *78*, 69–82.

72 Schmidt, B., Wandrey, C., Freitag., R., *J Chromatogr* **2003**, *1018*, 155–167.

73 Ghose, S., Cramer, S. M., *J Chromatogr* **2001**, *928*, 13–23.

74 Gerstner, J. A., *BioPharm* **1996**, *9*, 30.

75 Jen, S. C. D., Pinto, N. G., *React Polym* **1993**, *19*, 145–161.

76 Jen, S. C. D., Pinto, N. G., *J Chromatogr* **1990**, *519*, 87–98.

77 Antia, F. D., Fellegvári, I., Horvázth, C., *Ind Eng Chem Res* **1995**, *34*, 2796–2804.

78 Shukla, A. A., Sunasara, K. M., Rupp, R. G., Cramer, S. M., *Biotechnol Bioeng* **2000**, *68*, 672–680.

79 Sunasara, K. M., Xia, F., Gronke, R. S., Cramer, S. M., *Biotechnol Bioeng* **2003**, *82*, 330–339.

80 Ruaan, R.-C., Yang, A., Hsu, D., *Biotechnol Prog* **2000**, *16*, 1132–1134.

81 Freitag, R., Horvath, C., *Adv Biochem Eng Biotechnol* **1995**, *53*, 17–59.

82 Peterson, E. A., *Anal Biochem* **1978**, *90*, 767–784.

83 Gerstner, J. A., Cramer, S., *Biotechnol Prog* **1992**, *8*, 540–545.

84 Gerstner, J. A., Cramer, S., *BioPharm* **1992**, *Nov/Dec*, 42–45.

85 Gadam, S. D., Cramer, S. M., *Chromatographia* **1994**, *39*, 409–418.

86 Jayaraman, G., Li, Y.-F., Moore, J. A., Cramer, S. M., *J Chromatogr* **1995**, *702*, 143–155.

87 Jen, S. C. D., Pinto, N. G., *J Chromatogr Sci* **1991**, *29*, 478–484.

88 Chen, G., Scouten, W. H., *J Mol Recognition* **1996**, *9*, 415–425.

89 Kundu, A., Cramer, S. M., *Anal Biochem* **1997**, *248*, 111–116.

90 Kundu, A., Vunnum, S., Cramer, S. M., *J Chromatogr* **1995**, *707*, 57–67.

91 Barnthouse, K. A., Trompeter, W., Jones, R., Inampudi, P., Rupp., Cramer, S. M., *J Biotechnol* **1998**, *66*, 125–136.

92 Luo, Q., Andrade, J. D., *J Chromatogr B* **2000**, *741*, 23–29.

93 Vunnum, S., Gallant, S., Cramer, S. M., *Biotechnol Prog* **1996**, *12*, 84–91.

94 Shukla, A. A., Barnthouse, K. A., Su Bae, S., Moore, J. A., Cramer, S. M., *J Chromatogr* **1998**, *814*, 83–95.

95 Shukla, A. A., Bae, S. S., Moore, J. A., *J Chromatogr* **1998**, *827*, 295–310.

96 Tugcu, N., Park, S. K., Moore, J. A., Cramer, S. M., *Ind Eng Chem Res* **2002**, *41*, 6482–6492.

97 Torres, A. R., Edberg, S. C., Peterson, E. A., *J Chromatogr* **1987**, *389*, 177–182.

98 Torres, A. R., Peterson, E. A., *Anal Biochem* **1983**, *130*, 271–282.

99 Torres, A. R., Dunn, B. E., Edberg, S. C., Peterson, E. A., *J Chromatogr* **1984**, *316*, 125–132.

100 Torres, A. R., Peterson, E. A., *J Chromatogr* **1992**, *604*, 39–46.

101 Quintero, G., Vo, M., Farkas, G., *J Chromatogr* **1995**, *693*, 1–5.

102 Freitag, R., Vogt, S., Mödler, M., *Biotechnol Prog* **1999**, *15*, 573–576.

103 Kumar, A., Galaev, I. Y., Mattiasson, B., *J Chromatogr B* **2000**, *741*, 103–113.

104 Galaev, I. Y., Arvidsson, P., Mattiasson, B., *J Mol Recognition* **1998**, *11*, 255–260.

105 Patrickios, C. S., Gadam, S. D., Cramer, S. M., Hertler, W. R., Hatton, T. A., *Biotechnol Prog.* **1995**, *11*, 33–38.

106 Kundu, A., Vunnum, S., Jayaraman, G., Cramer, S. M., *Biotechnol Bioeng* **1995**, *48*, 452–460.

107 Freitag, R., Wandrey, C., EP 98810231.5-2109 (date of deposit: 18 March 1998), *Int Patent Appl PCT/IB 99/00455.*

108 Schmidt, B., Wandrey, C., Vogt, S., Freitag, R., Holzapfel, H., *J Chromatogr A* **1999**, *865*, 27–34.

109 Schmidt, B., Wandrey, Ch., Freitag, R., *J Chromatogr* **2002**, *944*, 149–159.

110 Mazza, C. B., Rege, K., Breneman, C. M., Sukumar, N., Dordick, J. S., Cramer, S. M., *Biotechnol Bioeng* **2002**, *80*, 60–72.

111 Rege, K., Ladiwala, A., Tugcu, N., Breneman, C. M., Cramer, S. M., *J Chromatogr* **2004**, *1033*, 19–28.

112 Rege, K., Ladiwala, A., Cramer, S. M., *Anal Chem* **2005**, *77*, 6818–6827.

113 Torres, A. R., Peterson, E. A., *J Chromatogr* **1990**, *499*, 47–54.

114 Kundu, A., Barnthouse, K. A., Cramer, S. M., *Biotechnol Bioeng* **1997**, *56*, 119–129.

115 Kalasz, H., Nagy, J., Bathori, M., *J Planar Chromatogr* **1989**, *2*, 39–43.

116 Kalasz, H., Kerecsen, L., Nagy, J., *J Chromatogr* **1984**, *316*, 95–104.

117 Kalasz, H., *J High Res Chromatogr Chromatogr Commun* **1983**, *6*, 49–50.

118 Kalasz, H., Bathori, M., Kerecsen, L., Toth, L., *J Planar Chromatogr* **1993**, *6*, 38–42.

119 Ramsey, R., Katti, A. M., Guiochon, G., *Anal Chem* **1990**, *62*, 2557–2565.

120 Frenz, J., *LC/GC Int* **1992**, *5*, 18–21.

121 Mhatre, R., Qian, R., Krull, I. S., Gadam, S., Cramer, S., *Chromatographia* **1994**, *38*, 349–354.

122 Frenz, J., Quan, C. P., Hancock, W. S., Bourell, J., *J Chromatogr* **1991**, *557*, 289–305.

123 Frenz, J., Bourell, J., Hancock, W. S., *J Chromatogr* **1990**, *512*, 299–314.

124 Vigh, G., Irgens, L. H., Farkas, G., *J Chromatogr* **1990**, *502*, 11–19.

125 Xiang, R., Horváth, C., Wilkins, J. A., *Anal Chem* **2003**, *75*, 1819–1827.

126 Wilkins, J. A., Xiang, R., Horvath, C., *Anal Chem* **2002**, *74*, 3933–3941.

127 Jozwiak, K., Haginaka, J., Moaddel, R., Wainer, I. W., *Anal Chem* **2002**, *74*, 4618–4626.

128 Maciuk, A., Renault, J.-H., Margraff, R., Trébuchet, P., Zéches-Hanrot, M.,

Nuzillard, J.-M., *Anal Chem* **2004**, *76*, 6179–6186.

129 Freitag, R., Vogt, S., *J Biotechnol* **2000**, *78*, 69–82.

130 Cramer, S. M., Horvath, C., *Prep Chromatogr* **1988**, *1*, 29–49.

131 Viscomi, G. C., Cardinali, C., Longobardi, M. G., Verdini, A. S, *J Chromatogr* **1991**, *549*, 175–184.

132 Jakobson, J., *ACS Symp Ser* **1993**, *529*, 77–84.

133 Kalghatgi, K., Fellegvari, I., Horvath, C., *J Chromatogr* **1992**, *604*, 47–53.

134 Vigh, G., Varga-Puchony, Z., Szepesi, G., Gadzag, M., *J Chromatogr* **1987**, *386*, 353–362.

135 Zammouri, A., Muhr, L., Grevillot, G., *Ind Eng Chem Res* **2000**, *39*, 2468–2479.

136 Kalasz, H., Horvath, C., *J Chromatogr* **1981**, *215*, 295–302.

137 Valko, K., Slegel, P., Bati, J., *J Chromatogr* **1987**, *386*, 345–351.

138 Subramanian, G., Phillips, M. W., Cramer, S. M., *J Chromatogr* **1988**, *439*, 341–351.

139 Ramanan, S., Velayudhan, A., *J Chromatogr* **1999**, *830*, 91–104.

140 Liao, A., Horvath, C., *Ann NY Acad Sci* **1988**, *589*, 182–191.

141 Torres, A. R., Peterson, E. A., In: *Separations for Biotechnology*, Verrall, M. S., Hudson, M. J. (Eds.). Ellis Horwood, Chichester, **1987**, pp. 176–184.

142 Luellau, E., von Stockar, U., Vogt, S., Freitag, R., *J Chromatogr* **1998**, *796*, 165–175.

143 Ghose, S., Mattiasson, B, *J Chromatogr* **1991**, *547*, 145–153.

144 Ghose, S., Mattiasson, B., *Biotechnol Tech* **1993**, *7*, 615–620.

145 Kim, Y. J., *Biotechnol Tech* **1995**, *9*, 417–422.

146 Kim, Y. J., *Biotechnol Tech* **1994**, *8*, 457–462.

147 Sunasara, K. M., Rupp, R. G., Cramer, S. M., *Biotechnol Prog* **2001**, *17*, 897–906.

148 Shukla, A. A, Hopfer, R. L., Chakravarti, D. N., Bortell, E., Cramer, S. M., *Biotechnol Prog* **1998**, *14*, 92–101.

149 Narahari, C. R., Strong, J. C., Frey, D. D., *J Chromatogr* **1998**, *825*, 115–126.

150 Vigh, G., Quintero, G., Farkas, G., *ACS Symp Ser* **1990**, *434*, 181–197.

151 Farkas, G., Irgens, L. H., Quintero, G., Beeson, M., Al-Saeed, A., Vigh, G., *J Chromatogr* **1993**, *645*, 67–74.

152 Irgens, L. H., Farkas, G., Vigh, G., *J Chromatogr A* **1994**, *666*, 603–609.

153 Quintero, G., Vo, M., Farkas, G., Vigh, G., *J Chromatogr A* **1995**, *693*, 1–5.

154 Qi, Y., Huang, J., *J Chromatogr* **2002**, *959*, 85–93.

155 Rathore, A. S., Horáth, C., *J Chomatogr* **1997**, *787*, 1–12.

156 Huang, S.-Y., Jin, J.-D., *Bioseparation* **1994**, *4*, 343–351.

157 el Rassi, Z., Horvath, C., *J Chromatogr* **1983**, *266*, 319–340.

158 Shukla, A. A., Deshmukh, R. R., Moore, J. A., Cramer, S. M., *Biotechnol Prog* **2000**, *16*, 1064–1070.

159 Tugcu, N., Deshmukh, R. R., Sanghvi, Y. S., Moore, J. A., Cramer, S. M., *J Chromatogr* **2001**, *923*, 65–73.

160 Tugcu, N., Deshmukh, R. R., Sanghvi, Y. S., Cramer, S. M., *React Polym* **2003**, *54*, 37–47.

161 Cramer, S. M., el Rassi, Z., Horvath, C., *J Chromatogr* **1987**, *394*, 305–314.

162 El Rassi, Z., Horvath, C., *J Chromatogr* **1983**, *266*, 319–340.

163 Bianchi, E., Del Guidice, G., Verdini, A. S., Pessi, A., *Int J Peptide Protein Res* **1991**, *37*, 7–13.

164 Viscomi, G. C., Lande, S., Horvath, C., *J Chromatogr* **1988**, *440*, 157–164.

165 Cardinalli, F., Ziggiotti, A., Viscomi, G. C., *J Chromatogr* **1990**, *499*, 37–45.

166 Torres, A. R., Krueger, G. G, Peterson, E. A., *Anal Biochem* **1985**, *144*, 469–476.

167 Camacho-Toralba, P. L., Beeson, M. D., Vigh, G., *J Chromatogr* **1993**, *646*, 259–266.

168 Sinibaldi, M., Castellani, L., Federici, F., Messina, A., *J Liq Chromatogr* **1993**, *16*, 2977–2992.

169 Horvath, C., Frenz, J., el Rassi, Z., *J Chromatogr* **1983**, *255*, 273–293.

170 Zhang, W.-N., Hu, Z.-X., Feng, Y.-Q., Da, S.-L., *J Chromatogr* **2005**, *1068*, 269–278.

171 Zhang, W.-N., He, H.-B., Feng, Y.-Q., Da, S.-L., *J Chromatogr* **2004**, *1036*, 145–154.

7

The Purification of Biomolecules by Countercurrent Chromatography

Ian J. Garrard

7.1
A Description of Countercurrent Chromatography

Countercurrent chromatography is a form of liquid–liquid separation technology. There is no solid stationary phase, as in the more "traditional" forms of chromatography such as high-performance liquid chromatography (HPLC) or flash chromatography. In countercurrent chromatography instruments, tubing is wound on a drum (called a bobbin) which is centrifugally rotated in planetary motion, i.e. revolving around the central axis of the sun gear while simultaneously rotating about its own axis at the same angular velocity (Fig. 7.1). A typical laboratory-size, preparatory instrument is shown in Fig. 7.2.

A two-phase solvent system is used for the separation. For most countercurrent chromatography separations, this two-phase system comes from an organic/aqueous solvent mix; however, alternatives more suited to labile biomolecules are described later in the chapter. A simple example would be heptane/water, but a more likely system might consist of heptane, ethyl acetate, methanol and water. This gives a two-phase system with an upper organic layer consisting mainly of heptane/ethyl acetate and a lower aqueous layer consisting mainly of methanol/water. The tubing is initially filled with the solvent phase intended to be stationary and, with the instrument spinning, the mobile phase is pumped through it. A small initial displacement of stationary phase occurs before an equilibrium is set up, with the planetary motion retaining up to 85% of the stationary phase in the coil and the mobile phase passing through, eluting at the far end without displacing any more stationary phase.

The planetary motion also sets up alternating zones of mixing and settling along the length of the tube that travel in waves synchronously with the rotation. Figure 7.3 shows a model of this mixing motion, which has been found to be similar to the "swish-swosh" motion that occurs when a tube of two liquids is tilted from side to side [2]. Samples injected with the mobile phase undergo many partitioning steps per minute (equal to the rotational speed in revolutions per minute) before they elute. This provides for high-resolution separations based on differen-

Bioseparation and Bioprocessing. Edited by G. Subramanian
Copyright © 2007 WILEY-VCH Verlag GmbH & Co. KGaA, Weinheim
ISBN: 978-3-527-31585-7

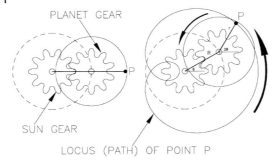

Fig. 7.1 The planetary motion of the J-type centrifuge [1].

Fig. 7.2 Brunel countercurrent chromatography – a J-type centrifuge that is typical of a modern laboratory-scale preparative instrument.

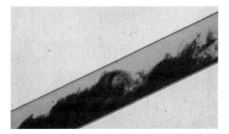

Fig. 7.3 A model of the wave mixing that occurs in the J-type centrifuge.

tial partitioning behavior as characterized by the distribution ratio ($D=$ concentration of solute in stationary phase/concentration of solute in mobile phase).

Countercurrent chromatography can be used in normal or reverse-phase mode, depending on which phase (upper or lower) is selected to be stationary. Furthermore, material that elutes very slowly can be recovered without any compound losses by pumping out the stationary phase while maintaining resolution [3].

As mentioned above, the use of aqueous/organic solvent two-phase systems is frequently unsuitable for the purification of biomolecules. An alternative is the aqueous/aqueous polymer systems described below. Unfortunately, it is harder to retain the liquid stationary phase of these systems than it is with most organic/ aqueous two-phase systems and this has an adverse effect on the resolution of the components being separated [4]. For this reason, a number of alternative centrifuge configurations have been investigated. The most commonly employed is the related technique of centrifugal partition chromatography [5] in which a drum, rotating simply about its own axis, contains a series of interconnecting chambers in which the stationary phase is retained. However, many alternatives have been created for the countercurrent chromatography concept, depending upon the relative orientation of the two rotational axes and the ratio of the two radii to the rotation speed. Figure 7.4 shows the various types and their mutual

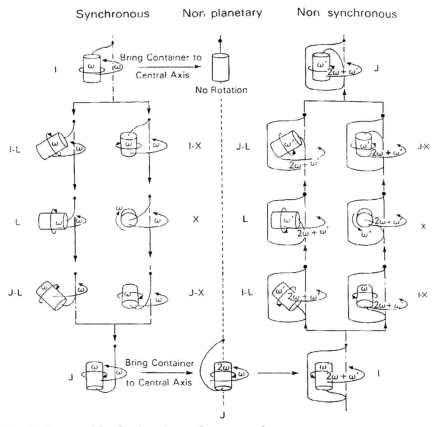

Fig. 7.4 Rotary seal free flowthrough centrifuge systems for performing countercurrent chromatography [6].

relationship. These have been classified as I, J, L and X schemes, as well as combinations of these [6]. Unlike the centrifugal partition chromatography centrifuge, all are configured to allow a connection to the pump without the need for a rotary seal.

There are three main classes, according to the type of planetary motion. With synchronous motion, one rotation of the holder corresponds to one revolution around the central axis of the apparatus. In the nonsynchronous series, the rates of rotation and revolution are independently adjusted. In the nonplanetary series, rotation and revolution of the holder share a common axis, resulting in either nonrotation or a single rotation around the central axis of the apparatus.

While a considerable amount of research has been done on the synchronous flow-through centrifuges, much less is known about the fields generated in the nonsynchronous systems. This is largely due to the substantially more complex mechanical design of apparatus required in order to make the rotation and revolution independently adjustable. However, it is possible that some significant future advances in countercurrent chromatography design will come from this series. By selecting a suitable combination of slow coil rotation and high-speed revolution, the system will be capable of stable retention of the stationary phase. This is particularly significant for the hydrophilic two-phase systems used for the partition of macromolecules and cell particles [7].

7.2
Countercurrent Chromatography Compared to Solid-phase Chromatography

Both countercurrent chromatography and solid-phase chromatography have the same ultimate aim – to separate a mixture of solutes into their component compounds in order to isolate one or more of them.

The manner in which separations are achieved is similar. Both techniques keep one phase stationary as a second phase passes through it. Furthermore, apart from the actual separation columns, much of the equipment is similar and shares the same technology, e.g. the pump, injector and detector apparatus.

However, there the similarities end. The most fundamental difference between solid-phase chromatography and countercurrent chromatography is that the former has a solid stationary phase and thus adsorption, ion exchange or hydrophobic interaction is the general mechanism of sample retention, whereas the latter has a liquid stationary phase and so liquid partitioning is the sample retention process.

7.2.1
The Advantages of Countercurrent Chromatography for the Purification of Biomolecules

The following is a list of the main advantages of countercurrent chromatography with reference to its use as a preparative instrument capable of separating a biomolecule sample into constituent fractions.

- Versatility
  - Lab-scale instruments can separate from a few milligrams to 10 g or more.
  - Many solvent systems are available to maximize α (the separation factor).
  - Can choose either phase as the mobile one (normal or reverse-phase chromatographic mode) or even switch over mid-run.
- Less likely cross-contamination
  - No contaminants are leached from adsorbents onto subsequent runs. Only compounds physically adhering to the coil internal walls can cause cross-contamination.
- Economical
  - The columns are virtually indestructible.
  - It uses significantly less solvent than comparatively sized solid-phase chromatography.
  - For preparative separations, the instruments are cheaper to purchase than the equivalent large-scale HPLC system.
- Crude samples tolerated
  - Dirty samples are accepted. Very little prior clean-up is necessary. Even particulates such as cells are tolerated.
- Predictable and reproducible
  - The retention time is very predictable from the distribution ratio D in the solvent system used. Furthermore, while solid-phase columns age and so retention can alter with time, with countercurrent chromatography a fresh column is made each column-fill, so the separation is always reproducible.
- High sample loading
  - Between 60 and 90% of the column volume is stationary phase – a large percentage compared to HPLC, which is often around 5% and may be below 1% volume of active sites (as opposed to the supporting matrix). This allows higher sample loadings.
- Total sample recovery
  - No irreversible adsorption or catalytic change, as is possible with solid supports.
  - The pH can be adjusted for maximum stability of solutes.
  - The total activity in bioassay-directed isolations can be recovered.

It is worth noting that countercurrent chromatography does not suffer from two of the biggest drawbacks that limit solid-phase separations, i.e. the presence of particulate matter and undesirable interactions with the stationary phase. As there is no solid support in countercurrent chromatography, the technique can cope with particulates. The only limitation to this is the physical blocking of the coil. In fact, countercurrent chromatography has successfully been used to separate particulates such as blood cells, organelles and other particles [7–9].

In addition, catalytic conversion of labile molecules is much less likely to occur than with the solid stationary phase. This makes countercurrent chromatography a useful technique for unstable compounds. Furthermore, total sample recovery is always possible with countercurrent chromatography. Since the column only contains liquids, blowing out the contents with a gentle nitrogen pressure will ensure that a complete mass balance analysis can always be performed and, since a fresh column can be created each run, there are no problems with excessively retained components leaching into subsequent analyses.

7.3
Solvent System Selection Process

The whole heart of a countercurrent chromatography separation is the solvent system. For the majority of countercurrent chromatography purifications, the two-phase system is created from a mixture of organic and aqueous solvents. In the past, developing a solvent system for countercurrent chromatography required lots of time, operator skill and experience. With biphasic systems coming from two-, three-, four- or even five or more-component solvents, this made the range of options almost limitless. On the plus side, this gave the technique the ability to separate compounds ranging from extremely polar to extremely nonpolar. The down side was that it was difficult to know where to begin.

Various systematic, step-by-step protocols have now been developed that take an inexperienced user logically through the process [10, 11]. The solvent selection part has even been fully automated using a liquid-handling robot and HPLC.

In brief, appropriate phase systems can be selected by screening a graded range of solvent systems to identify one with a suitable distribution ratio for the compound of interest (generally in the range $D=0.2$–5). A typical table might range from moderately polar (e.g. butanol/water) to moderately nonpolar (e.g. heptane/methanol). An example, taken from reference [11], is given in Tab. 7.1. Alternative tables exist for extremely polar or extremely nonpolar compounds and Tab. 7.2 shows a selection table for highly nonpolar compounds.

Using a liquid-handling robot (or a lab technician if a suitable robot is not available!), all systems are made up in small volume, e.g. 4 mL. A small amount of the crude mixture to be separated is added to the systems and mixed. After phase separation, samples are taken from the top and bottom phases, analyzed by an appropriate assay (frequently HPLC or gas chromatography), and D values calculated from the peak areas (or heights if not fully resolved) for each solvent system.

Tab. 7.1 Table for selecting a countercurrent chromatography solvent system graded from polar (1) to nonpolar (28) [11]

No.	Heptane	Ethyl acetate	Methanol	Butanol	Water
1	0	0	0	2	2
2	0	0.4	0	1.6	2
3	0	0.8	0	1.2	2
4	0	1.2	0	0.8	2
5	0	1.6	0	0.4	2
6	0	2	0	0	2
7	0.1	1.9	0.1	0	1.9
8	0.2	1.8	0.2	0	1.8
9	0.29	1.71	0.29	0	1.71
10	0.33	1.67	0.33	0	1.67
11	0.4	1.6	0.4	0	1.6
12	0.5	1.5	0.5	0	1.5
13	0.57	1.43	0.57	0	1.43
14	0.67	1.33	0.67	0	1.33
15	0.8	1.2	0.8	0	1.2
16	0.91	1.09	0.91	0	1.09
17	1	1	1	0	1
18	1.09	0.91	1.09	0	0.91
19	1.2	0.8	1.2	0	0.8
20	1.33	0.67	1.33	0	0.67
21	1.43	0.57	1.43	0	0.57
22	1.5	0.5	1.5	0	0.5
23	1.6	0.4	1.6	0	0.4
24	1.67	0.33	1.67	0	0.33
25	1.71	0.29	1.71	0	0.29
26	1.8	0.2	1.8	0	0.2
27	1.9	0.1	1.9	0	0.1
28	2	0	2	0	0

Quantities (in milliliters) required to make 4 mL of system.

The solvent systems highlighted in gray are multiples of 0.5 mL and are therefore particularly easy to make up. If doing the work by hand, testing just the highlighted systems first enables the approximate area of interest in the table to be determined to avoid having to make unnecessary systems. For example, if the results of testing just the highlighted systems show that 17 is too polar and 22 too nonpolar, then only the systems between 17 and 22 are tested to fine-tune the selection.

When analyzed, $D < 0.2$ will make the compound elute very quickly, not giving it much chance to separate from the other components. $D > 5$ will mean long elution times and possible band broadening. The degree of separation between the target compound and any of its contaminants can readily be calculated from α, the separation factor ($\alpha = D_2/D_1$, where $D_2 > D_1$). Once the performance of a particular instrument under various run conditions has been observed,

Tab. 7.2 Table for selecting a highly nonpolar countercurrent chromatography solvent system graded from nonpolar (28) to extremely nonpolar (40)

No.	Methanol	Acetonitrile	Toluene	Heptane
28	2	0	0	2
29	1.8	0.2	0	2
30	1.6	0.4	0	2
31	1.4	0.6	0	2
32	1.2	0.8	0	2
33	0.8	1.2	0	2
34	0.4	1.6	0	2
35	0	2	0	2
36	0	1.8	0.2	2
37	0	1.6	0.4	2
38	0	1.4	0.6	2
39	0	1.2	0.8	2
40	0	1	1	2

Quantities (in milliliters) required to make 4 mL of system.

it is relatively easy to predict the minimum α value necessary for a given purification.

7.4
Countercurrent Chromatography of Polar Biomolecules

For many biomolecules, a simple organic/aqueous solvent system is unsuitable. Biomolecules such as peptides and sugar derivatives are extremely water soluble, while compounds such as proteins, enzymes and nucleic acids can be readily denatured by contact with organic solvents.

Considering first the problem of extremely water-soluble molecules, several options are available. These include:

- pH adjustment of the aqueous layer.
- Addition of salts to the aqueous layer.
- pH zone-refining countercurrent chromatography.
- Addition of affinity ligands to the stationary phase.
- Room temperature ionic liquids.

If, however, the biomolecules to be purified are sensitive to organic solvents, e.g. most proteins, then an aqueous–aqueous polymer system must be developed. Each of these options will be considered in turn.

7.4.1
pH Adjustment of the Aqueous Layer

pH adjustment of the solvent system is a technique that will work best on ionizable biomolecules. Trifluoroacetic acid (TFA) is frequently used to acidify the
aqueous layer. For example, Knight and coworkers used TFA at concentrations of
1–5% in solvent systems of methylbutyl ether–butanol–acetonitrile for the purification of a range of peptides [12]. Likewise, Harada and coworkers, in the purification
of a novel, cyclic depsipeptide antibiotic known as WAP-8294A, found the compound to be soluble in water, but not in organic solvents, including ethanol and
butanol. However, when TFA was added at concentrations at or above 0.005 M,
the compound moved to the butanol layer of butanol–water (1 : 1) due to protonation of the carboxylic group [13].

Although protonation of acid functions is accepted as an explanation for the
effect of TFA acid in the aqueous phase, it is not only acid components which
show an effect. Working on the separation of colistin A and B (a peptide antibiotic),
Ikai and coworkers found the solvent system butanol–0.04 M aq. TFA considerably
more effective than simply butanol–water [14]. Furthermore, unlike the results of
Harada and coworkers on the antibiotic WAP8294A, they found that there was a
direct relationship between the TFA concentration and the distribution ratio, with
the partitioning of colistin components into the butanol phase increasing with
TFA concentration. Their explanation was that the free amino groups in the
peptide were being protonated by the TFA and subsequently forming neutral,
lipophilic ion pairs with the TFA anions. As a result, the hydrophobicity of colistin
components increased with the concentration of TFA.

7.4.2
Addition of Salts

Some varying results are found when searching the scientific literature for application examples where salts have been used to assist the partitioning of polar compounds. In 1991, Harada and coworkers investigated the effects of four different
salts on the distribution ratios of bacitracin (a peptide antibiotic) components. The
salts studied were sodium chloride, potassium chloride, ammonium chloride and
ammonium acetate [15]. All were dissolved at 10% (w/v) concentration, with
various organic solvents for the opposing layer. The authors concluded that the
salt addition had very little effect and finally selected a system consisting of chloroform, ethanol and water (5 : 4 : 3) for the separation.

Not everyone agrees with that conclusion, however. In separating a number of
water-soluble vitamins, Shinomiya and coworkers obtained effective results using
the polar salt system butanol–ethanol–0.15 M monobasic potassium phosphate
solution (8 : 3 : 8) [16]. This system proved to be more polar than the simpler
butanol–0.15 M potassium phosphate solution (1 : 1), i.e. the presence of the ethanol
improved the partitioning of the water soluble compounds into the butanol layer.

This salt concentration is fairly low compared to some other applications. For example, in a recent paper, Fahey and coworkers used saturated ammonium sulfate solution mixed with water in the ratio 1.2 : 1, giving a salt concentration of 31% (w/v). This was successfully used in the solvent system 1-propanol–acetonitrile–saturated aqueous ammonium sulfate–water (1 : 0.5 : 1.2 : 1) to separate a number of extremely similar and highly polar glucosinolates from a broccoli extract [17]. These compounds were also separated by the author and his colleagues at the Brunel Institute for Bioengineering on a preparative scale using the same solvent system [18]. However, the peak shape of the glucosinolate compounds was very strange – an upward-sloping triangle (Fig. 7.5).

This is unlikely to be due to volume overload (which in HPLC tends to produce a rectangular peak shape) or mass overload (which in HPLC tends to produce a triangular peak, but with a sharp initial rise and gradual downward slope). It is also unlikely to be due to ionic effects arising from operating with a solvent system pH that is too close to the molecule's pK_a value, since the pK_a of N-hydroxy sulfates is of the order of −9 and thus the molecules are ionized under nearly all conditions [19]. A possible explanation is that an ion pair is forming between the charged glucoraphanin molecule and the ammonium ion (GR-SO$_3^-$ ...NH$_4^+$) and it is this neutral species that is partitioning. If so, then this is the effect of the salt: to create ion pairs that are neutral and thus less hydrophilic.

An ion-pairing effect would make the nature of the salt highly significant. For example, the tetrabutyl ammonium–GR ion pair would be more hydrophobic than the ammonium–GR ion pair and this would change the distribution ratios. Unfortunately, at the time this was not investigated since the GR peak, although strange in shape, yielded pure material and the separation was deemed to be acceptable to achieve the goal of a preparative purification.

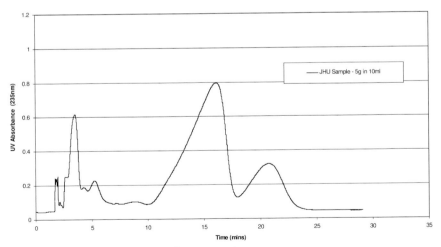

Fig. 7.5 Preparative countercurrent chromatography chromatogram of crude glucosinolate extract [18].

It should also be noted that the addition of salt is frequently necessary in order to create a two-phase system, i.e. without the salt, a single layer would be formed. For example, in the Fahey paper mentioned above [17], the solvent system consists of ethanol, acetonitrile, saturated salt solution and water. Since ethanol and acetonitrile are fully miscible with water, it is only the presence of the ammonium sulfate salt that makes the system form two layers and thus the addition of the salt allows more polar organic solvents to be used.

7.4.3
pH Zone-refining Chromatography

pH zone-refining countercurrent chromatography is a unique way of separating ionic molecules and was initially developed with the separation of organic acids. In brief, a short-chain organic acid, e.g. TFA (the "retainer"), is added to the sample solution or stationary phase, followed by isocratic elution with a basic mobile phase (the "eluter"). This results in the sharpening of peaks obtained from compounds containing acid groups. It follows that when basic solutes are to be separated, the peak sharpening agent is a base and the mobile phase is acidic.

The technique was discovered by Yoichiro Ito by accident [20]. A thyroxine analog was found to produce a broad, skewed peak in the solvent system hexane–ethyl acetate–methanol–15 mM ammonium acetate buffer (4 : 5 : 4 : 5) at pH 4, whereas when the ratios were slightly altered to (5 : 5 : 5 : 5) the peak became uncharacteristically sharp. It was found that the pH of the eluent made an abrupt rise which coincided with the elution of the sharpened peak. This was caused by the creation of an acid, bromoacetic acid, in the stationary phase, which acted as a "retainer" acid.

Although initially discovered and used for the separation of organic acids, pH zone refining has now been applied to a number of different compounds, some of which are listed in Tab. 7.3. As can be seen from Tab. 7.3, the technique has been successfully applied to quite a range of molecules, including synthetic mixtures, natural products, peptides and enzymes.

The biggest disadvantage of pH zone refining is that it only works with ionic compounds. Nonionizable analytes can be separated by a similar process using a suitable affinity ligand in the stationary phase with a displacer in the mobile phase as described in Section 7.4.4. Nevertheless, to operate as pH zone refining, the analytes to be separated should be ionizable with the difference in pK_a being, according to Ito, greater than 0.2. They should also be stable at a wide range of pH, preferably from 1 to 10, and the sample size should be at least 0.1 mmol and preferably over 1 mmol for each species, for an effective ionic solute equilibrium to be set up at the acid border [29].

pH zone refining bears a strong resemblance to displacement chromatography [30], the distinctive rectangular elution peaks being one of the most obvious similarities. However, there are some fundamental differences. In pH zone-refining countercurrent chromatography, the solutes are eluted in increasing order of their pK_a values and hydrophobicities. Furthermore, the concentration of each solute is

Tab. 7.3 A selection of separations achieved by pH zone-refining countercurrent chromatography

Sample	Solvent system	Retainer	Eluter	Reference
DNP amino acids	MtBE–ACN–H$_2$O (4:1:5)	TFA (200 μL/SS)	NH$_3$ (0.1%/MP)	21
Indole auxins	MtBE–H$_2$O (1:1)	TFA (0.04%/SP)	NH$_3$ (0.1%/MP)	21
Hydroxyxanthene dye: D&C Orange No. 5	Et$_2$O–ACN–10 mM AcONH$_4$ (4:1:5)	TFA (200 μL/SS)	MP	22
Alkaloids (natural product)	MtBE–H$_2$O (1:1)	HCl (10 mM/SP)	TEA (10 mM/MP)	23
Tetrachlorofluorescein (synthetic product)	Et$_2$O–ACN–10 mM AcONH$_4$ (4:1:5)	TFA (200 μL/SS)	MP	24
Carbobenzoxy dipeptides	MtBE–ACN–H$_2$O (2:2:3)	TFA (16 mM/SP)	NH$_3$ (5.5 mM/MP)	25
Carbobenzoxy tripeptides	nBuOH–MtBE–ACN–H$_2$O (2:2:1:5)	TFA (16 mM/SP)	NH$_3$ (2.7 mM/MP)	25
Curcuminoids (natural product)	MtBE–ACN–H$_2$O (4:1:5)	TFA (20 mM/SP)	NaOH (30 mM/MP)	26
Carminic acid (natural product)	nBuOH–MtBE–ACN–H$_2$O (3:1:1:5)	TFA (15 mM/SP)	NH$_3$ (15 mM/MP)	27
Enzymes	PEG 1540–15% ammonium sulfate (1:1)	Acetic acid (10 mM/SP)	NaOH (100 mM/MP)	28

SS: in sample solution; SP: in stationary phase; MP: in mobile phase.

mainly determined by that of the counterions in the mobile phase and the valency of the solute molecule. In displacement chromatography, the solutes' affinity to the stationary phase determines both the elution order and their concentration in the mobile phase [21].

7.4.4
Affinity Ligands

The optimization of the distribution ratio for a desired analyte in a two-phase solvent system has, up until this point, been achieved either by changing the relative hydrophobicity (Section 7.3) or for charged analytes the ionic strength (Section 7.4.2) or the pH of the system (Sections 7.4.1 and 7.4.3). This section adopts a new strategy – that of adding a complexing agent or ligand which selectively changes the distribution ratio of a particular group of compounds.

Affinity separations have been performed for many years in liquid chromatography for the purification of biomolecules. For example, various nucleic acids and proteins have been separated on affinity columns with a nucleotide-bonded solid support [31], and columns containing a chiral selector bonded onto the solid phase are in regular use for the separation of enantiomers [32]. Unfortunately, the difficulties of bonding the affinity ligand onto the stationary phase makes these columns expensive and thus limited to analytical use. Countercurrent chromatography, however, with its liquid stationary phase, has the potential to offer affinity separations on a preparative scale at an acceptable cost.

A World Patent entitled "Separation of polar compounds by affinity countercurrent chromatography" [33] makes the following claim:

> Separations of very hydrophilic compounds has now been
> achieved using countercurrent chromatography in which
> a ligand for the analytes of interest is used to enhance the
> partitioning of the polar species into the organic layer of an
> aqueous/organic solvent mixture. The compounds are
> separated according to their affinity for the ligand in the
> stationary organic phase. This method of affinity
> countercurrent chromatography can also be conducted
> in a pH zone-refining mode.

For the affinity ligand to be effective, it must remain in the stationary phase. Any leakage of the ligand from the end of the column may not affect the detector, but will contaminate the eluate. The ligand can be added to the stationary phase before filling the column, but the better approach is to first pump a given amount of ligand-free stationary phase into the column, followed by a known volume of ligand-containing stationary phase. This results in a calculated volume of ligand-free stationary phase at the end of the column, which acts as an absorbent to any ligand that may leak out [34]. However, it is still important to select a ligand with a strong affinity to the stationary phase of the solvent system being used.

With the organic phase as stationary, the ligand must be hydrophobic. A typical example is di(2-ethylhexyl) phosphoric acid (DEHPA), which can be purchased from companies like Sigma and has been used for the separation of organic amines [34], dipeptides [29], polar catecholamines [35] and heavy metals [36]. For the chiral separation of amino acids and peptides, the ligand N-dodecanoyl-L-proline-3,5-dimethylanilide (DPA) has been used [34, 37]. Unfortunately, this cannot be purchased directly and has to be synthesized according to the procedure described by Oliveros and coworkers [38]. More recently, Shinomiya and coworkers have used the compound 1-octane sulfonic acid to act as an ion-pair reagent (in effect, the same as an affinity ligand) for the separation of water-soluble vitamins [39]. The concentration of affinity ligand in the stationary phase is generally around 10–50 mM.

The positive effect of an affinity ligand in a successful separation can be quite striking. In a paper by Ma and Ito, four isomeric pairs of dipeptides were run on countercurrent chromatography using the system methylbutyl ether–0.1 M KH_2PO_4 (1 : 1) without any ligand and just a single peak was obtained, i.e. no separation occurred. However, when the same run was done with the addition of 10% DEHPA to the organic stationary phase, the peaks of all eight dipeptides can be seen [40]. Similarly impressive results can be found in other papers.

Although the potential for affinity ligands to separate extremely polar compounds is there, much more research is required before this can be confirmed. At present, the applications are limited to just a few and the choice of ligands is even smaller, greatly reducing the current usefulness of the technique.

7.4.5
Room Temperature Ionic Liquids (RTILs)

Ionic liquids have been used for many years as high-temperature solvents in inorganic chemistry [41], generally referred to, perhaps more recognizably, as molten salts. However, if a salt has a melting point below about 30 °C, then it will be a liquid at, or close to, normal room temperatures and is thus known as a RTIL. One of the first discovered was ethyl ammonium nitrate (an explosive), which has a melting point around 12 °C. However, in recent years, more RTILs have been discovered and, according to Berthod and Carda-Broch, they have the potential for being useful in countercurrent chromatography for the separation of ionic compounds [42].

In appearance, ionic liquids can look very like viscous, organic solvents. However, they are actually molten salts made of ions free to move within the liquid. In Section 7.4.2, the addition of salts to the aqueous phase of a solvent system was considered. Such a salt solution will contain both ionic and molecular interactions between both the solvent molecules (usually water) and the salt ions. An ionic liquid, on the other hand, will simply contain strong electrostatic interactions in a similar manner to the solid salt crystals. Furthermore, while an aqueous salt solution has to be mixed with suitable organic solvents in order to make a two phase system, RTILs are frequently immiscible with water, and a two-phase system can consist of just the RTIL and water. Unfortunately, the viscosity of RTILs is generally too high for use in most countercurrent centrifuges, creating too much back pressure. Thus, the RTIL has to be thinned with an organic solvent, such as acetonitrile or methanol, before it can be used in countercurrent chromatography.

A typical RTIL will consist of a bulky organic cation that contains nitrogen or phosphorous and either an organic or inorganic anion. For example, cations such as *N*-alkylpyridinium and 1-alkyl-3-methylimidazolium may be combined with inorganic anions such as Cl^-, NO_3^-, PF_6^- and BF_4^-. Organic anions are less common in RTILs, but may include trifluoroacetate, pentafluoropropanate, bis(trifluoromethanesulfonyl) imide $(CF_3SO_2)_2N^-$ and trifluoromethanesulfonate $CF_3SO_3^-$. With these various combinations, a range of RTILs can be created. For example, if the cation 1-butyl-3-methyl imidazolium (BMIM), is combined with the anion PF_6, the melting point is −8 °C, but if combined with the anion BF_4, it is −82 °C, a massive 74 °C difference. Furthermore, if combined with Cl^-, the melting point is +65 °C and thus the salt BMIM-Cl⁻ is a solid rather than a RTIL. However, the viscosity of BMIM-BF₄ is 233 cP and BMIM-PF₆ is 312 cP, both at 20 °C, hence the need to mix the RTIL with a thinning solvent. (All data obtained from Ref. [42].)

Unfortunately, although ionic liquids are new solvents with unique properties and are potentially useful in countercurrent chromatography for the separation of ionic or polar mixtures, the testing of these unusual liquids is only just beginning. The critical question of whether they actually work in countercurrent chromatography, or have any advantages over the other techniques described in this chapter, has yet to be answered. Much more research is required on this subject; however, if RTILs fulfill their potential, a whole new branch of countercurrent chromatography may be created.

7.5
Aqueous Polymer Systems

Aqueous polymer systems were introduced by Albertsson in the early 1960s [7, 43] and are predominantly employed in countercurrent chromatography for the purification of proteins [44, 45]. However, they have also been successfully employed in the purification of enzymes [46], chiral amino acids [47, 48], DNA and its binding proteins [49, 50], antibiotics [51], and polysaccharides [52].

Some of these separations were performed on the planetary motion X-type centrifuge [53]. This is because the X-type creates a gentler mixing between the phases and yields better retention of the stationary phase when compared to the J-type, especially when viscous, low interfacial tension systems are used. However, its complicated motion results in a tortuous path for the connecting leads and machine complexity. Prototypes of the X-type exist and have been successfully applied [45, 52], but they are not likely to lead to robust, commercially reliable instruments.

Many separations using aqueous polymer systems have been performed using centrifugal partition chromatography. However, in recent years the design of the J-type centrifuge has improved significantly and successful experiments have been performed using aqueous–aqueous polymer systems on the J-type [49]. The simplicity and robust nature of the J-type centrifuge make it likely that further research will be done using these systems on this type of instrument in the future.

Although aqueous polymer systems are polar systems designed for the separation of highly polar molecules, the primary reason for their use is that they can be buffered and made isotonic to provide a gentle environment for cells, subcellular particles and macromolecules. The traditional organic–aqueous phase systems frequently cause denaturation of these molecules, whereas it is possible to recover near-complete biological activity with aqueous polymer systems. Separation of fragile biological particles can be based on the charge density [54], lipid composition [55] or ligand affinity [56, 57].

In the past, many different types of polymer have been tried [58] including polypropylene glycol, polyvinyl alcohol, methylcellulose, polyvinylpyrrolidone, hydroxypropyl starch and maltodextrin. However, at the current time nearly all systems use poly(ethylene glycol) (PEG). The two most common systems are:

- PEG–potassium phosphate systems. These are generally used for the separation of macromolecules, e.g. proteins. The system is adjusted by changing the molecular weight of the PEG used and/or the pH of the phosphate buffer.
- PEG–dextran systems, frequently PEG 8000–Dextran 500. May be used for proteins with low solubility in the PEG–phosphate systems or for the separation of mammalian cells. The system is adjusted by changing the osmolarity and pH with electrolytes.

PEG–phosphate systems have a characteristic feature in that low-molecular-weight compounds are frequently partitioned exclusively in the upper or the lower phase, while macromolecules are more evenly distributed between the phases. This means that, when the optimized system is run on countercurrent chromatography, mixtures containing macromolecules can yield high purity fractions free from the low-molecular-weight impurities that either elute immediately with the mobile phase or remain permanently in the stationary phase in the column [59].

For example, Shinomiya and coworkers investigated a number of systems for the general purification of proteins and concluded that most of them could be fractionated on the system 12.5% (w/w) PEG 1000–12.5% (w/w) potassium phosphate solution [44]. Similar PEG–phosphate systems were used to separate the enzymes, chiral amino acids, DNA proteins and polysaccharides listed in the references above.

PEG 1000, which has a number average molecular weight around 1000, is the most commonly used polymer in PEG–phosphate systems. However, some separations were achieved using the higher-molecular-weight PEG 8000 [60] or PEG 6000 [61], or the moderate weight PEG 3350 [62]. The salt concentration is generally around 12.5% (w/w), although variations can be found. For example, Shibusawa and coworkers used 16% (w/w) dibasic potassium phosphate for the separation of high- and low-density lipoproteins from human serum [63], and likewise Chao and coworkers used the same salt concentration for the purification of polysaccharides from a plant root [64]. In the isolation of plasmid DNA, 18% (w/w) phosphate was employed by Kendall and coworkers [49]; for the purification of glycoprotein components from a fermentation media, a strong 25% (w/w) potassium phosphate solution was used [60]. However, these concentrations appear to be the exception rather than the rule, with 12.5% (w/w) covering the majority of the separations found in the literature.

Very rarely is a salt other than potassium phosphate used in these systems. In the paper mentioned above on the general purification of proteins, Shinomiya and coworkers tested eight different salt solutions with PEG 1000: potassium phosphate, sodium phosphate, potassium citrate, sodium citrate, sodium carbonate, sodium sulfate, magnesium sulfate and ammonium sulfate [44]. As mentioned, they concluded that most proteins could be purified on the 12.5% (w/w) PEG

1000–12.5% (w/w) potassium phosphate system, although they found that cyto-chrome *c* and apotransferrin performed better on the system 12.5% (w/w) PEG 1000–24% (w/w) potassium citrate.

However, wherever a salt other than potassium phosphate has been used, it is generally ammonium sulfate. Examples include a 17% (w/w) ammonium sulfate solution in Ref. [50] for the purification of DNA binding protein and 17.5% (w/w) ammonium sulfate used for the recovery of the antibiotic cephalosporin C from a fermentation broth [51]. Interestingly, in this latter reference, Lin and Chu found that the separation of cephalosporin C and its desacetyl derivative was enhanced by the addition of solvents, e.g. 5% acetone, or neutral salts, e.g. 1.45% potassium thiocyanate (KSCN).

As might be expected, the resolution of proteins and similar analytes is highly dependent upon the pH of the aqueous system [45, 65]. With PEG–phosphate systems, the pH can be adjusted by altering the ratio of monobasic to dibasic phosphate in the salt solution. The change that this effects can be seen in Tab. 7.4.

The strong influence of pH on the distribution ratio of proteins in aqueous polymer systems has resulted in some effective separations using the pH peak focusing countercurrent chromatography technique [28]. This is related to pH zone refining described previously.

The second class of aqueous polymer systems are the PEG–dextran systems. These are much less frequently encountered in the literature because they yield considerably lower peak resolution than the PEG–salt systems [66]. However, some proteins, such as globulins and histones, have poor solubility in the high-salt PEG–phosphate systems, so require the low-salt PEG–dextran systems for purifica-tion. Purifications of proteins have been achieved using an entirely aqueous polymer system of, for example, 4.0% PEG 8000–7.0% Dextran T500 (weight average molecular weight of 500 000) [67]. However, in the Shinomiya paper quoted above, a number of PEG–Dextran systems were investigated and the authors concluded that the best separations were obtained with the system 4.0% (w/w) PEG 8000–5.0% (w/w) Dextran T500 in a 5 mM potassium phosphate buffer (pH 7.0) and 3 M sodium chloride [66].

Tab. 7.4 The effect of the ratio of monobasic to dibasic phosphate on the pH of a PEG–phosphate system [59]

Concentration (w/w) (%)			pH
PEG 1000	K_2HPO_4	KH_2PO_4	
16.0	12.5	0	9.2
16.0	10.4	2.1	8.0
16.0	8.33	4.17	7.3
16.0	6.25	6.25	6.8

7.6
Conclusion

Whereas countercurrent chromatography is regularly employed, in both academia and industry, for the separation of natural product extracts, synthetic mixtures and intermediate-polarity compounds, the purification of biomolecules by countercurrent chromatography is a technique still in its infancy and thus its full potential has yet to be realized. A number of different approaches exist for the purification of compounds such as peptides, proteins, hormones, lipids, sugar derivatives and enzymes, but currently none have been completely investigated and the full benefits of the technique in this field of research are still to be determined.

However, some impressive examples of biomolecule purifications exist in the scientific literature, many where labile or otherwise difficult compounds were purified by countercurrent chromatography when other techniques failed. With its inherent advantages over solid-phase chromatography, such as high loading, total sample recovery and the ability to accept particulates, countercurrent chromatography looks set to become an important process in the field of bioseparations. Furthermore, the ever-increasing understanding of the processes that affect the separation and the development of easy-to-follow protocols for the selection of solvent systems combine to make countercurrent chromatography a technique that has a definite future in the bioseparations business.

References

1 Wood, P. L., *The hydrodynamics of countercurrent chromatography in J-type centrifuges, PhD Thesis*, Brunel University, **2002**.

2 König, C. S., Sutherland, I. A., *J Liq Chromatogr Rel Technol* **2003**, *26*, 1521–1535.

3 Berthod, A., Ruiz-Angel, M. J., Carda-Broch, S., *Anal Chem* **2003**, *75*, 5508–5517.

4 Berthod, A. (Ed.), *Countercurrent Chromatography: The Support Free Liquid Stationary Phase (Comprehensive Analytical Chemistry XXXVIII: Countercurrent Chromatography)*. Elsevier, Amsterdam, **2002**.

5 Foucault, A. (Ed.), *Centrifugal Partition Chromatography*. Marcel Dekker, New York, **1995**.

6 Ito, Y., Review, *J Chromatogr* **1991**, *538*, 3–25.

7 Albertsson, P. A., *Partition of Cell Particles and Macromolecules*, 3rd

edn. Wiley-Interscience, New York, **1986**.

8 Ito, Y., Bowman, R. L., *Anal Biochem* **1974**, *61*, 288–291.

9 Sutherland, I. A., in *Countercurrent Chromatography: Theory and Practice*, Mandava, N. B., Ito, Y. (Eds.). Marcel Dekker, New York, **1988**.

10 Friesen, J. B., Pauli, G. F., *J Liq Chromatogr Rel Technol* **2005**, *28*, 2777–2806.

11 Garrard, I. J., *J Liq Chromatogr Rel Technol* **2005**, *28*, 1923–1936.

12 Knight, M., Fagarasan, M. O., Takahashi, K., Geblaoui, A. Z., Ma, Y., Ito, Y., *J Chromatogr A* **1995**, *702*, 207–214.

13 Harada, K.-I., Suzuki, M., Kato, A., Fujii, K., Oka, H., Ito, Y., *J Chromatogr A* **2001**, *932*, 75–81.

14 Ikai, Y., Oka, H., Hayakawa, J., Kawamura, N., Harada, K.-I., Suzuki, M., Nakazawa, H., Ito, Y., *J Liq Chromatogr Rel Technol* **1998**, *21*, 143–155.

15 Harada, K.-I., Ikai, Y., Yamazaki, Y., Oka, H., Suzuki, M., *J Chromatogr* **1991**, *538*, 203–212.

16 Shinomiya, K., Komatsu, T., Murata, T., Kabasawa, Y., Ito, Y., *J Liq Chromatogr Rel Technol* **2000**, *23*, 1403–1412.

17 Fahey, J. W., Wade, K. L., Stephenson, K. K., Chou, F. E., *J Chromatogr A* **2003**, *996*, 85–93.

18 Fisher, D., Garrard, I. J., van den Heuvel, R., Chou, F. E., Fahey, J. W., Sutherland, I. A., *J Liq Chromatogr Rel Technol* **2005**, *28*, 1913–1922.

19 Prestera, T., Fahey, J. W., Holtzclaw, W. D., Abeygunawardana, C., Kachinski, J. L., Talahay, P., *Anal Biochem* **1996**, *239*, 168–179.

20 Ito, Y., Shibusawa, Y., Fales, H. M., Cahnmann, H. J., *J Chromatogr* **1992**, *625*, 177–181.

21 Ito, Y., Shinomiya, K., Fales, H. M., Weisz, A., Scher, A. L., *ACS Symp Ser* **1995**, *593*, 156–183.

22 Weisz, A., Scher, A. L., Shinomiya, K., Fales, H. M., Ito, Y., *J Am Chem Soc* **1994**, *116*, 704–707.

23 Yang, F., Quan, J., Zhang, T., Ito, Y., *J Chromatogr A* **1998**, *822*, 316–320.

24 Shinomiya, K., Weisz, A., Ito, Y., *ACS Symp Ser* **1995**, 218–230.

25 Ma, Y., Ito, Y., *J Chromatogr A* **1995**, *702*, 197–206.

26 Patel, K., Krishna, G., Sokoloski, E., Ito, Y., *J Liq Chromatogr Rel Technol* **2000**, *23*, 2209–2218.

27 Degenhardt, A., Winterhalter, P., *J Liq Chromatogr Rel Technol* **2001**, *24*, 1745–1764.

28 Shibusawa, Y., Misu, N., Shindo, H., Ito, Y., *J Chromatogr B* **2002**, *776*, 183–189.

29 Ito, Y., Ma, Y., *J Chromatogr A* **1996**, *753*, 1–36.

30 Horváth, C., Nahum, A., Frenz, J. H., *J Chromatogr* **1981**, *218*, 365–393.

31 Schott, H., *Affinity Chromatography (Chromatographic Science Series 27)*. Marcel Dekker, New York, **1984**.

32 Lindner, W., *J Chromatogr A* **1994**, *666*, 3–53.

33 Ma, Y., Ito, Y., Separation of polar compounds by affinity countercurrent chromatography, *WIPO Patent WO 97/07396*, **1997**.

34 Ma, Y., Ito, Y., *Anal Chem* **1996**, *68*, 1207–1211.

35 Ma, Y., Sokoloski, E., Ito, Y., *J Chromatogr A* **1996**, *724*, 348–353.

36 Kitazume, E., Sato, N., Ito, Y., *J Liq Chromatogr Rel Technol* **1998**, *21*, 251–261.

37 Ma, Y., Ito, Y., Foucault, A., *J Chromatogr A* **1995**, *704*, 75–81.

38 Oliveros, L., Franco Puertolas, P., Minguillon, C., Camacho-Frias, E., Foucault, A., Le Goffic, F., *J Liq Chromatogr* **1994**, *17*, 2301.

39 Shinomiya, K., Yoshida, K., Kabasawa, Y., Ito, Y., *J Liq Chromatogr Rel Technol* **2001**, *24*, 2615–2623.

40 Ma, Y., Ito, Y., *Anal Chim Acta* **1997**, *352*, 411–427.

41 O'Donnell, T. A., *Eur J Inorg Chem* **2001**, *1*, 21–34.

42 Berthod, A., Carda-Broch, S., *J Liq Chromatogr Rel Technol* **2003**, *26*, 1493–1508.

43 Albertsson, P. A., *Partition of Cell Particles and Macromolecules*. Almqvist and Wiksell, Stockholm, **1960**.

44 Shinomiya, K., Kabasawa, Y., Ito, Y., *J Liq Chromatogr Rel Technol* **1998**, *22*, 1727–1736.

45 Shibusawa, Y., Yamaguchi, M., Ito, Y., *J Liq Chromatogr Rel Technol* **1998**, *21*, 121–133.

46 Kroner, K. H., Virkajarvi, I., Peltola, P., Hellberg, K., Hustedt, H., Countercurrent aqueous two phase extraction of enzymes [abstract], in *Proc 6th Int Conf on Partition in Aqueous Two Phase Systems*, Assmannshausen, **1989**, p. 16.

47 Hollander den, J. L., Stribos, B. J., van Buel, M. J., Luyben, K. C. A. M., Van Der Wielen, L. A. M., *J Chromatogr B* **1998**, *711*, 223–225.

48 Shinomiya, K., Kabasawa, Y., Ito, Y., *J Liq Chromatogr Rel Technol* **1998**, *21*, 135–141.

49 Kendall, D., Booth, A. J., Levy, M. S., Lye, G. J., *Biotechnol Lett* **2001**, *23*, 613–619.

50 Shibusawa, Y., Ino, Y., Kinebuchi, T., Shimizu, M., Shindo, H., Ito, Y., *J Chromatogr B* **2003**, *793*, 275–279.

51 Lin, P. C., Chu, I. M., *Biotechnol Tech* **1995**, *9*, 549–552.

52 Chao, Z., Shibusawa, Y., Shindo, H., Ito, Y., *J Liq Chromatogr Rel Technol* **2003**, *26*, 1895–1903.

53 Ito, Y., Zhang, T. Y., *J Chromatogr* **1988**, *449*, 135–151.

54 Walter, H., Tamblyn, C. H., Levy, E. M., Brooks, D. E., Seaman, G. V. F., *Biochim Biophys Acta* **1980**, *598*, 193–199.

55 Eriksson, E., Albertsson, P. A., *Biochim Biophys Acta* **1978**, *507*, 425–432.

56 Flanagan, S. D., Barondes, S. H., Taylor, P., *J Biol Chem* **1976**, *251*, 858–863.

57 Johanssen, G., Gysin, R., Flanagan, S. D., *J Biol Chem* **1981**, *256*, 9126–9135.

58 Diamond, A. D., Hsu, J. T., *Adv Biochem Eng Biotechnol* **1992**, *47*, 89–135.

59 Shibusawa, Y., Ito, Y., *Prep Biochem Biotechnol* **1998**, *28*, 99–136.

60 Wei, Y., Zhang, T., Ito, Y., *J Chromatogr A* **2001**, *917*, 347–351.

61 Chen, J., Ma, G. X., Li, D. Q., *Prep Biochem Biotechnol* **1999**, *29*, 371–383.

62 Qi, L., Ma, Y., Ito, Y., Fales, H. M., *J Liq Chromatogr Rel Technol* **1998**, *21*, 83–92.

63 Shibusawa, Y., Chiba, T., Matsumoto, U., Ito, Y., *ACS Symp Ser* **1995**, *593*, 119–128.

64 Chao, Z., Shibusawa, Y., Shindo, H., *Zhongguo Yaoxue Zazhi* **1999**, *34*, 444–446.

65 Shibusawa, Y., Hosojima, T., Nakata, M., Shindo, H., Ito, Y., *J Liq Chromatogr Rel Technol* **2001**, *24*, 1733–1744.

66 Shinomiya, K., Hirozane, S., Kabasawa, Y., Ito, Y., *J Liq Chromatogr Rel Technol* **2000**, *23*, 1119–1129.

67 Shibusawa, Y., Ito, Y., *Am Biotechnol Lab* **1997**, *16*, 8–1.

8

Continuous Chromatography in the Downstream Processing of Products of Biotechnological and Natural Origin

Michael Schulte, Klaus Wekenborg, and Jochen Strube

8.1
Introduction

Intensive work has been done to improve single chromatographic units for the separation of small molecules. In particular, for difficult separations with low selectivities and larger production amounts in the multi-ton range, continuous chromatographic concepts like the simulated moving bed (SMB) have proven their ability to improve the process performance and thus are accepted techniques applied to an industrial scale. The time has come to bring the SMB technology in the pharmaceutical industry to the next level, approaching more complex separations under difficult constraints.

One of the biggest challenges in today's chemical and pharmaceutical industries is the economic production of macromolecular products. These products might be of synthetic nature (e.g. peptides and oligonucleotides), extracted from natural sources (e.g. plant extracts, blood serum), or recombinant products from yeast, fungi, bacterial or mammalian sources. Regardless from which source the target substance is isolated, the feedstock to be processed always contains a multitude of substances with different properties with respect to size, charge and hydrophobicity. Some of the impurities differ from the main product only in very small parameters related to the overall size of the molecules. In addition, the product itself is rather unstable in most cases, creating the need for short processing times, and specific conditions in terms of temperature, pH and ionic strength.

In addition to other separation techniques like liquid–liquid extraction as well as filter or membrane units, chromatography perfectly serves the need for mild conditions in combination with very specific interactions and high resolution power. However, it sometimes lacks productivity, scalability and cost-effectiveness.

When designing a complete downstream process, which often consists of three or more chromatographic steps, it is not only the choice of single chromatographic principles (e.g. affinity, ion-exchange or size-exclusion) and their interconnection that secures the success of a project. The mode of operation might also contribute

Bioseparation and Bioprocessing. Edited by G. Subramanian
Copyright © 2007 WILEY-VCH Verlag GmbH & Co. KGaA, Weinheim
ISBN: 978-3-527-31585-7

and improve the overall process performance. Apart from the classical batch operation in a single column, continuous-process modes have been available for a long time, but their potential has not yet been realized in macromolecule production. In this chapter we will show where and how the continuous chromatographic purification of macromolecular pharmaceutical compounds will deliver a valuable contribution in the race towards new and economically manufactured drug products.

8.2
SMB Chromatography

8.2.1
Basic Principle

Continuous countercurrent chromatography in its technical form of SMB chromatography was invented in the 1960s and was successfully transferred to a large scale in the area of xylol separation from petroleum fractions [1]. A second field of application arose in the 1970s in the sugar-refining industry for the separation of monosaccharides, e.g. glucose and fructose. To enter the pharmaceutical industry the technology had to overcome several obstacles. It was in 1993 when the small French company Separex presented the first laboratory-scale SMB unit, which fulfilled the needs for drug purification in preclinical and clinical stages. Due to a clever design of the valve switching process the influence of the system dead volume could be significantly reduced, allowing high purities even with small-scale SMB units. At the same time as the first pharmaceutical SMB units were developed, chiral stationary phases for chromatographic separations of enantiomers were substantially improved and made commercially available at the ton scale. The first enantiopure drug substance introduced to the market produced by SMB chromatography was Levetiracetam (Keppra®) by the Belgian company UCB Pharma. Today, UCB is operating several large-scale SMB units with column sizes of up to 1 m in diameter at two different production sites [2]. In total, six drugs on the market are manufactured with at least one chromatographic SMB production step, i.e. beside Keppra, a precursor of Pfizer's Zoloft® (Sertralin), Lundbeck's Escitalopram (Cipralex®), UCB's Escetiricin (Xusal®), Cephaoln's Armodafinil (Nuvigil®) and an intermediate (DOLE) for a cholesterol-lowering drug produced by Daicel. In 2005, these drug substances amounted a total of 1500 tons of active pharmaceutical ingredient produced by SMB technology [3].

SMB technology is characterized by two main principles: it is continuous and it is operated in a countercurrent mode. The theoretical form of a continuous countercurrent operation is the true moving bed (TMB) as schematically displayed in Fig. 8.1. In TMB chromatography the feed is introduced in the middle of a column, where the solid and fluid phases are transported in opposite directions. The less-adsorbed component B is transported with the mobile phase in the direction of the raffinate port, while the well-adsorbed component A is carried with the

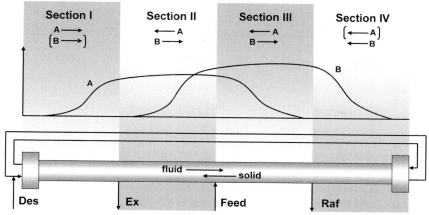

Fig. 8.1 TMB chromatography with internal concentration profile.

stationary phase towards the extract port. As well as the feed line and the two product streams, a fourth stream of fresh eluent is present.

These streams divide the column into four sections or zones with different flow rates of the liquid phase, while each of these sections plays a specific role in the process. In this way both the solid and the liquid phase can be recycled to section IV and I, respectively. In summary, the four sections have to fulfill the following tasks:

Section I   Desorption of the more retained component
Section II   Desorption of the less retained component
Section III   Adsorption of the more retained component
Section IV   Adsorption of the less retained component

With a proper choice of all individual internal flow rates in sections I–IV and the velocity of the stationary phase, complete separation of the feed mixture can be realized. This leads to a distribution of the fluid concentrations as displayed in the axial concentration profile in the upper part of Fig. 8.1. Since the TMB process reaches a steady state, it can be seen from the diagram that pure component A can be withdrawn with the extract stream. On the other hand, the raffinate line contains the pure component B only.

Unfortunately, the movement of a particulate solid phase is rather difficult to realize and, consequently, only a few examples can be found. One of them is the "hypersorption" process for the recovery of ethylene from a gas stream [4, 5]. Activated carbon, the solid phase, slowly moves down in vertical tubes, while the gaseous fluid phase is transported in the opposite direction. Other approaches presented for liquid chromatographic separations employ a continuous conveyor belt transported through different sections. Kuhn and Martin had already demonstrated the enrichment of mandelic acid ester enatiomers on a wool fiber in 1941 [6]. A continuous nylon belt with immobilized affinity ligands was used by Niven and Scurlock in order to purify trypsin [7].

Even though these examples of the application of TMBs can be found, the breakthrough for countercurrent operation of chromatographic processes came with the invention of the simulated movement of the solid phase [1]. Nowadays, a standard SMB set-up is an interconnection of packed chromatographic columns (Fig. 8.2). In a typical SMB system up to 12 columns are connected in a circle, while the introduction and withdrawal ports are located in between the packed beds.

The mobile phase passes all columns in one direction. The countercurrent flow of both phases is achieved by moving the columns periodically downstream, in the opposite direction to the liquid flow. Actually, in modern SMB units the columns are not shifted, but all ports are moved in the direction of the liquid flow by means of valves, obeying the same effect of countercurrent contact between solid and fluid. This countercurrent character of the process becomes more obvious when the relative movement of the packed beds to the inlet and outlet streams during several switching intervals is observed (Fig. 8.3).

In order to demonstrate the advantages of the SMB technology, internal as well as outlet concentration profiles, obtained by dynamic simulation, are compared to those of a discontinuous batch separation (Fig. 8.4). For identical conditions in terms of isotherms, feed concentrations, etc., both processes were optimized to realize complete separation (purity and yield of both components greater than 99.9%). In the case of batch chromatography, these constraints lead to a baseline separation, while the throughput has been maximized by touching band operation as depict in Fig. 8.4(b). The corresponding internal profile (Fig. 8.4a) reveals that only minor parts of the stationary phase are utilized for the separation of the two components with rather low concentration of the components in the column. This is completely different for the continuous SMB process. As can be seen from Fig. 8.4(c), large parts of the adsorbents are used for separation purposes, while the substances to be purified are present at quite high concentrations. The latter fact results in rather concentrated product streams as visualized in Fig. 8.4(d).

Fig. 8.2 Columns in a laboratory-scale SMB unit.

switching interval 1

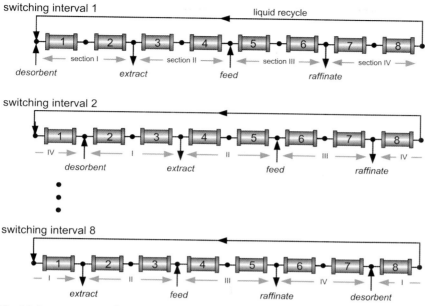

Fig. 8.3 Demonstration of the countercurrent nature in a SMB unit.

Batch chromatography

SMB chromatography

Internal concentration profiles

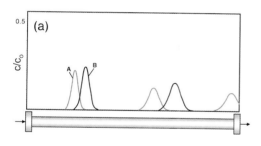

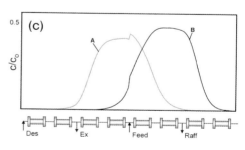

Outlet concentration profiles

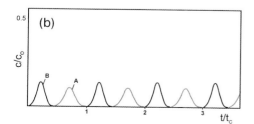

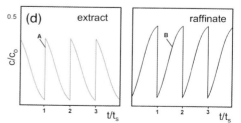

Fig. 8.4 Comparison between internal and outlet concentration profiles for batch and SMB systems.

The "sawtooth" profiles result from the dynamic port-switching characteristic for every SMB unit. One special feature that only the SMB technology can offer is the direct recycling of the liquid solvent.

8.2.2
SMB Equipment

To simulate the countercurrent character of the SMB unit, the columns have to be moved relative to the ingoing and outgoing process streams. This can be realized in general by two different constructional approaches. In the first possibility, the columns, mounted on a carousel, are really moving, while the fluid distribution is enabled by one center multiport valve. This center valve offers a great variety of possible process set-ups and operation is not necessarily countercurrent. A detailed description of possible operating modes of this device is given in Section 8.4.

In the second design concept the columns remain at fixed positions, but the process streams are switched by mean of valves in the direction of the fluid stream. To ensure a proper movement of the mobile phase in one direction only, the implementation of a so-called recycle pump is required. In the standard set-up, this pump is located at a fixed position between two columns as displayed in Fig. 8.5(a). Since the columns and the recycle pump is moving from one section to the next, the recycle pump has to deliver changing flow rates depending on the current section. In addition to that, a fixed recycle pump produces a rather large dead volume. This problem can be solved by applying an asynchronous shifting strategy of the process stream along the recycle line [8].

In alternative approaches the recycle pump is not located at a fixed position, but it performs a similar movement as the inlet/outlet streams do (Fig. 8.5b). The pump can be placed somewhere in the circuit of columns, but it is preferably located near the desorbent line between sections I and IV. Here, only the pure

(a) fixed recycle pump

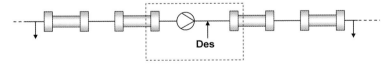

(b) moving recycle pump

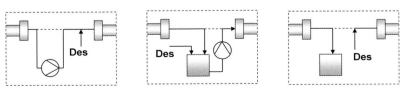

Fig. 8.5 Different recycle pump concepts in SMB systems.

solvent is present, and contact between recycle pump and the target components can be avoided. Direct recycling of the mobile phase is not necessarily required. "Open" concepts with an extra buffer tank or even without any solvent recycling can be considered. For these the number of pumps can be reduced, and process control issues regarding pressure and flow rates can be simplified.

8.2.3
Layout and Optimization

It has been proven several times that the SMB concept is able to improve the performance of a chromatographic separation in terms of productivity, solvent consumption and product concentration. These advantages, however, are achieved by a higher process complexity with respect to operation and layout. Since a purely empirical approach is simply not feasible, model-based strategies have been developed to handle layout issues.

An established method to generate operating parameters is based on the ideal or local equilibrium model of a TMB process, neglecting band-broadening effects like axial dispersion and mass transfer limitations as well as the dynamic SMB behavior due to the port switching. With the dimensionless flow rate ratios m_j as the ratio between the flow rate of the fluid ($Q_{j,TMB}$) and the solid phase (Q_{ads}) in every section j a first layout is possible [9]:

$$m_{i,j} = \frac{\text{net fluid flow}}{\text{net solid flow}} = \frac{Q_{j,TMB} - Q_{ads} \cdot \varepsilon_p}{Q_{ads} \cdot (1 - \varepsilon_p)}. \tag{1}$$

In the case of linear isotherms, where the components isotherm slope is the Henry constant H_i, a complete separation of a two component feed mixture can be realized when the following constraints are fulfilled:

$$\text{Section I:} \quad H_A < m_I \tag{2}$$

$$\text{Sections II and III:} \quad H_A < m_{II} < m_{III} < H_B \tag{3}$$

$$\text{Sections IV:} \quad H_B < m_{IV}. \tag{4}$$

According to these constraints, the dimensionless flow rate ratios can be chosen. To visualize all possible operating points for sections II and III, the dimensionless flow rate ratio m_{III} is plotted versus m_{II} in the so-called "triangle" diagram. For linear isotherms, all possible operating points are within the triangle shown in Fig. 8.6(a). However, in order to achieve the best performance with respect to productivity, the difference between m_{II} and m_{III} should be as high as possible, giving the highest possible feed flow rate. The flow rate of fresh desorbent and, therefore, the eluent consumption can be minimized by choosing m_I as low and m_{IV} as high as possible.

When the isotherms are no longer linear, things become complicated since the migration of the components strongly depends on the fluid concentration. Strate-

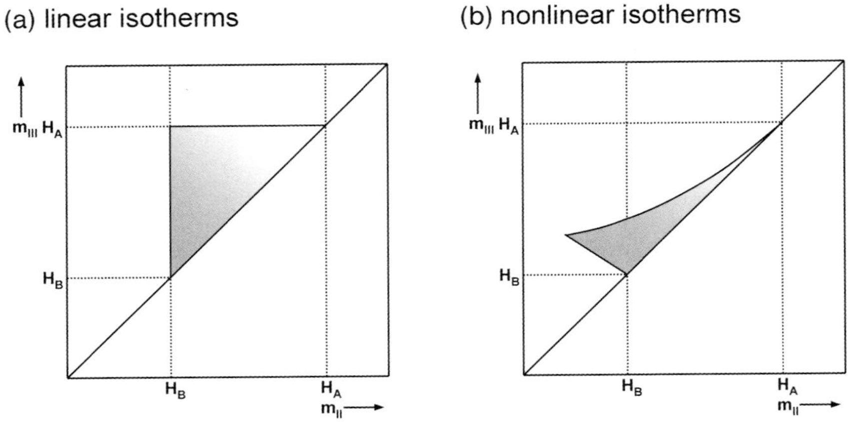

(a) linear isotherms

(b) nonlinear isotherms

Fig. 8.6 Operating diagrams for systems with linear (a) and nonlinear (b) isotherms.

gies for parameter determination are available for standard nonlinear isotherm models such as different kinds of Langmuir isotherms [10–15]. The impact of the isotherm's nonlinearity on the operating diagram is schematically displayed in Fig. 8.6(b). Even though the triangle is no longer of a rectangular shape, a set of operating parameters for the separation region can be found that will give a complete separation. As the ideal model does not take any band-broadening effects into account, safety margins have to be considered when applying this method to the layout of SMB separations. Further optimization of operating parameters can be done on the running process or applying experimentally verified, more detailed models [16].

8.2.4
SMB Modifications

8.2.4.1 Operation 1: Isocratic Modifications

The classical SMB, with its constant switching times and section flow rates, offers a great variety of possible modifications. Many approaches have been made to investigate the impact of different SMB operating modes on objective functions like productivity, eluent consumption, etc.

On the one hand, there is the group of isocratic process modifications with constant eluent composition but variable switching times and/or section flow rates. The most recognized and applied one at industrial scale is the VARICOL process [17]. Using the same set-up as in classical SMB, during the VARICOL operation the process lines are no longer switched downstream in a synchronous manner. According to the process needs every single port can be switched individually. The main benefit of this concept is a reduction of the total number of columns needed.

Slightly different ideas are realized in the PowerFeed concept introduced by Kloppenburg and coworkers [18] and Zhang and coworkers [19] or the PartialFeed approach by Zang and coworkers [20]. For these concepts, the switching times remain constant, but the in- and outgoing and thus the internal flow rates during one switching interval are used as optimization variables. Finally, there is the additional possibility to vary the feed concentration as it realized in the so-called ModiCon SMB proposed by Schramm and coworkers [21].

8.2.4.2 Operation 2: Gradient Modifications

In addition to design (number of sections, columns) or operating parameters (flow rates, shifting times), thermodynamic aspects that influence the chromatographic separation can be utilized to gain improved process performance. In general, conditions for increased desorption in section I, combined with a better adsorption environment in section IV, will reduce the need for fresh solvent. A similar gradient in sections II and III will have a positive impact on product purity and/or feed throughput.

Depending on the method applied to the separation (ion exchange, affinity, etc.), different conditions for adsorption as well as desorption can be realized. In solvent-gradient SMB, the fresh solvent stream is fed to the system at a higher solvent strength than the feed stream. Due to the circulation of the fluid in the SMB unit together with the mixing at the desorbent and the feed port, a two-step gradient can be realized as displayed in Fig. 8.7. If, for example, ion exchange is used for the separation, the solvent strength can be adjusted either by the ionic strength or the pH [22, 23]. For other chromatographic systems, such as the reversed-phase system, further information regarding layout and optimization is given in Refs. [24–28].

Even though the solvent gradient concept requires no extra modification of the SMB plant compared to the classical set-up, the gradient is limited to a two-step gradient. When, for instance, the temperature is used to influence the adsorption equilibrium, a multistep gradient can be realized. Starting with a high temperature level in section I for improved desorption of strongly adsorbed components, the temperature can be lowered from section to section or even for every single

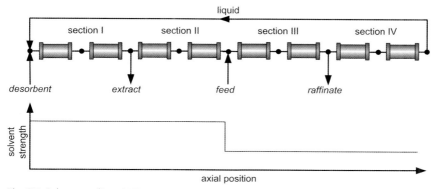

Fig. 8.7 Solvent gradient SMB process.

column [29]. This approach might be especially beneficial with temperature-responsive adsorbents, e.g. based on N-isopropylacrylamide [30]. Another gradient can be realized, when supercritical fluid chromatography is applied for the separation, as the solvent strength can be controlled by the pressure level. Various examples are given in Refs. [31–36].

8.2.4.3 Design: Number of Columns/Sections

Apart from the four-section SMB set-up with its eight columns as described before, a multitude of variations with respect to plant design are possible. First of all the number of columns in every section can be adjusted to the process needs [37].

A three-section SMB unit, without section IV, might be considered when recycling of the solvent is no longer required. This set-up might be useful in purely adsorption processes, e.g. capturing steps. On the other hand, a countercurrent SMB process is not limited to four sections. Five-section processes with internal recycle streams or an intermediate third fraction have been proposed [38].

When feed streams of biological origin have to be processed, a large number of impurities are generally present. Some of these substances might block the column due to their size; others might bind very strongly to the stationary phase, precipitate or denature during the separation process. The introduction of an additional cleaning-in-place (CIP) section is thus required as displayed in Fig. 8.8. The basic set-up is similar to a concept proposed by Hotier and coworkers [39] or Voigt and coworkers [40] in order to withdraw a third pure fraction.

After the extract component or fraction has been removed from the first column in section I, this bed together with all irreversibly bound substances is moved to the regeneration section. With an appropriate buffer, Des*, strongly adsorbed substances are collected as a third or waste fraction. In a subsequent CIP step, the columns are prepared for the reuse in the main SMB unit.

Before the cleaned column can be returned to the main SMB cycle, a readjustment of the solvent is generally necessary. This equilibration could be done in the same section as cleaning and regeneration are performed within one switching interval. On the other hand, an additional section can be implemented for the solvent readjustment only. Details regarding layout and optimization as well as experimental results for the purification of nucleosides have been presented by Abel and coworkers [41] and Paredes and coworkers [42]. The same concept has been used to purify plasmid DNA from RNA by size-exclusion chromatography [43].

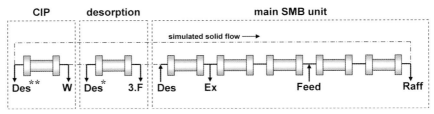

Fig. 8.8 SMB process with extra sections for desorption and CIP.

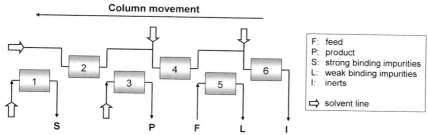

Fig. 8.9 Schematic representation of the multicolumn countercurrent solvent gradient purification operation.

In addition to systems with different numbers of zones, a new multicolumn scheme was introduced by the group of Morbidelli [44, 45]. The so-called multi-column countercurrent solvent gradient purification (µCSGP) divides the whole column cycle into an interconnected column lane and a batch lane (Fig. 8.9). While the columns in the batch lane are eluted using single pumps, the mixture in the interconnected lane is transported in the cycle to the next segment. With an appropriate choice of the single flow rates and solvent strengths, this set-up allows purifying a mixture of compounds into several fractions under continuous operation.

The laboratory-scale system, consisting of three ÄKTA high-performance liquid chromatography systems and a multitude of multiport switching valves, has been demonstrated to be able to purify a peptide drug from 46% starting purity to over 98% in a single continuous operation. During this operation impurities eluting in front of the main compound as well as those eluting later than the target compound were removed under gradient operation. A second example showed the purification of three isoforms of a monoclonal antibody by applying weak cation exchange.

8.2.5
Application to Biochromatography

Even though the continuous countercurrent operation of chromatographic processes has become an accepted tool in various industries, it has not yet made its way to the biopharmaceutical industry. Nevertheless, examples of successful implementation of this technique into downstream processes of products from biological origin are available and exhibit a great potential. The examples given in this chapter are divided according to the chromatographic method chosen to perform the separation.

Size-exclusion chromatography or gel filtration is a widely applied method for desalting or buffer exchange as well as for separation purposes. In the latter case, the technique suffers from several drawbacks which restrict its large-scale use in the biopharmaceutical industry. The first and most severe drawback is linked to the low loading capacity and long bed length. Both parameters are needed to achieve a sufficient separation efficiency, which is solely based on the partitioning of the feed compounds in the pore system and lacking the differentiation power of selective adsorption. Therefore, rather large columns with long packing length have to be used, operated with low linear velocities to achieve an adequate resolution. A good packing quality of these production-scale columns with column length up to 1 m can only be achieved with rigid polymeric resins, e.g. Fractogel® BioSEC.

Size-exclusion chromatography, which is a *per se* isocratic process, can easily be transferred to standard SMB equipment. The most problematic issue of stabilizing the raffinate front in an SMB system, where the compounds show no adsorption, was solved in 1998 by Merck researchers [46]. Fractogel BioSEC was packed into eight columns of 15 cm length and 2.6 cm in diameter giving a total column volume of 637 mL. The separation was carried out using a Licosep® Lab SMB system in a four-zone configuration with two columns in each section. As a demonstration example, the casein fraction from skim milk was separated from all other proteins and ingredients (Tab. 8.1). Casein and, especially, its aggregates are the largest molecules present in skim milk, and thus elute first (as the raffinate fraction) in a chromatographic size-exclusion system. Using the described SMB system, it was possible to isolate the two compound groups at high purities. The calculation of specific productivity and buffer consumption for the batch and the SMB system shows the tremendous optimization effect for the continuous countercurrent operation.

In a second example as reported by Wang [47] and Xie and coworkers [48], size-exclusion chromatography was used to separate insulin from zinc chloride and a high-molecular-weight protein on a laboratory scale. Since the target product insulin elutes as the intermediate fraction, a tandem SMB consisting of two subsequent units was applied. The high-molecular-weight protein was removed with the raffinate stream in a first SMB unit, while the extract was collected and processed in a second unit. Compared to the batch process, insulin could be recovered at almost the same high purity (above 99.9%) while the yield was increased by 10% (from 90 to 99%). Additional improvements were observed with respect to productivity, which could be increased by 400% in combination with 72% lower eluent consumption. The same set-up and model system has been used to investigate the residence time as well as the residence time distribution

Tab. 8.1 Specific process parameters for the isolation of a casein fraction by SMB chromatography

Parameter	Batch	SMB
Productivity (g_{Feed} day^{-1} $l_{Column\ volume}^{-1}$)	1.99	189.9
Buffer consumption (ml_{Buffer} g_{Feed}^{-1})	723	18

of the target product [49]. It has been found that a smaller residence time of the target product can be realized when a strategy according to the partial feed concept is applied.

The separation of bovine serum albumin (BSA) and myoglobin on Sepharose Big Beads® has been investigated by Houwing and coworkers [50]. In this study the main focus was on the influence of mass transfer effects.

The continuous purification of plasmid DNA from a clarified cell lysate has been successfully demonstrated by Paredes and coworkers [43]. The specific feature of the presented process is an online "CIP" section in combination with an additional re-equilibration of the stationary phase as described before.

In order to remove lactose from a mixture of human milk oligosaccharides, Geisser and coworkers used SMB technology operated in two different modes [51]. On the one hand, the separation was performed in a size-exclusion mode, while in the second case ligand-exchange chromatography has been used. In both cases lactose could be withdrawn in the raffinate fraction in high purity. The remaining lactose in the value stream of oligosaccharides eluting as the extract was reduced to 0–20% for the ligand exchange mode and 0–4% for the size-exclusion mode respectively.

Apart from the separation of components due to their size as done in gel filtration, ion-exchange chromatography exploits the different charges of substances for purification purposes. In the classical, discontinuous batch mode, the ion exchange is preferably operated under gradient conditions. The loading interval is therefore followed by a linear or stepwise increase of the solvent strength, which can be achieved by changing the ionic strength or the pH value of the buffer solution.

Experimentally determined equilibrium parameters for alpha-lactalbumin and β-lactoglobulin on an ion-exchange resin served Lucena and coworkers as a starting point for numerical studies of an isocratic SMB process [52]. The purification of BSA from a yeast protein [53] and myoglobin [22] has been reported by Houwing and coworkers. In both cases stepwise solvent gradients were implemented within the SMB unit, where the concentration of the counterions was varied. In these special cases, the test systems exhibit azeotropic behavior, as the selectivity between the two components can be reversed for different salt concentrations in the buffer solution.

Another example of solvent gradient operation of an SMB unit is the separation of β-lactoglobulins A and B on a strong anion exchanger [23]. Again, the elution strength was adjusted by the concentration of the counterions, while the pH was held constant at a value of 7. Before experiments were carried out, suitable operating parameters were generated using a transport dispersive SMB model [54], while the steric mass action (SMA) model [55] was used to describe the ion-exchange equilibrium. As the result of a simulation study, operating conditions were found, leading to an internal axial concentration profile at the end of one switching interval as displayed in Fig. 8.10.

The SMB experiments were carried out on a pilot-scale SMB unit (LICOSEP® LAB; Novasep, France) with eight HPLC columns (87 × 10 mm) arranged in a

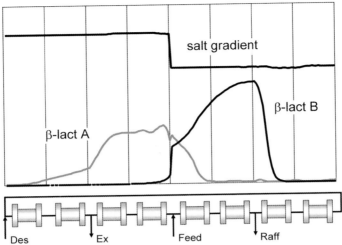

Fig. 8.10 Internal concentration profile for the separation of the β-lactoglobulins A and B by solvent-gradient SMB.

2–2–2–2 configuration. In this plant one recycle pump and a set of detectors are located at a fixed position between columns 8 and 1.

The experimental results as well as the corresponding simulations are displayed in Fig. 8.11. The comparison between the measured and calculated salt concentration is given in Fig. 8.11(a). One can observe the two regimes of different salt concentrations characteristic of the nonisocratic operation of the SMB process. The second part displays the development of the protein concentration. Apart from the measured total concentration, the corresponding profile obtained by simulation is shown, while this curve results from the summation of the calculated single protein concentrations. In the last part of Fig. 8.11 the composition of samples taken during the SMB experiment is visualized and compared with the expected composition obtained by simulation.

In order to show the potential of the solvent-gradient operation of SMB processes, a case study based on the equilibrium theory has been performed. The ionic strength was held constant in the first two sections, while it was decreased stepwise for the remaining parts of the process. Operating diagrams (Fig. 8.12) were calculated for a first approximation of the process parameters like productivity and eluent consumption. With an increasing gradient step, the productivity can be increased by a factor of almost 3 compared to the corresponding isocratic process. At the same time, the eluent consumption undergoes a reduction of nearly 90%.

Due to its highly specific interactions, affinity chromatography can offer for the purification of proteins, it is an established method in downstream processing. One example for continuous operation has been presented by Huang and coworkers for the purification of trypsin [56]. According to the solvent-gradient concept, two regimes with different pH values were realized in a six-column unit.

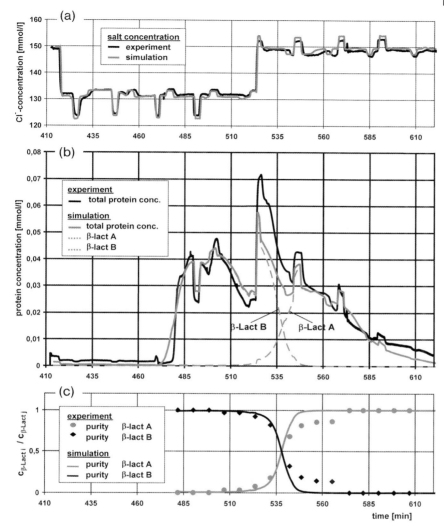

Fig. 8.11 Comparison of experiments and simulation or the separation of β-lactoglobulins and B by SMB chromatography: (a) salt concentration, (b) protein concentration and (c) collected samples.

A similar approach for the enrichment of bovine alpha-chymotrypsin on immobilized trypsin inhibitors was investigated by Gottschlich and coworkers [57]. The SMB process without solvent recycling consisted of two sections with different pH values for adsorption and desorption, respectively. It has to be pointed out that these two sections were not directly interconnected, as is generally the case in the classical SMB set-up. Due to the direct transfer of the loaded columns from one

(a) operating diagrams   **(b) process parameter**

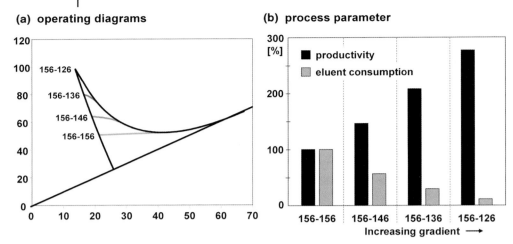

Fig. 8.12 Process performance for the solvent-gradient SMB.

section to the other, a pollution of the extract (the product) was observed. For that reason a reduced recovery had to be accepted to realize an appropriate purity. To overcome this problem additional purge zones between adsorption/desorption steps were introduced [58]. This improved concept was applied to the recovery of monoclonal antibodies from a cell culture supernatant. The target product could be isolated with a yield higher than 90% while more than 99% of contaminating proteins were removed.

8.2.5.1 Adsorption Chromatography in Normal Phase and Reversed-phase Mode

One of the first multicomponent separations of a product of biotechnological origin was patented in 1997 by Arzneimittelwerk Dresden, Merck and Novasep [59]. The product to be isolated was a cyclosporine A from a defatted *Trichoderma polysporum* extract. In the original process, other cyclosporines, i.e. U, D, L, G and B, were removed by batch chromatography. In the normal phase system used for the separation, the target product elutes in the middle of the surrounding impurities. Due to low recoveries and a lot of low-purity fractions that had to be reworked, this process was considered to be inefficient. An alternative three-step approach was therefore chosen, where two fractions were prepared in a first batch pre-purification step, which were suitable for subsequent SMB separations. The first fraction with cyclosporine A eluting as the last compound was submitted to a reversed-phase system to change the elution order and to collect the target compound in the raffinate stream, which gives favorable results in terms of productivity and eluent composition [60]. Compared to batch chromatography, a 2-fold increase in productivity could be shown for the normal phase system and a 10-fold increase for the reversed-phase step.

Other applications of SMB technology in adsorption chromatography of multi-component separations have been reported for the production of the cancer drug paclitaxel (Taxol®). Paclitaxel is extracted from different parts of the yew tree. In the extract, the target product is associated with several very close related compounds, such as 10-deacetyl 7-epi-taxol and cephalomanin. Using a silica sorbent (LiChrospher® Si60, 10 μm) Taxol could be isolated in closed-loop recycling mode as well as SMB mode [61]. In 2004, the Canadian company Bioxel announced that they had developed a commercial-scale Taxol separation system in cooperation with the Californian company Aerojet, which offers extensive SMB services [62]. In addition to Taxol, the Canadian yew tree also serves as source of Taxol precursors such as 10-deacetylbaccatin III and 13-acetyl-9-dihydrobaccatin III.

The purification process of an ascomycin derivate, an anti-inflammatory drug substance, has been developed by Küsters and coworkers [63]. The starting material ascomycin is produced by fermentation and chemically modified afterwards to obtain the desired product. The first isolation includes a crude silica gel filtration followed by crystallization. For the final purification, in order to achieve a purity of the target product greater than 98%, two SMB steps were applied. This alternative process concept finally replaced an old batch elution process step, allowing continuous operation at higher productivities combined with lower solvent consumption.

A new therapeutic principle is currently being developed with antisense oligonucleotides. The complex DNA or RNA molecules with a chain length between 20 and 40 nucleotide units are synthesized via solid-phase synthesis protocols. During the repeated synthetic operations, deletion sequences or other imperfect molecules can be formed. Different purification strategies for this type of molecules have been used [64]. One of these purification strategies to isolate the final drug molecule is based on the adsorption of the 4,4'-dimethoxytrityl (DMT) protecting group on reversed-phase silica materials. The DMT group should be present only at the full-sequence oligonucleotides and not at the malformed molecules. The separation between DMT-on and DMT-off molecules is therefore basically a binary separation suitable for SMB chromatography. One of the problems of the oligonucleotide separation is the rather different adsorption behavior of DMT-on and DMT-off oligonucleotides. In particular, the stabilization of the DMT-off molecules in the raffinate zone is of importance. In 2005, Schulte and Lühring demonstrated the possibility of an SMB separation of antisense oligonucleotides on a reversed-phase silica sorbent [65]. With their different elution behavior, the separation between DMT-on and DMT-off oligonucleotides could be a good candidate for transfer to a gradient SMB separation, which has been published for other drug substances.

Jensen and coworkers reported on continuous gradient separations of products from biological origin by reversed-phase chromatography [28]. In the first example, nystatin was purified from its main contaminant in aqueous methanol solutions, whereas in the second bovine insulin was separated from its porcine analog in acetonitrile/water. In both cases the solvent gradient SMB concept was investigated. From theoretical studies it was found that the eluent consumption could be reduced by at least 50% with a simultaneous 2-fold enrichment of the products.

8.2.6
Countercurrent Fractional Aqueous Two-phase Extraction

Aqueous two-phase extraction systems, e.g. based on poly(ethylene glycol) and dextran, have often demonstrated their power in the separation of biological molecules. The necessary efficiencies can be achieved in two ways: either by fixation of one of the phases onto a solid support and operating the system as a liquid–liquid partitioning chromatography [66] or by applying multistage countercurrent systems. Various countercurrent chromatographic systems have been reported in the literature and applied for analytical purposes [67]. Different approaches have been made to optimize the systems and to cope with the high centrifugal forces needed for high separation efficiencies. Until recently all systems used for this technology lacked the necessary robustness and scale-up possibility for preparative separations. In addition, they were all operated in batch mode.

One of the first attempts to operate a fractional extraction in a continuous manner was described in the PhD thesis of Kabasawa in 1977 [68]. His countercurrent fractional extraction with continuous sample feeding consists of two segmented glass columns rotating around their longitudinal axis. The columns are filled with the two phases and held at an inclined position. The lower phase is pumped down to the lower exit and simultaneously the upper phase is forced up the column toward the upper exit. The sample is fed continuously into the middle between the two columns (Fig. 8.13). Even though separations were performed successfully, the efficiency of this apparatus with only a few dozen theoretical plates was rather poor.

A new approach for continuous two-phase extraction was recently presented by the small French engineering company Armen Instrument. They developed a preparative-scale centrifugal partitioning chromatography system consisting of single 12.5-L units with good performance in terms of efficiency. These rotators can be operated in two directions in order to keep either the upper or the lower phase stationary. The connection of two centrifugal partitioning chromatography units to one central multiport valve allows the continuous operation, where the direction of the flow is reversed periodically. The components to be separated are introduced with the upper or the lower phase depending on the actual flow direction. The goal of process development for such systems is to select

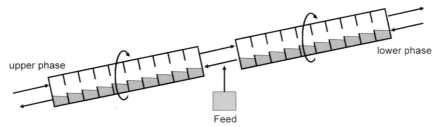

Fig. 8.13 Schematic diagram of a continuous fractional extractor.

appropriate two-phase solvent systems and the corresponding operating conditions, to cut chromatograms at exactly the right position to get the desired separation.

The described two-zone sequential-centrifugal partitioning chromatography mode is preferably used for the fast continuous separation of a mixture into two subgroups, e.g. in capturing steps. In addition, a four-zone system is available, giving a total column volume of 50 L [69]. The size of the system, which has recently shown its reliability in a long-term experiment, allows for the first time its use in the production of biopharmaceutical compounds. The full potential of this new technology is still to be exploited, taking into account the known drawbacks of aqueous two-phase extraction systems, e.g. re-extraction, cost, quality and re-use of the solvent systems.

8.3
Continuous Annular Chromatography

8.3.1
Basic Principle

Continuous feed introduction has been a goal in preparative chromatography for a long time. In 1949, Martin suggested a system for continuous feed introduction that operates in an annular mode [70]. The idea was to pack the selective sorbent into an annulus, which slowly rotates around a central axis, while fresh eluent is introduced over the whole cross-section at the top of the bed. This mode of operation results in a cross-current movement of solid and fluid phase. The feed, containing the components to be separated, is continuously introduced at a fixed position (Fig. 8.14a). Depending of the affinity of the different substances towards the solid adsorbent, they can be collected and purified at different positions at the

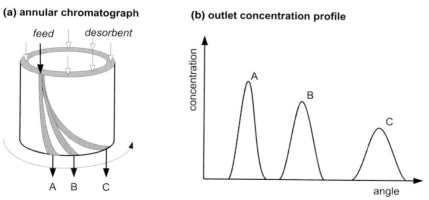

Fig. 8.14 Schematic principle of a annular chromatograph (a) and the concentration profile at the outlet (b).

column outlet. The more a molecule is retained, the larger its elution distance is, relative to the initial feed point. This results in a chromatogram as displayed schematically in Fig. 8.14(b). This appears very similar to what one knows from classical batch chromatography, just plotted against the elution angle instead of the retention time. In fact, results from simple batch experiments can easily be transferred to an annular process.

In the mid-1950s, two different systems where constructed to make Martin's dream come true. In 1955, Svensson built a machine where the annulus was formed out of a multitude of single columns [71]. In Svensson's system, the annulus consisted of 36 columns with an internal diameter of 11 mm. As this system is close to the ISEP (Ion Exchange Separator) principle, we will show its technical implications in Section 8.4. The other apparatus was constructed by the Swiss engineer Solms [72]. He folded a paper sheet for paper chromatography to form a cylinder and put one end into a reservoir with a mobile phase. Due to capillary forces, the annulus is constantly fed with fresh eluent. At a fixed point the feed is introduced through a capillary. Both systems have not been extremely successful for real separations, but the basic principle of continuous cross-current chromatography has been demonstrated.

The first practical system for solid–liquid chromatography was constructed by Fox [73–75]. The system consisted of two acrylic cylinders forming an annulus of 2.58 L with a width of 3/8 in and 100 outlet ports. The uniform flow rate over the annulus was achieved by a constant-level solvent reservoir. Even with this first simple system, protein separations were carried out. Protein, salt, lactose and riboflavin from skim milk have been separated on Sephadex® G-25. In a second example, the separation of myoglobin from hemoglobin in beef heart extracts on Sephadex G-75 was performed. Achieving a uniform bed as well as avoid clogging of the sorbent turned out to be challenging tasks at that time. Improved systems allowing gradient elution and reducing the problem of back-mixing were introduced by Scott in 1976 [76] and Canon in 1978 [77]. In principle, it is not necessarily the annulus to be rotated, but the feed and eluent introduction as well as the complete fraction collection device at the column outlet instead. Systems following this concept have been reported by Dunnill and Lilly [78] and Goto [79]. Unfortunately, these systems lacked the required reliability and robustness. The first real preparative applications of continuous annular chromatography for bioseparations on commercial systems were later performed on the so-called "P-CAC" systems manufactured by Prior Separation Technology (Götzis, Austria) [80]. The "P" stands for preparative, pilot-scale and/or pressurized. They constructed and marketed laboratory-scale systems with an annulus volume of a few hundred milliliters as well as production-scale systems in stainless steel. For their system, they used a pump to supply the eluent over the entire column head. The biggest hurdle to be overcome was the packing of the annulus with small particle sorbents (for biochromatography), because a uniform flow rate throughout the stationary phase of the annular column is an absolute prerequisite for efficient and successful operation of an annular chromatographic system. By a combination of an advanced packing procedure using high rotational velocities (up to 5000 R h^{-1})

at simultaneously maximum main eluent speed, it was possible to achieve stable beds with sufficient column efficiency [81]. Unfortunately, the company had to discontinue its operation in 2004 and for the time being no continuous annular chromatography system is readily available on the market.

8.3.2
Application to Biochromatography

With the advent of modern large-scale biopharmaceutical production processes, mainly for recombinant proteins, alternative downstream processes have been investigated by several academic groups for more than a decade. A variety of examples using different chromatographic methods have been reported, which are summarized in Tab. 8.2.

Although the applicability and ease of operations has been demonstrated, continuous annular chromatography has found only limited interest in industrial applications so far. The Austrian blood fractionation company Octapharma in collaboration with the University of Agricultural Sciences, Vienna, reported on the use of an annular chromatograph in the size-exclusion mode [89]. An intravenous immunoglobulin concentrate with an IgG concentration of 2 mg mL^{-1} was separated into the functional IgG and its nonfunctional aggregates using Superdex® 200 prep grade as stationary phase. The authors found a higher efficiency in the annular mode compared to their batch columns. They claim that the lower plate height [height equivalent of a theoretical plate (HETP)] is linked to reduced extra-column effects in the annulus. This is in contrast to results by Giovannini and Freitag [90], who compared the performances of analytical and preparative batch columns with an annular column when using 40-μm particles. They found the highest efficiencies in the preparative batch column due to the low influence of the wall effects.

The most advanced industrial continuous annular chromatography process has been reported by Bayer Biotechnology, Berkeley [91, 92]. Bayer produces the recombinant blood coagulation Factor VIII (Kogenate®) in a continuous perfusion

Tab. 8.2 Examples for continuous annular chromatography in biochromatography

Substrate	Separation/elution mode	Reference
BSA/myoglobin/vitamin B$_{12}$	size-exclusion chromatography	82
BSA/hemoglobin/cytochrome *c*	ion-exchange chromatography, isocratic	83
Myoglobin/hemoglobin	ion-exchange chromatography, step gradient	84
BSA/salt	size-exclusion chromatography	85
Lipase	size-exclusion chromatography	86
BSA/IgG	size-exclusion chromatography	87
IgG (Chinese hamster ovary cell culture supernatant)	affinity and hydroxyapatite	88

fermenter using a baby hamster kidney cell line. The product feed is continuously filtered over a 0.2- µm filter and clarified in a disk-stack centrifuge. If the first chromatographic capture step is performed in a batch mode, the unpurified product has to be cooled and stored. This treatment results in severe loss in activity of the desired product. It was therefore the goal of the Bayer researchers to extend the continuous operation to the capture step. Since large volumes of clarified supernatant had to processed, the feed for the continuous annular chromatography column was introduced over the whole annulus, while the elution buffer was introduced at one fixed location only. With this "reversed" set-up, the bound target product could be desorbed using small volumes of desorption buffer, resulting in a rather concentrated product fraction. Experiments have been performed using Poros® DE 2 (PerSeptive Biosciences), SP Sepharose® FF (GE Healthcare) and Fractogel EMD DEAE 650 M (Merck) as chromatographic stationary phases. Ideally, the flow rate is equally distributed over all 90 outlet lines, while in fact deviations of ±20% were observed. The authors claim that this effect is due to uneven ("sidewards") fluid flow through the annulus, leading to an effect they call "peak wobbling". The withdrawal of the product fraction can no longer be realized with one outlet port. The target compound is distributed over a circle segment of approximately 20° instead, resulting in a higher dilution. This effect is strongly related to the packing quality of the stationary phases used. It was shown that for the rigid Fractogel sorbent the effect is much less pronounced than for the "soft" Sepharose material. By using Fractogel DEAE the product outlet could be focused on one outlet port, instead of six outlet ports for Sephadex [93].

The application of a P-CAC system for the size-exclusion separation of fructans (fructo-oligosaccharides and inulin) has been demonstrated by Numico Research (Friedrichsdorf, Germany) [94]. An annular chromatograph with an annulus width of 1 cm and a bed height of 40 cm was packed with Toyopearl® HW 40 (S) (Tosoh Biosep, Stuttgart, Germany) by a slurry-packing method using a peristaltic pump. The feed with a total concentration of 100 g L^{-1} was separated at 60°C into 90 fractions (corresponding to the 90 outlet lines of the P-CAC), which were used to determine the physiological effect of oligosaccharides with different chain lengths afterwards. For fructan separation, different rotational speeds as well as various feed concentrations were examined. The angular velocity profile became unstable, particularly for very concentrated feed mixtures and higher rotational speeds. Due to the high viscosity of the feed, the pure solvent breaks through the feed band close to the top of the annulus, reducing the separation efficiency further down the chromatographic bed. In additional studies the resolution of their P-CAC (with a maximum bed height of 60 cm) was found to be lower compared to a batch separation where a bed length of 220 cm was used. This is not very surprising, since different process concepts and column geometries are compared with each other. For such comparisons, the productivity (amount of product produced per time and amount of stationary phase), specific eluent consumption and final product concentration should be used.

Continuous annular chromatography is a production method that allows continuous feed introduction. In combination with continuous perfusion fermentation, the concept should show its strongest advantages as a continuous capture step. After the technical problems have been solved, the system might be used in combination with rigid packing materials for production processes, especially when the feed is introduced over the complete annulus to bind the product and to drastically reduce the volume to be further processed. The process performance in terms of productivity, solvent consumption and product dilution of a continuous annular chromatography should be in the range of what one could achieve by batch chromatography, provided that a good packing quality can be realized. Nevertheless, there is the need for a system manufacturer for the technology be more widely applied.

8.4
ISEP

8.4.1
Basic Principle

As already demonstrated by Svensson in 1955 [71] with his annular chromatographic system consisting of 36 single columns, the annulus can be divided into single columns and still be operated as continuous annular chromatography. However, the constellation of a carrousel of columns can be used much more flexible in co-, cross- or countercurrent mode. The most important part of such a column carrousel is the multiport valve, which distributes the different solvent flows into the single columns. Most systems reported in the literature or available on the market consist of an upper ring with a multiport valve to which the solvent lines from different pumps are connected. The column carrousel and a lower ring with another multiport valve realize the fraction connection or re-injection into the next column. In principle, the instrument can be operated with rotating columns or with rotating solvent lines. The two most prominent systems on the market (CSEP® or ISEP® from Calgon Carbon, Pittsburgh, PA and SMB from Knauer, Berlin, Germany) are both constructed with rotating columns. The ISEP system was developed and introduced into the market in 1990 by Advanced Separation Technologies (Lakeland, Florida; now within Calgon Carbon). Since then, more than 300 systems have been installed covering 40 different applications, according to the companies website. The installed systems are considerably large with 30 columns of a column diameter up to several meter. The upper and lower distributor valves have 20 fixed ports each and can withstand a valve inlet operating pressure of up to 2 bar. The most pressure critical situation is the shift of the flow inlet from one column to the next one. To avoid a complete blockade at a certain point, the fluid from one port enters one column for 70% of the time, while it enters two columns for the remaining 30% of the time. As many of the applications are in the food or basic chemical industries, a certain leakage of the valve is not that dramatic as it would be in pharmaceutical productions. While Calgon Carbon recently decided to focus more on their traditional business and less on chromatography, an improved version of the ISEP technology was introduced in the market by SepTor Technologies, a daughter company of the Finish company Outokumpu Technology Minerals. The main feature of the SepTor system is the distribution valve in the lower part of the system. Together with other valve improvements, this reduces the internal leakage between the column segments inside the system.

Advanced Separation Technologies, as well as the German company Wissenschaftliche Gerätebau Dr. Knauer, Berlin, also developed laboratory-scale systems with stainless steel rotary valves, which can withstand higher pressure drops up to 100 bar. Those systems incorporate up to 16 columns and can be used in different open- and closed-loop modes. A special set-up for eight-zone SMB operation has been introduced recently. The laboratory-scale systems show a rapid column carrousel movement to avoid partial leakage from one column to another, as it is a constructive detail in the large-scale systems.

The biggest advantage of the multiport systems is their flexibility in connecting the different lines to the single columns. The two boundary conditions of the systems are the all-parallel connection, resulting in an annular chromatographic system where the annulus is divided into single columns (thus corresponding to Svensson's system from 1955) and an all-serial connection, which is a classical SMB configuration.

The most widely used set-up is an operation of several zones, e.g. adsorption, rinse, regeneration/backwash and desorption zone, in which the columns are connected in series or in parallel, according to the need of high volume processing (parallel connection) or high separation efficiency (serial connection). An example for a process realized for purification of an amino acid from a fermentation broth is shown schematically in Fig. 8.15.

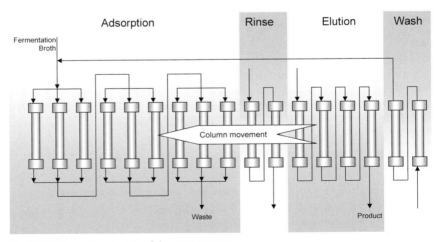

Fig. 8.15 Schematic principle of the ISEP process.

Using this flexible approach, several zones can be realized in one system with partly cocurrent, partly countercurrent movement of the stationary and mobile phase. Independent from the internal line connections the systems always have the advantage of continuous feed introduction. There is also another very important aspect to be considered for such chromatographic plants – in classical batch processes, columns are loaded only up to a dynamic binding capacity of 30–50% of their theoretical equilibrium as a result of limiting mass transfer resistances. With the given concept it is possible to completely load single columns before they are passed over to the following elution section.

8.4.2
Application to Biochromatography

The main application areas of the ISEP systems can be found in the food industry. Sugar separations are carried out in multi-ton scales, e.g. to obtain fructose, xylose, mannitol and betaine, as well as different amino acids. For smaller amounts of special sugars, such as raffinose, arabinose, galactose, mannose and others, pilot plants have been realized.

To show the advantages of this continuous operation mode, Andersson and coworkers used a C920 system from Calgon Carbon to separate lactoperoxidase and lactoferrin on Streamline® SP [95]. Their system consisted of 20 columns with a volume of 10 mL each (1.1 cm column diameter) arranged in a six-section set-up.

The zone configuration was as follows:
- Loading the feed over two series of four columns in parallel.
- Washing with conditioning buffer (two columns).
- Elution of lactoperoxidase (early eluting compound, three columns in series).
- Elution of lactoferrin (late eluting compound, three columns in series).
- Regeneration with NaOH (two columns in series).
- Equilibration with buffer to adjust column pH (two columns in series in backflush mode).

Compared to a conventional single-column batch process, the productivity was increased by 48%, while the buffer consumption was 4.8 times lower. Additionally, the final concentration of the target protein was 6.5 times higher.

A similar system was recently published by BiogenIdec in a filed patent [96]. An antibody was purified using a Protein A sorbent in a four-zone process of adsorption, wash, elution and re-equilibration. Different column configurations have been used to capture the antibody in the first step of a classical downstream cascade. A CIP zone using first 1% phosphoric acid and afterwards 4 M urea is integrated into the continuous purification process.

8.5
Conclusions and Outlook

Continuous chromatography has found an attractive niche in the production of small-molecule pharmaceutical compounds with several production systems running and a multitude of systems in research and development for the preparation of first-kilogram amounts. With the lessons learned in small-molecule production, continuous chromatographic processes are now ready to make the next step into biopharmaceutical manufacturing.

The continuous operation itself might be improvement enough as it fits well into continuous fermentation systems, where it helps to avoid the intermediate storage of labile products. The countercurrent operation, e.g. in SMB mode, will show its benefits in difficult separations, e.g. in size-exclusion chromatography, where otherwise long columns have to be packed to achieve the necessary column efficiency. SMB chromatography in size-exclusion chromatography mode will show much higher productivities and will help size-exclusion chromatography to gain more impact in biopharmaceutical production. New modes of countercurrent operation will be developed and fitted to the needs of biochromatography. These modes will combine the advantages of continuous and countercurrent operations with the flexibility in mobile phases, which are typically used in biochromatography.

Some obstacles still have to be overcome to show the full potential of continuous countercurrent chromatography in biopharmaceutical manufacturing. The availability of suitable stationary phases will be no problem as today rigid and pressure-stable phases for all operation modes (e.g. ion exchange, size exclusion chromatography, hydrophobic interaction) have been developed and shown to be easily packed into large-scale production columns.

Major concerns regarding SMB, ISEP and production-scale continuous annular chromatography are related to the contact of the sometimes labile product with a multitude of stainless steel columns, tubes, pumps and valves. The residence time of the product inside the systems is also an issue [49] as is the cleanability of the system. Appropriate CIP and sterilization-in-place protocols have to be developed under production conditions to show the equivalence of the new chromatographic methods.

Modeling and simulation of the new processes is still a major task as the separation problems are becoming more and more complex, with a multitude of very different impurities and changing mobile-phase compositions during the elution. Together with the better understanding of the chromatographic processes it is absolutely necessary to teach the new concepts and complex operation modes so that the operators in biopharmaceutical production feel safe in the daily use of the new systems.

Nevertheless, continuous and, in particular, countercurrent chromatographic operations are ready to play their role in modern biopharmaceutical production, helping to fulfill the production demands of modern biopharmaceutical drug compounds. Two promising areas of application can be seen in the field of industrial biotechnology, where productive and economic downstream processes are needed to isolate complex value compounds, e.g. from oligo saccharide feed streams. The other potential application arises with the advent of proteins produced in the milk of transgenic animal [97]. In-line downstream processing of proteins from animal milk with continuous biochromatographic systems is both a challenge and a dream. This dream might come true some day, with farmers herding transgenic cattle and being as familiar with SMB chromatography for product isolation as they are today with continuous centrifugation for defatting milk.

References

1 Broughton, D. B., Gerhold, C. G., Continuous sorption process employing fixed bed of sorbent and moving inlet and outlets. *US Patent 2985589*, **1961**.

2 Hammende, M., Case study in production-scale multicolumn continuous chromatography. In: *Preparative Enantioselective Chromatography*, Cox, G., *Blackwell Publishing*, **2005**, pp. 2533.

3 Blehaut, J., Ludemann-Hombourger, O., Use of green solvents and complete integration of solvent recycling dramatically reduce emissions of VOCs and waste in large scale chromatography separation processes. Presented at: Achema 2006, Frankfurt, **2006**.

4 Berg, C., Hypersorption process for separation of light gases. *Trans Am Inst Chem Eng* **1946**, *42*, 665–680.

5 Kehde, H., Fairfield, R. G., Frank, J. C., Zahenstecher, L. W., Ethylene recovery. Commercial Hypersorption operation. *Chem Eng Prog* **1948**, *44*, 575–581.

6 Martin, H., Kuhn, W., Multiplikationsverfahren zur Spaltung von Racematen. *Z Elektrochem* **1941**, *47*, 216–220.

7 Niven, G. W., Scurlock, P. G., A method for continuous purification of proteins by affinity chromatography. *J Biotechnol* **1993**, *31*, 179–190.

8 Hotier, G., Nicoud, R. M., Separation by simulated moving bed chromatography with dead volume correction by desynchronization of periods. *EP 688589A1*, **1995**.

9 Storti, G., Mazzotti, M., Morbidelli, M., Carrà, S., Robust design of binary countercurrent adsorption separation processes. *AIChE J* **1993**, *39*, 471–492.

10 Mazzotti, M., Storti, G., Morbidelli, M., Robust design of countercurrent adsorption separation processes: 2. Multicomponent systems. *AIChE J* **1994**, *40*, 1825–1842.

11 Mazzotti, M., Storti, G., Morbidelli, M., Robust design of countercurrent adsorption separation processes: 3. Nonstoichiometric systems. *AIChE J* **1996**, *42*, 2784–2796.

12 Mazzotti, M., Storti, G., Morbidelli, M., Optimal operation of simulated moving bed units under nonlinear chromatographic separation. *J Chromatogr A* **1997**, *769*, 3–24.

13 Storti, G., Baciocchi, R., Mazzotti, M., Morbidelli, M., Design of optimal conditions of simulated moving bed adsorptive separation units. *Ind Eng Res* **1995**, *34*, 288–301.

14 Gentilini, A., Migliorini, C., Mazzotti, M., Morbidelli, M., Optimal operation of simulated moving bed units under nonlinear chromatographic separation, II. Bi-Langmuir isotherm. *J Chromatogr A* **1998**, *805*, 37–44.

15 Migliorini, C., Mazzotti, M., Morbidelli, M., Robust design of countercurrent adsorption separation processes: 3. Nonconstant selectivities. *AIChE J* **2000**, *46*, 1384–1399.

16 Susanto, A., Wekenborg, K., Epping, A., Jupke, A., Model Based Design and Optimization. In: *Preparative Chromatography of Fine Chemicals and*

Pharmaceutical Agents. Schmidt-Traub, H., VCH-Wiley, Weinheim, **2005**, pp. 313–366.

17 Ludemann-Hombourger, O., Bailly, M., Nicoud, R.-M., The VARICOL process: a new multicolumn continuous chromatographic process. *Sep Sci Technol* **2000**, *35*, 1829.

18 Kloppenburg, E., Gilles, E. D. A., A new concept for operating simulated moving bed processes. *Chem Eng Technol* **1999**, *22*, 10, 813–817.

19 Zhang, Z., Mazzotti, M., Morbidelli, M., Power-feed operation of simulated moving bed units: changing flow rates during the switching interval. *J Chromatogr A* **2003**, *1006*, 87–99.

20 Zang, Y., Wankat, P. C., SMB operation strategy – partial feed. *Ind Eng Chem Res* **2002**, *41*, 2504–2511.

21 Schramm, H., Kienle, A., Kaspereit, M., Seidel-Morgenstern, A., Improved operation of simulated moving bed processes through cyclic modulation of feed flow and feed concentration. *Chem Eng Sci* **2003**, *58*, 5217–5227.

22 Houwing, J., van Hateren S. H., Billiet H. A. H., Van der Wielen, L. A. M., Effect of salt gradients on the separation of dilute mixtures of proteins by ion-exchange in simulated moving beds. *J Chromatogr A* **2002**, *952*, 85–98.

23 Wekenborg, K., Susanto, A., Schmidt-Traub, H., Modeling and validated simulation of solvent-gradient simulated-moving bed processes for protein separation. *Eur Symp Comput Aided Process Eng* **2005**, *15*, 313–318.

24 Abel, S., Mazzotti M., Morbidelli, M., Solvent gradient operation of simulated moving beds I. Linear isotherms. *J Chromatogr A* **2002**, *944*, 23–39.

25 Abel, S., Mazzotti M., Morbidelli, M., Solvent gradient operation of simulated moving beds II. Langmuir isotherms. *J Chromatogr A* **2004**, *1026*, 47–55.

26 Belscheva, D., Hugo, P., Seidel-Morgenstern, A., Linear two-step gradient counter-current chromatography. Analysis based on a recursive solution of an equilibrium stage model. *J Chromatogr A* **2003**, *989*, 31–45.

27 Antos, D., Seidel-Morgenstern, A., Two-step solvent gradients in simulated moving bed chromatography. Numerical study for linear equilibria. *J Chromatogr A* **2002**, *944*, 77–91.

28 Jensen, T. B., Reijns, T. G. P., Billiet H. A. H., van der Wielen, L. A. M., Novel simulated moving-bed method for reduced solvent consumption. *J Chromatogr A* **2000**, *873*, 149–162.

29 Migliorini, C., Wendlinger, M., Mazzotti, M., Morbidelli, M., Temperature gradient operation of a simulated moving bed unit. *Ind Eng Chem Res* **2001**, *40*, 2606–2617.

30 Gao, J., Wu, C., The "coil-to-globule" transition of poly(N-isopropylacrylamide) on the surface of a surfactant-free polystyrene nanoparticle. *Macromolecules* **1997**, *30*, 6873–6876.

31 Clavier, J.-Y., Nicoud, R. M., Perrut, M., A new efficient separation process: the simulated moving bed with supercritical eluent. In: *High Pressure Chemical Engineering*, von Rohr, Ph, R., Trepp, Ch. (Eds.). Elsevier Science, London, **1996**, pp. 429–434.

32 Depta, A., Giese, T., Johannsen, M., Brunner, G., Separation of stereoisomers in a simulated moving bed supercritical fluid chromatography plant. *J Chromatogr A* **1999**, *865*, 175–186.

33 Denet, F., Hauck, W., Nicoud, R. M., DiGiovanni, O. D., Mazzotti, M., Jaubert, J. N., Morbidelli, M., Enantioseparation through supercritical fluid simulated moving bed (SF-SMB) chromatography. *Industrial & Engineering Chemistry Research*, **2001**, *40*, 4603–4609.

34 DiGiovanni, O., Mazzotti, M., Morbidelli, M., Denet, F., Hauck, W., Nicoud, R. M., Supercritical fluid simulated moving bed chromatography: II. Langmuir isotherm. *J Chromatogr A* **2001**, *919*, 1–12.

35 Rajendran, A., Mazotti, M., Morbidelli, M., Enantioseparation of 1-phenyl-1-propanol on Chiralcel OD by supercritical fluid chromatography I. Linear isotherm. *J Chromatogr A* **2005**, *1076*, 183–188.

36 Rajendran, A., Peper, S., Johannsen, M., Mazotti, M., Morbidelli, M., Brunner, G., Enantioseparation of 1-phenyl-1-propanol by supercritical fluid-simulated moving bed chromatography. *J Chromatogr A* **2005**, *1092*, 55–64.

37 Nicoud, R. M., Simulated moving bed (SMB): some possible applications for biotechnology. In: *Bioseparation and Bioprocessing*. Subramanian, G. (Ed.). Wiley-VCH, Weinheim, **1998**, pp. 3–38.

38 Ganetsos, G., Barker, P. E., *Preparative and Production Scale Chromatography*. Marcel Dekker, New York, **1993**.

39 Hotier, G., Toussaint, J.-M., Terneuil, G., Lonchamp, D., Continuous process and apparatus for chromatographic separation of a mixture of at least three constituents into three purified effluents using two solvents. *US Patent 005093004A*, **1992**.

40 Voigt, U., Kinkel, J., Hempel, R., Nicoud, R.-M., Chromatographic process for obtaining highly purified cyclosporin A and related cyclosporins. *US Patent 006306306B1*, **2001**.

41 Abel, S., Bäbler, M. U., Arpagaus, C., Mazzotti, M., Morbidelli, M., Two-fraction and three-fraction continuous simulated moving bed separation of nucleosides. *J Chromatogr A* **2004**, *1043*, 201–210.

42 Paredes, G., Abel, S., Mazzotti, M., Morbidelli, M., Stadler, J., Analysis of a simulated moving bed operation for three-fraction separations (3F-SMB). *Ind Eng Chem Res* **2004**, *43*, 6157–6167.

43 Paredes, G., Mazzotti, M., Stadler, J., Makart, S., Morbidelli, M., SMB operation for three-fraction separations: purification of plasmid DNA. *Adsorption* **2005**, *11*, 1, 841–845.

44 Aumann,L., Morbidelli,M., *European Patent 05405327.7, 05405421.8*, **2005**.

45 Ströhlein, G., Aumann, L., Mazzotti, M., Morbidelli, M., A continuous, counter-current multi-column chromatographic process incorporating modifier gradients for ternary separations. *J Chromatogr A* **2006**, *1126*, 338–346.

46 Britsch, L., Schulte, M., Strube, J., Continuous method for separating substances according to molecular size. *European Patent 1140316B1*, **1998**.

47 Wang, N.-H. L., Xie, Y., Mun, S., Kim, J.-H., Hritzko, B. J., Insulin purification using simulated moving bed technology. *World Patent 01/87924A2*, **2001**.

48 Xie, Y., Mun, S., Kim, J., Wang, N.-H. L., Standing wave design and experimental validation of a tandem simulated moving bed process for insulin purification. *Biotechnol Prog* **2002**, *18*, 1332–1344.

49 Mun, S., Xie, Y., Wang, N.-H. L., Residence time distribution in a size-Exclusion SMB for insulin purification. *AIChE J* **2003**, *49*, 2039–2058.

50 Houwing, J., Billiet H. A. H., van der Wielen, L. A. M., Mass-transfer effects during separation of proteins in SMB by size exclusion. *AIChE J* **2003**, *49*, 1158–1167.

51 Geisser, A., Hendrich, T., Boehm, G., Stahl, B., Separation of lactose from human milk oligosaccharides with simulated moving bed chromatography. *J Chromatogr A* **2005**, *1092*, 17–23.

52 Lucena, S. L., Rosa, P. T. V., Furlan, L. T., Sanatana, C. C., Separation of alpha-lactalbumin and beta-lactoglobulin by preparative chromatography using simulated moving beds. *Eng Manufact Biotechnol* **2001**, 325–337.

53 Houwing, J., Billiet H. A. H., van der Wielen, L. A. M., Optimization of azeotropic protein separations in gradient and isocratic ion-exchange simulated moving bed chromatography. *J Chromatogr A* **2002**, *944*, 189–201.

54 Michel, M., Epping, A., Jupke, A., Modeling and determination of model parameters. In: *Preparative Chromatography of Fine Chemicals and Pharmaceutical Agents*. Schmidt-Traub, H., VCH-Wiley, Weinheim, **2005**, pp. 215–307.

55 Brooks, C. A., Cramer, S. M., Steric-mass-action ion exchange: displacement profiles and induced salt gradients. *AIChE J* **1992**, *38*, 1969–1978.

56 Huang, S. Y., Lin, C. K., Chang, W. H., Lee, W. S., Enzyme purification and concentration by simulated moving bed chromatography: an experimental study. *Chem Eng Commun* **1986**, *456*, 291.

57 Gottschlich, N., Weidgen, S., Kasche, V., Continuous biospecific affinity purification of enzymes by simulated moving-bed chromatography. Theoretical description and experimental results. *J Chromatogr A* **1996**, *719*, 267–274.

58 Gottschlich, N., Kasche, V., Purification of monoclonal antibodies by simulated moving-bed chromatography. *J Chromatogr A* **1997**, *765*, 201–206.

59 Voigt, U., Kinkel, J., Hempel, R., Nicoud, R.-M., Chromatographisches Verfahren zur Gewinnung von hochgereinigtem Cyclosporin A und verwandten Cyclosporinen. *Patent 09734918*, **1997**.

60 Schulte, M., Britsch, L., Strube, J., Continuous preparative liquid chromatography in the downstream processing of biotechnological products. *Acta Biotechnol* **2000**, *29*, 345.

61 Jordan, G., Präparative Trennung von Mehrkomponentengemnischen durch simulierte Gegenstromchromatographie. Diploma Thesis. University of Applied Sciences, Nürnberg, **1999**.

62 Bioxel. Press Release. Sainte-Foy, Quebec, **2004**.

63 Küsters, E., Heuer, Ch., Wieckhusen, D., Purification of an ascomycin derivative with simulated moving bed chromatography. A case study. *J Chromatogr A* **2000**, *874*, 155–165.

64 Schulte, M., Sanghvi, Y., Therapeutic oligonucleotides: state-of-the-art in purification technologies. *Curr Opin Drug Discov Dev* **2004**, *7*, 765–776.

65 Schulte, M., Lühring, N., Keil, A., Sanghvi, Y., Purification of oligo-DMT-on by simulated moving bed (SMB)-chromatography. *Org Proc Res Dev* **2005**, *9*, 212–215.

66 Müller, W., LLPC – Liquid–Liquid Partition Chromatography of Biopolymers. GIT-Verlag, Darmstadt, **1988**.

67 Sutherland, I. A., Heywood-Waddington, D., ITo, Y., Counter-current ChromaTography – applications To The separation of biopolymers, organelles and cells using either aqueous–organic Or Aqueous–aqueous phase systems. *J Chromatogr* **1987**, *384*, 197–207.

68 Kabasawa, Y., Counter-current fractional extraction with continuous sample feeding (CFE). PhD Dissertation. University of Tokyo, **1977**.

69 Foucault, A., Legrand, J., Marchal, L., Couillard, F., A new concept of centrifugal partition chromatography for high performance continuous purification. Presented at: SPICA, Aachen, **2004**.

70 Martin, A. J. P., Summarizing paper on a general discussion on chromatography. *Discuss Farad Soc* **1949**, *7*, 332–336.

71 Svensson, H., Agrell, C.-E., Dehlen, S.-O., Hagdahl, L., An apparatus for continuous chromatographic separation. *Sci Tools* **1955**, *2*, 17–21.

72 Solms, J., Kontinuierliche Papierchromatographie. *Helv Chim Acta* **1955**, *38*, 5.

73 Fox, J. B., Calhoun, R. C., Eglington, W. J., Continuous chromatographic apparatus I. Construction. *J Chromatogr* **1969**, *43*, 48–54.

74 Fox, J. B., Continuous chromatographic apparatus II. Operation. *J Chromatogr* **1969**, *43*, 55–60.

75 Nicholas, R. A., Fox, J. B., Continuous chromatographic apparatus III. Application. *J Chromatogr* **1969**, *43*, 61–65.

76 Scott, C. D., Spence, R. D., Sisson, W. G., Pressurized, annular chromatograph for continuous separations. *J Chromatogr* **1976**, *126*, 381–400.

77 Canon, R. M., Sisson, W. G., Operation of an improved continuous annular chromatograph. *J Liq Chromatogr* **1978**, *1*, 427.

78 Dunnill, P., Lilly, M. D., *Biotechnol Bioeng Symp* **1972**, *3*, 97.

79 Goto, M., Goto, S., *J Chem Eng Jpn* **1987**, *20*, 598.

80 Bart, H. J., Messenböck, R. C., Byers, C. H., Prior, A., Wolfgang, J., Continuous chromatographic separations of fructose, mannitol and sorbitol. *Chem Eng Process* **1996**, *35*, 459–471.

81 Giovannini, R., Freitag, R., Continuous separation of multicomponent protein mixtures by annular displacement chromatography. *Biotechnol Prog* **2002**, *18*, 1324–1331.

82 Hashimoto, K., Continuous Seperation of Bio-Compounds by Rotating Annular Chromatography. *Prep Chromatogr* **1989**, *1*, 163.

83 Bloomingburg, G. F., Bauer, J. S., Carta, G., Byers, C. H., Continuous protein separation by annular chromatography. *Ind Eng Chem Res* **1991**, *30*, 1061–1068.

84 Takahashi, Y., Goto, S., Continuous separation and concentration of proteins using an annular chromatograph. *J Chem Eng Japan* **1992**, *25*, 403.

85 Reissner K., Prior A., Wolfgang J., Bart H. J., Byers C. H., Preparative desalting of bovine serum albumin by continuous annular chromatography. *J Chromatogr A* **1997**, *763*, 49–56.

86 Genest, P. W., Field, T. G., Vasudevan, P. T., Palekar, A. A., Continuous purification of porcine lipase by rotating annular size-exclusion chromatography. *Appl Biochem Biotechnol* **1998**, *73*, 215.

87 Uretschläger, A., Jungbauer, A., Scale-down of continuous protein purification by annular chromatography. Design parameters for the smallest unit. *J Chromatogr A* **2000**, *890*, 53–59.

88 Giovannini, R., Freitag, R., Isolation of a recombinant antibody from cell culture supernatant: continuous annular versus batch and expanded bed chromatography. *Biotechnol Bioeng* **2001**, *73*, 522–529.

89 Iberer, G., Schwinn, H., Josic, D., Jungbauer, A., Buchacher, A., Improved performance of protein separation by continuous annular chromatography in the size-exclusion mode. *J Chromatogr A* **2001**, *921*, 15–24.

90 Giovannini, R., Freitag, R., Continuous isolation of plasmid DNA by annular chromatography. *Biotechnol Bioeng* **2002**, *77*, 445–454.

91 Vogel, J. H., Pritschet, M., Wolfgang, J., Wu, P., Konstantinov, K., Continuous Isolation of rFVIII from mammalian cell culture. In: *Animal Cell Technology: From Target to Market*, Lindner-Olsson, E., Chatzissavidou, N., Luellau, E. (Eds.). Kluwer, Dordrecht, **2001**, pp. 313–317.

92 Vogel, J. H., Nguyen, H., Pritschet, M., Van Wengen, R., Konstantinov, K., Continuous annular chromatography: general characterization and application for the isolation of recombinant protein drugs. *Biotechnol Bioeng* **2002**, *80*, 559–568.

93 Schmidt, S., Wu, P., Konstantinov, K., Kaiser, K., Kauling, J., Henzler, H.-J., Vogel, J. H., Kontinuierliche Isolierung von Pharmawirkstoffen mittels annularer Chromatographie. *Chem Ing Tech* **2003**, *75*, 302–305.

94 Finke, B., Stahl, B., Pritschet, M., Facius, D., Wolfgang, J., Boehm, G., Preparative continuous annular chromatography (P-CAC) enables the large-scale fractionation of fructans. *J Agr Food Chem* **2002**, *50*, 4743–4748.

95 Andersson, J., Mattiasson, B., Simulated moving bed technology with a simplified approach for protein purification. Separation of lactoperoxidase and lactoferrin from whey protein concentrate. *J Chromatogr A* **2005**, *1107*, 88–95.

96 Thömmes, J., Sonnenfeld, A., Pieracci, J., Conley, L., Method of purifying polypeptides by simulated moving bed-chromatography. *Patent* 2004/024284, **2002**.

97 Stix, G., The land of milk and honey. *Sci Am* **2005**, *11*, 72–75.

9
Continuous Annular Chromatography
Frank Hilbrig and Ruth Freitag

9.1
Introduction

Liquid chromatography is often the first and sometimes the only choice for the isolation of biological macromolecules such as proteins from complex multicomponent mixtures, including typical feeds from the biotechnology industry such as culture supernatants or cell lysates. This continued popularity is due to a number of factors – chromatography is universally applicable, versatile, has high-resolution capabilities and can be operated under "physiological" conditions. Biomolecules (proteins, antibodies, nucleic acids, etc.) can be isolated using a number of interaction *modi* ranging from electrostatic and hydrophobic interactions (ion-exchange, respectively, reversed-phase and hydrophobic-interaction chromatography) to specific biological interaction (affinity chromatography). If necessary, several orthogonal separation principles can be used in series. In addition, the relative simple separation by size (size-exclusion chromatography, gel filtration) is often applied. The development of a chromatographic separation is also aided by the ever-increasing selection of dedicated chromatographic stationary phases that become commercially available. At the preparative scale, however, the capacity/loadability of the column and thus the throughput of the chromatographic separation frequently presents a limiting factor for such operations. Scale-up of the typical batch column is simplest done by increasing the cross-sectional area (diameter) of the column, while keeping the column length constant. Under such circumstances the resolution should stay comparable. However, this approach is limited by the maximum diameter, which is still compatible with packing a uniform and stable bed. Even with radial and axial compression, such columns will rarely be much more than a 1 m wide. The column length (bed height), on the other hand, is limited by the applicable pressure. In most production environments, high pressure is not an option.

The idea of combining the adaptation of the chromatographic separation process to large-scale with the introduction of continuous feed injection and product withdrawal has therefore been discussed for more than 50 years as a means to make

Bioseparation and Bioprocessing. Edited by G. Subramanian
Copyright © 2007 WILEY-VCH Verlag GmbH & Co. KGaA, Weinheim
ISBN: 978-3-527-31585-7

chromatography more competitive in the industrial sector in terms of reduced residence time for the feed and simplified process control if the continuous system reaches the steady-state, for example. One approach to continuous chromatography, i.e. the simulated moving bed (SMB) approach is eminently suited to the separation of two-component/two-fraction mixtures. However, many separations, especially in recombinant biotechnology, require multicomponent separations at several stages, while the scale of the operation (kilogram amounts) is often below that most suited to the SMB. In such cases another concept for continuous chromatography, i.e. continuous annular chromatography, may be considered.

9.2
Continuous Annular Chromatography – The Basic Principle

In modern continuous annular chromatography systems the "stationary" phase is packed into a slowly rotating concentric cylinder. The feed is introduced from a fixed inlet point, while the eluting substances can be collected from fixed outlet points. The annulus is continuously perfused from top to bottom with liquids such as eluent(s), column regeneration buffer or cleaning-in-place (CIP)/ sanitization-in-place (SIP) solutions. Solid and liquid phases thus move in a cross-current manner. The separation of a substance mixture occurs in the axial direction in a manner analogous to that found in standard batch column chromatography, while the resolution in time ("elution time") of the batch column is transformed into a resolution in space ("elution angle") in continuous annular chromatography. Figure 9.1 elucidates this principle in a schematic manner, which also may serve to explain the principle of continuous annular chromatography in general [1].

If we consider first a set of seven batch columns (Fig. 9.1a), which are at various stages of separating a binary mixture by isocratic elution, we find the first column at $T=0$, i.e. at the moment of feed injection, while columns 2–5 (2: $T=1$; 3: $T=2$; 4: $T=3$; 5: $T=4$) are at various points of separation. Column 6 ($T=5$) is at the stage of the elution of the first compound, while column 7 ($T=6$) has reached the stage of elution of the second compound. If we now imagine these columns arranged into a slowly rotating circular array (Fig. 9.1b), in which they are continuously perfused with eluent (all columns are assumed to have identical flow resistance), we already approach a set-up for a separation by continuous cross-current chromatography. For that purpose we now require a fixed inlet point from which the feed is introduced to one column after the other as the revolver of "batch" columns slowly rotate past it. Once a column has passed the injector, it is again flushed with eluent and the separation starts along the column axis (hence the cross-current movement of the liquid and the solid phases). After a given time, the separated substances appear at the column outlet. Depending on the retention time, this will – in the case of the column carrousel – correspond to a certain rotation and hence a certain distance (angle) in relation to the injection point. This elution

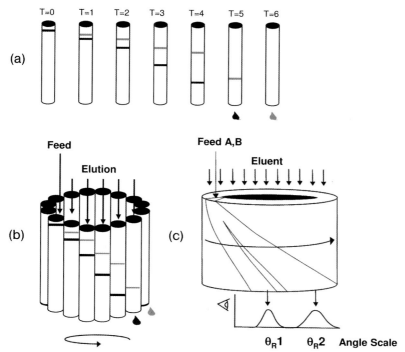

Fig. 9.1 Schematic presentation of the principle of a
two-component separation by continuous annular
chromatography: (a and b) column-based separation and
(c) annular column.

angle is therefore fixed in relation to the injection point and – given a reproducible
column performance – the separated components can be collected at their fixed
exit angles, once steady-state conditions have been reached. The fixed elution times
of the batch column have thus been transformed into fixed elution angles in the
rotating column carrousel. It is important to realize that in spite of the rotation
of the columns in the carrousel, the chromatographic separation within a
given column is in no way different from that taking place in an ordinary batch
column.

Up to now we have only considered a simple isocratic elution, where we have
only two liquids entering a column, i.e. the feed and the eluent. This set-up would
already correspond to a quasi-continuous gel filtration, for example. However,
depending on the number of available inlet ports for the entry of liquid into the
columns, we could also realize step elution (several inlets for eluents of different
strength placed in judiciously chosen distances from each other), or continuous
integrated column regeneration by introducing a suitable liquid, or even CIP/SIP
steps. Other than in the batch column approach these liquids would all be applied
simultaneously to all columns in the carrousel.

If we now transfer the continuous separation from a carrousel of columns to a continuous bed of chromatographic phase particles packed between two concentric cylinders of different diameter, i.e. an annular bed/chromatographic column (Fig. 9.1c), the involved phenomena will be similar. In particular, the chromatographic separation will occur in the axial direction as the sample/feed components move from column top to bottom. Due to the continuous nature of the chromatographic bed, however, the overall effect is the resolution of the feed components into a number of helical bands (Fig. 9.1c). This sometimes leads to the assumption of a radial component in the separation. However, just as in conventional batch chromatography, the separation in continuous annular chromatography occurs strictly in the axial direction. Some diffusion in the radial direction may occur in the annular column due to the presence of radial concentration gradients. However, within the typical timescale of a chromatographic separation, this peak broadening in radial direction should not contribute significantly to the observed zone development.

9.3
Development of Instrumentation and Operation

Continuous annular chromatography may be performed on a continuous (annular) bed or be approximated using a carrousel of columns. Both approaches have been realized in the past, and both have their distinct advantages and disadvantages. The homogeneous packing of the involved chromatographic bed(s) is of primary importance in both cases, as differences in flow resistance will obviously bias the separation. In the case of a column carrousel, the number of columns limits the separation. Introduction of feed and eluent is technically involved, and becomes more so as the number of columns increases. Fraction collection at distinct points, on the other hand, can be done by a simple fraction collector. In the case of an annular column, the chromatographic bed is continuous. However, the number of distinct collectable fractions will be limited by the number or outlet points. The application of pressure to such a column requires the design of a dedicated "head", while the seal between the rotating column and the fixed outlet points may present a weak point in column design.

The idea of using a cross-current movement of solid and liquid phase for chromatographic separations is by no means new. The theoretical possibility of using such an approach to render chromatography continuous, which should in principle have been applicable to most chromatographic operation modes known at that time, was already discussed in 1949 by Martin [2]. In this context, he already envisaged both possible manifestations of the approach, i.e. the true annular column and the realization in quasi-continuous form as a carrousel of individual columns. In his opinion, using a carrousel of columns should make no essential difference to the operation of the system, but would perhaps present less practical difficulty. While this was certainly true at that point in time, modern continuous annular chromatography systems tend to work with annular beds instead of column assemblies (see below).

In 1955, Svensson and coworkers realized a system that consisted of 36 columns and which was mainly of academic interest [3]. The first true continuous annular chromatograph was constructed by Fox in 1969 [4]. The device consisted of a rotating annulus with an outer diameter of 294 mm, a width of 9.6 mm and a height of 300 mm. In this continuous annular chromatography apparatus the feed was pumped continuously onto the column through a fixed nozzle located just above the adsorber bed. A specially designed "leveling plough" kept the bed surface absolutely flat during operation. The apparatus was designed in such a way that the chosen eluent could be filled in a space on top of the annulus. By keeping the liquid level constant, a uniform, gravity-enforced flow of eluent through the entire continuous annular chromatography column was achieved. After the continuous separation had reached steady-state, stationary helical component bands developed from the feed inlet point to the foot of the column. This is clearly shown in the corresponding paper (reproduced here as Fig. 9.2) for the separation of two colored metal ion complexes. The slope of an individual helical band was found to depend on the eluent velocity, the rotational speed of the column, and the distribution coefficients of the components between the fluid and adsorbent phases, i.e. the retention factors. At steady-state, the component bands formed regular helices between the feed sector at the top of the bed and the individual fixed exit points (100) at the bottom of the annular bed, where the separated components could continuously be recovered. As long as the conditions remained constant, the angular displacement of each component band from the feed point was also found to remain constant.

Fig. 9.2 Continuous annular chromatograph as designed by Fox and coworkers [4] (reprinted with permission).

This rotating annular chromatograph thus allowed realizing a truly continuous, steady-state chromatographic process that retained the ability for multicomponent separations and the flexibility, which is typical for chromatography in general. No information on process stability over time was given. However, two examples of isocratic biomolecule separation have been reported with the apparatus [5]. This includes a separation of proteins, salt, lactose and riboflavin from skim milk using Sephadex G-25 as chromatographic phase, and the separation of myoglobin from hemoglobin in beef heart extracts on Sephadex G-75 (1.2 L column volume). In the latter case, the feed was introduced at a flow rate of 0.4 mL min^{-1}, the eluent flow rate was 15 mL min^{-1} and the annular column rotated with a rotation speed of 180° h^{-1}. The myoglobin was recovered with 97% purity. According to the authors, the continuous annular chromatography separation was comparable in terms of product quality to an otherwise similar separation using a (discontinuous) batch column (8.3 × 30.0 cm, 1.6 L) available in the laboratory, but it was superior with regard to time and effort.

In 1976, Scott and coworkers [6] introduced the first prototype of a gas-pressurized continuous annular chromatography device, which allowed applying pressures of up to 1.8 bar. In addition, an optical monitoring device was included for an estimation of the substance concentration in the helical bands. For this purpose the band was illuminated through the outer glass wall of the apparatus and the reflected light was measured by a photo detector. Scott and coworkers also suggested the use of segmented spacers at the feed entry to allow gradient elution. Applications were shown for gel-filtration and ion-exchange chromatography of metal ion complexes. A simple theoretical plate model was used to describe the separation, which was shown to approximate the separation of cobalt and nickel complexes satisfactorily. The continuous annular chromatography apparatus suggested by Scott and coworkers was subsequently redesigned by Canon and Sisson [7] with the aim of improving chromatographic performance and the consistency. The inlet was redesigned to allow for more flexibility and manifolds were added for gradient formation. The feed system was changed to a two-nozzle system in order to reduce concentration variation in the radial direction. Instead of interfacing or even burying the feed nozzle in the adsorber bed, the nozzle was placed in a layer of glass beads, which allowed the feed band to spread until its velocity matched that of the main eluent.

A noteworthy contribution of this group was the introduction of a dedicated column exit unit consisting of 180 holes in the bottom plate, which were interfaced to 180 discrete exit tubes. This significantly reduced the circumferential mixing near the exit. The application of the system was again for the separation of metal ion complexes. Instead of rotating the annulus, it is in principle also possible to rotate the inlet and the outlet unit of a continuous annular chromatography unit and keep the annulus stationary. Such systems have, for example, been reported by Dunnill and Lilly [8] as well as by Goto [9]. However, this approach seems to be mechanically more involved; generally, these systems were regarded as lacking in reliability and robustness.

9.4
Modern Continuous Annular Chromatography Systems

Since 1999, continuous annular chromatography instruments have been commercially available under the name "P-CAC" from Prior Engineering (Götzis, Austria). The "P" in stands for preparative, pilot-scale and/or pressurized. The principle difference of the P-CAC to the units developed earlier is the design of the stationary distribution head (Fig. 9.3). Instead of using gas pressure, the entire

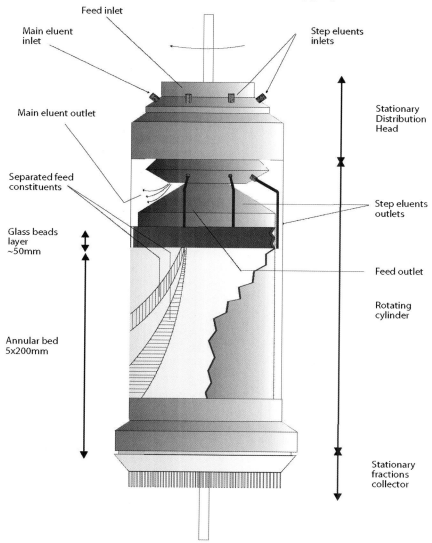

Fig. 9.3 Schematic drawing of the P-CAC system commercialized by Prior Engineering (Götzis, Austria).

head and the annulus are flooded with the main eluent at defined pressure (flow rate) provided by a pump. The chromatographic bed (annulus) is packed between an inner cylinder made of a polymeric material and an outer cylinder made of glass in the standard system. The diameter of the inner cylinder in the standard system is 14 cm; the inner diameter of the glass cylinder is 15 cm. The typical annulus thickness is therefore 10 mm. A somewhat wider inner cylinder allowing the packing of an annulus with only 5 mm thickness is available upon request. The annular bed is toped by a layer of glass beads averaging 150–250 μm in diameter. These glass beads serve to stabilize the bed surface and act as flow distributors allowing the incoming feed/eluent streams to spread until their velocities match the velocity of the main eluent, thereby helping to overcome the problem of defining the initial bandwidth of the feed/eluent.

Up to six fixed inlets nozzles are available for the introduction of the feed and additional eluents. These nozzles reach deep into the layer of the glass beads without touching the adsorber bed surface. At the bottom, the column is flanged to a bottom plate containing 90 exit ports below the annulus and a porous filter membrane supporting the resin bed. This bottom plate glides on a stationary plate, which contains 90 fraction collection ports in which tubings are fixed. Instead of straight tubings, tubings forming a siphon are used to collect the eluted fractions, which has the advantage of maintaining homogeneous flow conditions around the annular column. At present, the eluting fractions are analyzed off-line, but prototypes of a fiber optic multichannel detector have already been developed that eventually will allow the continuous monitoring of the eluting substance zones [10].

P-CAC instruments fabricated by dedicated polymer and steel materials are available [11]. Normally the material of the outer cylinder is glass for operation under visual inspection, which will withstand pressure of up to 3 bar. Alternatively, the outer cylinder can also be made from stainless steel, then allowing for operation with up to 10 bar. The rotation of the annulus can be controlled between 50 and 5000° h^{-1} by a speed drive system. A crucial point for the performance of a continuous annular chromatography column is the quality of the bed packing. Depending on the type and the stability of the chosen chromatographic phase, either automatic (by pump) or manual slurry packing can be used. During packing the column is rotated at the maximum rotation rate or in the (optional) "fast rotation mode" at a rotation rate of up to 10 rpm. In order to avoid uneven phase sedimentation the use of the "variable shaking modus" is recommended, thereby avoiding entrapment of gas bubbles, especially in the exit tubings [12]. Otherwise, uneven flow distribution in the exit ports up to the complete blockage of some outlets and overcompensated flow in adjacent outlets and thus deteriorated resolution in the chromatogram will occur. Several relevant separation applications have been demonstrated using the various P-CAC devices, some of which will be discussed below (see also Tab. 9.1).

Tab. 9.1 Overview of published biomolecule (pressurized) continuous annular chromatography separations

Substrate	Separation mode	Elution mode	Reference
Amino acids			
glutamic acid/valine/leucine	displacement	isocratic	22
aspartic acid/glutamic acid/glycine	ion exchange	isocratic	52
glutamic acid/glycine/valine	ion exchange	isocratic/step elution	53
glutamic acid/valine	ion exchange	isocratic	41
glutamic acid/valine	ion exchange	isocratic	26
glutamic acid/valine	ion exchange	isocratic	39
glutamic acid/valine/leucine	ion exchange	step elution	40
Proteins			
BSA/hemoglobin/cytochrome c	ion exchange	isocratic	19
myoglobin/hemoglobin	ion exchange	step elution	54
BSA/hemoglobin	ion exchange	step elution	21
IgG (Chinese hamster ovary cell culture supernatant)	affinity and hydroxyapatite	step elution	31
whey proteins/soybean trypsin inhibitor	displacement	isocratic	30
rFVIII	ion exchange	step elution	12
Factor IX	ion exchange	step elution	38
BSA/myoglobin/vitamin B12	size exclusion		25
BSA/salt	size exclusion		17
lipase	size exclusion		55
Nucleic acids			
pDNA (bacterial lysate)	hydroxyapatite	step elution	28

9.5
Theory and Modeling of Continuous Annular Chromatography

Predicting the elution profiles and changes thereof upon changes of the operational parameters is an important tool in the set-up and optimization of chromatographic separations. In continuous annular chromatography the space and time coordinates typically used in the modeling of conventional batch chromatographic separation are replaced by the space and annular displacement coordinates of the rotating bed. Two-dimensional, steady-state separations in continuous annular chromatography are therefore analogous to one-dimensional, time-dependent separations in conventional fixed-bed elution chromatography. Elution time and angle are related by [13]:

$$t_R = \alpha_R / \omega, \tag{1}$$

where t_R is the elution time, α_R is the elution angle and ω is the rotational velocity.

Assuming an isothermal, steady-state separation process with uniform concentration and fluid profiles in the radial direction, and local equilibrium between the fluid and solid phase, the mass balance describing the development of the concentration profiles for a component i in an annular chromatograph can be formulated by a coupled set of N_c two-dimensional convection diffusion equations [14]. The general form of these equations is given by:

$$\omega \frac{\partial c_i}{\partial \theta} + \omega F \frac{\partial q_i}{\partial \theta} + u \frac{\partial c_i}{\partial z} = D_z \frac{\partial^2 c_i}{\partial z^2} + D_\theta \frac{\partial^2 c_i}{\partial \theta^2} \quad \text{in } \Omega; \, i = 1, N_c. \tag{2}$$

In this equation the solid-phase concentration q_i and the liquid phase concentration c_i are related through the adsorption isotherm. The coefficients D_z and D_θ quantify band broadening by dispersion and diffusion in the axial (z) and the angular (θ) directions, respectively. The phase ratio F is given by $F = (1 - \varepsilon)/\varepsilon$, with ε being the porosity of the column. ω corresponds to the rotation velocity and u is the axial flow velocity.

According to Eq. (2), the partitioning of a solute between the liquid phase and the solid phase together with changes in concentration by convection equals the diffusion terms. If angular dispersion is neglected ($D_\theta = 0$), the continuity Eq. (2) is a term-by-term equivalent transformation ($t_R \rightarrow \alpha_R/\omega$, with α_R corresponding to the center of the elution angle) of the corresponding unsteady state equation for conventional elution chromatography [13].

If typical continuous annular chromatography conditions apply, i.e. no eluate backmixing, continuity of the outgoing flow and periodicity for the angular direction, the numerical solution of Eq. (2) with the assumption of nonlinear isotherms (e.g. of the Langmuir type) for the two components, the separation of a binary mixture can be simulated and used to demonstrate possible effects of a change in D_θ on the peak shape. Seidel-Morgenstern and coworkers [14] have simulated in this context the effect of D_θ on peak height, elution region and peak shape. They found that when D_θ is increased, the peak maxima decrease, while the elution regions (peak width) increase and the peak shapes become more symmetrical. In addition, their simulations predict that an increase of the rotation rate should increase and shift the elution range of the components. Concomitantly, smaller peak maxima, broader outflow intervals and, in a certain range, better separations should manifest themselves under these conditions. Increasing the feed's concentration, on the other hand, should lead to more asymmetric peaks with maxima at smaller elution angles. Stronger adsorption of the components serves to increase the elution angles and to reduce the peak heights, while a more pronounced nonlinearity of the adsorption isotherm reduces the elution angle. All of these effects correspond exactly to similar phenomena known from conventional fixed-bed chromatography.

Carta and coworkers [15] have used a linear two-film mass transfer model to simulate continuous annular chromatography elution profiles of a sugar mixture where the fluid–particle mass transfer rate equation (Eq. 3) is written in terms of a global interphase mass transfer coefficient $k_o a$:

$$\omega(1-\varepsilon)\frac{\partial q_i}{\partial \theta} = k_o a \left(c_i - \frac{q_i}{K}\right), \tag{3}$$

where K is the distribution coefficient of a given substance between the stationary and the mobile phase.

Under the assumptions that (i) the feed arc is infinitely small, (ii) the number of transfer units n $[n=k_o a(z/u)] > 5$, (iii) axial and angular dispersion are negligible, (iv) the adsorption equilibrium is described by a linear isotherm, and (v) concentration and fluid velocity gradients in the radial direction can be ignored, an approximate analytical solution can be derived for the chromatographic response $c(z,\tau)$ in an isocratic continuous annular chromatography operation for appropriate boundary conditions:

$$c_i(z,\tau) = \frac{Q_I}{2\pi^{0.5}}\left\{\frac{(k_o a)^2}{u^3 z\tau[(1-\varepsilon)K]^3}\right\}^{0.25} \exp\left\{-\left[\sqrt{\frac{k_o az}{u}} - \sqrt{\frac{k_o a\tau}{(1-\varepsilon)K}}\right]^2\right\}, \tag{4}$$

where $\tau = \dfrac{\alpha_R}{\omega} - \dfrac{\varepsilon z}{u}$ and $Q_I = \dfrac{c_F u Q_F}{Q_T}\dfrac{360°}{\omega}$.

Q_I is the quantity of solute injected with the feed mixture per unit cross-sectional area, c_F is the solute concentration in the feed mixture, Q_F is the feed flow rate and Q_T is the total flow rate of fluid through the annular bed.

The authors subsequently attempted to predict the isocratic continuous annular chromatography elution profiles of a glucose–fructose model mixture assuming that the time required for establishing the equilibrium is much longer than the time required for the separation. For this, K, n and ε (and thus $k_o a$) were determined from fixed-bed chromatographic experiments. Comparison of the theoretical results with the experimental ones obtained during an isocratic continuous annular chromatography separation of the sugar mixture at low feed concentration and modest column loading showed acceptable agreement. In particular, the separation was achieved using a calcium-loaded cation-exchange adsorbent, i.e. Dowex 50W-X8 (particle diameter 50–60 µm), in a continuous annular chromatograph (27.9 cm outer diameter, 1.27 cm annulus width) at 66.7 mL min^{-1} eluent flow, 1.0 mL min^{-1} feed flow and 240° h^{-1} column rotation rate.

For this particular separation the angular dispersion could indeed be neglected, as the peak spreading was largely determined by intraparticle diffusion as the major resistance to mass transfer. The comparison between numerical and experimental results confirmed expected trends in the profiles upon changes of the operation parameters. The peak position shifted to higher elution angles and the peak broadened when the rotation rate was increased. In the range of 0.5–2.5 mL min^{-1} (under these conditions the feed may be considered as infinitesimal), the feed flow rate had no effect on the resolution, while increasing the rotation speed of the continuous annular chromatography column did increase the resolution. Increasing the axial flow rate of the liquid phase or the feed concentration had a detrimental effect on the resolution, at least under experimental conditions. The calculations, on the other hand, predicted no change in resolution with increasing feed concentration.

Qualitatively, it was observed that under the given operating conditions the experimentally determined resolution of the fructose–glucose mixture with the continuous annular chromatography column was about 20% lower than the value calculated based on the parameters obtained from the fixed-bed experiments. The deviation between experimental results and calculations was attributed to the presence of calcium ions in the eluent and to a generically greater dispersion in the continuous annular chromatography compared to the fixed-bed column. Possible reasons for the latter phenomenon, whose extent can be characterized by a determination of the Peclet number as a function of the Reynolds number, are the annulus packing method and putative wall effects. Recently, Bart and coworkers [16] as well as Wolfgang and coworkers [17] repeated these experiments and calculations for a modified sugar mixture consisting of fructose, mannitol and sorbitol under nearly identical experimental conditions. Using Dowex 50W-X8 as column packing material, they found an excellent agreement between the measured continuous annular chromatography profiles and those predicted by Eq. (4) under a number of operation conditions.

It can thus be stated with some confidence that for molecules that are linearly and noncompetitively adsorbed, and for cases where axial and angular dispersion can be neglected, continuous annular chromatography elution profiles can be calculated in the simplified manner described above and, particularly, by Eq. (4). At high loading the feed occupies a significant sector of circumference (according to Ref. [18] this is the case for a feed sector occupying more than 6°) and Eq. (4) no longer applies. In this case, the analytical solution derived for the periodic application of a finite feed volume to a chromatographic annulus under the assumption of negligible axial and angular dispersion can be used [18]:

$$
\frac{c_i(z, \theta)}{c_F} = \frac{\theta_F}{\theta_F + \theta_E} + \frac{2}{\pi} \sum_{j=1}^{\infty} \left\{ \frac{1}{j} \exp\left[-\frac{j^2 k_o az}{(j^2 + r^2)u} \right] \times \right.
$$

$$
\left. \sin\left[\frac{j\pi\theta_F}{\theta_F + \theta_E} \right] \cos\left[-\frac{j\pi\theta_F}{\theta_F + \theta_E} + \frac{2j\pi\theta_F}{\theta_F + \theta_E} - \frac{2j\pi z\omega\varepsilon}{u(\theta_F + \theta_E)} - \frac{jrk_o az}{(j^2 + r^2)u} \right] \right\}, \quad (5)
$$

where θ_F is the arc over which feed is applied to the column, θ_E is the elution arc and r is given by:

$$
r = \frac{k_o a(\theta_F + \theta_E)}{2\pi(1 - \varepsilon)K\omega}. \quad (6)
$$

This equation can be used when a finite feed sector is used, but also for multiple interfering feeds. Integration of Eq. (5) between two feed angles yields the average product concentration collected in this outlet section of the column:

$$\frac{c_i(z)}{c_F} = \frac{\theta_F}{\theta_F + \theta_E} + \frac{2(\theta_F + \theta_E)}{\pi(\theta_2 + \theta_1)} \sum_{j=1}^{\infty} \left\{ \frac{1}{j^2} \exp\left[-\frac{j^2 k_o a z}{(j^2 + r^2)u} \right] \times \sin\left[\frac{j\pi\theta_F}{\theta_F + \theta_E} \right] \right.$$

$$\left. \sin\left[\frac{j\pi(\theta_2 + \theta_1)}{\theta_F + \theta_E} \right] \cos - \left[\frac{j\pi\theta_F}{\theta_F + \theta_E} + \frac{j\pi(\theta_1 + \theta_2)}{\theta_F + \theta_E} - \frac{2j\pi z\omega\varepsilon}{u(\theta_F + \theta_E)} - \frac{jrk_o a z}{(j^2 + r^2)u} \right] \right\}.$$

$$(7)$$

This approach was used to model the separation of industrial sugar feedstocks under overloading conditions, in particular a feed sector of up to 55°. Good agreement was found between the experimental results and the predictions based on the linear adsorption isotherm model, although the experimental peaks were higher than expected from calculations based on parameters determined from fixed-bed experiments. Interestingly, and in spite of the overall similarity of the applied experimental conditions, the global mass transfer coefficient $k_o a$ for fructose, derived in this series of publications [15–18] from fixed-bed experiments showed significant differences. The value given by Byers and coworkers [18] is 61% lower than that determined by Howard and coworkers [15].

As long as the separation is in the linear range, i.e. no overloading, simulation and modeling of the separation tend to be rather insensitive to the exact nature of the mass transfer mechanism assumed in the model [19]. In the case of large molecules such as proteins and a porous chromatographic resin, intraparticle diffusion can usually be assumed to present the main resistance to mass transfer. For instance, when a continuous annular chromatography column packed with S-Sepharose was used for the isocratic separation of a mixture of bovine serum albumin (BSA), hemoglobin and cytochrome c at modest sample loading, and under essentially linear adsorption conditions ([Na⁺] > 100 mM), the separation could be adequately described by a particle diffusion model, whose linear equilibrium constants as well as the values for the diffusivities and the mass transfer parameters were obtained from fixed-bed experiments. The calculations also showed that the continuous annular chromatography behaved almost ideally, which *inter alia* indicated the virtual absence of hydrodynamic and extra column nonidealities.

A model for the simulation of the continuous annular chromatography in step gradient elution was published by Carta and coworkers in 1989 [20]. The example discussed was the separation of Fe^{3+} and Cr^{3+} ions using the Dowex 50W-X8 resin and a step gradient of increasing ammonium sulfate concentration for elution. In order to correctly predict the performance of the continuous annular chromatograph, an additional conservation term for the eluent had to be included into the continuity equation. The numerical solution of the continuous step elution equation showed, in good agreement with the experimental results, that the ammonium sulfate concentration is the most important variable. As the ammonium sulfate concentration increases, the dispersion coefficients decrease rapidly. As a result the peaks are sharpened and exit the annulus at smaller elution angles than in the corresponding isocratic experiment, albeit at the cost of an overall reduced resolution.

In addition, Bloomingburg and Carta [21] have developed a model that by numerical solution of the differential mass balance equations accurately predicted the elution profile of BSA and hemoglobin in a continuous NaCl step gradient elution separation (pH 6.5) using a continuous annular chromatography column packed with a cation-exchanger material (S-Sepharose). The model is based on the assumptions that (i) BSA is not adsorbed to the resin and is retained only in the pore volume, while concomitantly being unaffected in its progress along the column axis by the presence of salt and hemoglobin, (ii) NaCl is in large molar excess, is retained only in the pore volume and its propagation through the column bed is not affected by the presence of the proteins, (iii) hemoglobin is adsorbed by the resin and its progress through the column bed hence depends on the local salt concentration, (iv) the intraparticle mass transport of all compounds can be described by the linear driving force approximation, and (v) axial and angular dispersion can be neglected. Equilibrium isotherms and constants (Langmuir type) and mass transfer parameters were determined by independent experiments. The model predictions agreed well with the experimental results as long as the different hemoglobin variants were not resolved. If this was not the case, the model could be used to design and optimize the separation based on batch column parameters.

DeCarli and coworkers used a simple mathematical equilibrium stage model to describe approximately the continuous displacement separation of a dilute mixture of three amino acids by NaOH using a cation-exchange resin [22]. When the plate number N was set to be equal to 50 and assuming constant selectivity values as well as a sharp displacer front, this model predicted concentration and pH profiles, which were in general agreement to the experimental (batch column and continuous annular chromatography) results. Based on their results, the authors stated that under such conditions continuous annular chromatography suffers little or no additional peak dispersion relative to conventional batch chromatography.

In general, continuous annular size-exclusion chromatography can qualitatively be predicted by using the plate theory (e.g. Ref. [23]), developed already by Scott and coworkers [6]. As demonstrated for the separation of Blue Dextran from $CoCl_2$ using Sephadex G-15, given a known theoretical plate height, experimental and predicted results were in excellent agreement. More stringent mathematical models for continuous annular size-exclusion chromatography were published by Dalvie and coworkers [24] and Hashimoto and coworkers [25].

Another model, this time considering the separation of 2 amino acids by ion-exchange chromatography, was proposed by Kitakawa and coworkers [26]. This model considers also the dissociation of the amino acids and the eluent buffer composition. In addition, Nernst–Planck-type intraparticle ionic transport equations and a nonlinear ion-exchange equilibrium term based on ion-exchange selectivity are used.

9.6
Issues of Continuous Annular Chromatography Application in Bioseparation and Bioprocessing

Several examples for the separation of biological molecules by continuous annular chromatography have been published [1] (see also Tab. 9.1). These applications demonstrate that continuous annular chromatography is a viable option for continuous separation even of complex feedstreams. A number of practical aspects of continuous annular chromatography application in such (bio)separations are discussed below.

9.6.1
The Annular Column

Bed instabilities are known to be a problem in large-scale preparative chromatography columns. As Giddings [27] recognized already in 1962, many industrial-scale packed columns exhibit flow nonuniformities due to the bed instabilities that manifest themselves at larger diameters. By using an annular column it is possible to pack a bed of similar height and total cross-sectional area as the corresponding (already instable) batch column, yet to keep the width of the annulus small enough to avoid bed instabilities and flow inhomogeneities. Given the fact that the method transfer from a small batch column to an annular column is usually straightforward, a chromatographic separation can thus be scaled up from a column of a few milliliters to several liters without loss of resolution. This was recently investigated in detail for a purification of plasmid DNA (pDNA) from clarified bacterial lysates by continuous annular chromatography using 40-µm hydroxyapatite particles as the chromatographic phase [28]. The commercially available P-CAC system (Prior Engineering) was used in this as in most other recently published research reports on continuous annular chromatography.

In this investigation, the height equivalent of a theoretical plate (HETP) of the continuous annular chromatography system was determined in comparison to a small-diameter batch column of similar width (4 mm column diameter versus 5 mm annulus width) and height (250 mm column bed height versus 210 mm annulus bed height) as well as to a larger diameter batch column of similar cross-sectional area (50 mm diameter and 210 mm bed height) and volume under otherwise identical conditions. It was found that the 50-mm diameter batch column consistently had lower HETPs (by a factor of 3) than either the small batch column or the continuous annular chromatography. This was ascribed to the wall effect, which has a dual effect on the column bed. It stabilizes the bed, but it is also known to decrease the packing density (and consequently the plate number) over a distance of up to 30–50 times the particle diameter from the wall [29]. As 40-µm particles were used in the above-mentioned study, this adverse effect of the wall exerted an influence over 50% of the bed in case of the 4-mm diameter batch column, but also the annular system, compared to only 5% in the case of the 50-mm diameter column.

Chromatographic columns are typically scaled-up by increasing the column diameter while keeping the bed height and, hence, the separation distance constant. In conventional batch columns this causes a rapid increase in column diameter, and concomitantly the already mentioned bed instabilities and flow inhomogeneities. In the case of a continuous annular chromatography, a bed volume that would already cause concern in the case of a conventional batch column could be packed as an annulus of only a few tenths of a centimeter, which would easily be stabilized by the adjacent walls. By using the continuous annular chromatography approach for scale-up, the column volume could therefore be considerably increased compared to conventional preparative batch columns before bed instabilities become a problem.

A second observation made in the pDNA separation study [28] concerned the relationship between the plate height and the rotation speed of the column. As shown in Fig. 9.4, at a given flow rate, the experimentally observed column efficiency appears to increase with the rotation speed of the annulus until for a speed of approximately 520° h^{-1} the plate height approaches the one measured as for the 4-mm diameter batch column. Since the chromatographic events in the continuous annular chromatography and the batch column should be identical, this behavior was attributed to an artifact caused by the design of the P-CAC fraction

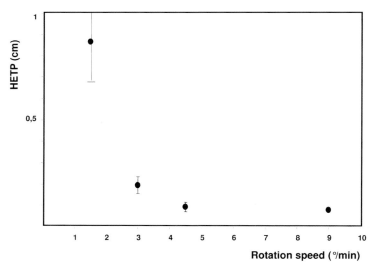

Fig. 9.4 HETP as a function of the rotation speed for the P-CAC (•) with a 5-mm thick annulus and a small batch column (inner diameter 4 mm) at the same loading factor. The volumetric flow rate and the linear flow rate was kept constant in all cases. P-CAC conditions were as follows: feed flow rate (2% acetone in 10 mM phosphate buffer, pH 6): 0.2 mL min^{-1}; eluent flow rate: 35.85 mL min^{-1}; rotation rates: 90° h^{-1} (LF=0.102), 180° h^{-1} (LF=0.051), 270° h^{-1} (LF=0.034) and 540° h^{-1} (LF=0.017). Batch column conditions were as follows: volumetric flow rate: 0.24 mL min^{-1} (linear flow rate: 1.6 cm min^{-1}); injected volumes: 52.8 µL (LF=0.017), 105.6 µL (LF=0.034), 160.8 µL (LF=0.051), 319.2 µL (LF=0.102). (Adapted from Giovannini and Freitag [28]; reprinted with permission.)

collection system. Whereas the outflow of the batch column is continuously monitored, in continuous annular chromatography the "chromatogram" is represented
by the (steady-state) composition of effluent from the column's 90 outlets. As a
result, each outlet represents an elution angle of 4° over which the concentrations
are averaged. Especially at low rotation speed, where a given peak elutes only over
a few outlets, each outlet averages over a substantial part of the "peak" zone and
this contributes considerably to peak broadening. At higher rotation speeds, the
approximation of the true peak shape becomes better, the broadening effect is less
pronounced and the efficiency approaches that of a conventional column of similar
bed width. Vogel and coworkers [12] have recently confirmed this argumentation
and called this hardware artifact "peak oscillation". According to Vogel and coworkers this "peak oscillation" together with the "peek wobbling" effect (see below)
present the limiting factors for the chromatographic performance of P-CAC *in
praxis.* Due to these effects, the experimentally observed concentration and separation factors for protein isolation by continuous annular chromatography will
always remain below the ones to be expected from the results of the corresponding
small batch column scouting separations.

A prerequisite for any high-performance separation by continuous annular chromatography is the uniform packing of the annular column. Experience with the
commercially available P-CAC system shows that relatively uniform packing can
be achieved by slurry packing at maximum rotation rate (5000° h^{-1} for the P-CAC)
and maximum main eluent flow for bed compression during packing [30]. Independent of the type of stationary material some inhomogeneities of the packed
bed are unavoidable and therefore the outlet flows must be expected to vary somewhat around the column. According to the investigation by Vogel and coworkers
[12], even for a well-packed P-CAC, deviations of ±20% from the average flow are
unavoidable. In addition to this deviation in liquid flow it must be remembered
that in continuous annular chromatography the column effluent is not continuously collected, but fractionated over the 90 outlets and that the substance zones
are typically collected over a certain angular region usually represented by several
outlets. Therefore, a deviation of the center of angular displacement ("peak
maximum") by ±4° must be expected even at constant rotation rate as well as some
variation of the asymmetry factor. This effect has been called "peek wobbling" by
Vogel and coworkers [12]. It is a normal occurrence in continuous annular chromatography separations. True inhomogeneities in the annular bed ("bed warping"),
on the other hand, are rare and easily recognizable, e.g. by the fact that at constant
rotation rate the center of elution angle of an inert tracer shifts by more than one
outlet position.

9.6.2
Method Transfer and Scale-up

According to theoretical considerations, but also many of the experimental results
gained with small molecules, the chromatographic events are identical in batch
and continuous annular chromatography. Moreover, save for the peak oscillation

and peak wobbling effects discussed above, an annular column can be expected to be of similar efficiency as an analytical column with a diameter similar to the annulus width under otherwise identical conditions (particle type and diameter, liquid flow rate, bed height). The transfer of a method developed using such an analytical column to the continuous annular chromatography system should therefore be easy and requires solely a transfer from the one-dimensional, time-dependent system in the case of the batch to the two-dimensional steady-state system of the annular column. In principle, this relationship is given by Eq. (1) that, as mentioned before, links the retention time t_R in the batch column to the rotation speed ω and the elution angle α_R in continuous annular chromatography [13]. Assuming an isothermal, steady-state separation process with uniform concentration and fluid profiles in radial direction and local equilibrium between fluid and solid phase, we expect:

$$\alpha_R = t_R \times \omega. \tag{7}$$

It was demonstrated for an isolation of a recombinant human antibody from a Chinese hamster ovary cell culture supernatant that the direct method transfer from a small 3 mL (4 × 250 mm) batch column to a 0.5 L (5 × 200 mm) continuous annular chromatography is possible taking this approach [31]. This paper was also among the first discussing continuous annular chromatography of proteins by other than gel filtration, as both hydroxyapatite and Protein A-affinity chromatography were used. Concomitantly, the compatibility of the continuous annular chromatography with complex cell culture supernatants as feed streams, step gradient elution and continuous column regeneration was demonstrated for the first time in this paper. Similar yields (between 75 and 85% in case of the rProtein A columns) and purification factors (50 and 52 for the rProtein A continuous annular chromatography and batch columns, respectively) were obtained, and 99% of the nucleic acids were removed from the product zone in the process by both columns. A direct comparison showed, however, that the chromatograms recorded for the continuous annular chromatography and the batch column were not identical. Instead the peaks were shifted in the continuous annular chromatography separation. Additional experiments showed that this was not only due to the difference in bed height.

The same group later showed that the effect is due to discontinuities in the development of the eluent steps around the annulus [28]. In the P-CAC (pressure limit 3 bar) as well as in most other continuous annular chromatography systems discussed in the pertinent literature, feed, gradient steps and column regeneration/cleaning agents are introduced from fixed inlets, albeit by individual pumps, the flow rate can therefore be adjusted. As these liquids enter the column (or, more precisely, the glass bead layer) their zone broadens until the linear flow rate in the corresponding volume stream equals that of the average liquid flow in the annular column. It is therefore possible that a zone of main eluent establishes itself between two "adjacent" gradient steps. This can even be used advantageously to avoid buffer incompatibilities in continuous annular chromatography separations. However, the phenomenon of "gradient interruption" has no equivalent in batch column chromatography, where the eluents enter the column consecutively in time. Some groups have suggested the use of adjustable inlet nozzles to avoid this problem. As demonstrated by Giovannini and Freitag, however, a very simple concept, i.e. that of the "loading factor" (LF), may be used for the direct transfer of conditions between a time resolved batch column separation and a space-resolved continuous annular chromatography separation [28].

The LF is well known in batch column separations, where it is used to set the injected sample volume in relation to the total column volume and whenever columns of different sizes are to be compared. In this case, the LF may be calculated by:

$$LF[-] = \frac{Q_F[cm^3\,min^{-1}]\,t_I[min]}{H[cm]\,S[cm^2]}, \tag{8}$$

where Q_F is the volumetric flow during injection, t_I is the time required for injection, H is the bed height and S is the cross-sectional area.

For continuous annular chromatography, the authors proposed the following equation:

$$LF[-] = \frac{Q_L[cm^3\,min^{-1}]\,360[°]}{H[cm]\,S[cm^2]\,\omega[°\,min^{-1}]}, \tag{9}$$

where Q_L is the volumetric flow of the liquid of interest, H is again the bed height, S is the cross-sectional area and ω is the rotation speed of the annular column.

For method transfer the LF is calculated for each step and used for the direct transfer of a given separation method from batch to continuous annular chromatography column or *vice versa*. Fig. 9.5 shows the overlay of a separation of plasmid DNA from a clarified lysate by continuous annular chromatography (150 mm outer diameter, 5 mm annulus thickness, 215 mm bed height) on hydroxyapatite and a separation performed on an analytical column (4 × 250 mm) using otherwise identical chromatographic conditions and using the loading factor concept for method transfer (see Tab. 9.2) [28].

The general applicability of the loading factor concept has most recently been demonstrated in the direct transfer of a protein separation in the displacement mode [30]. In this application, separations of mixtures containing up to three standard proteins (two whey proteins, soybean trypsin inhibitor) were developed and optimized using a small (4 × 250 mm) batch column filled with strong anion-exchanger phases. Glyoxal bisulfite (GBS) served as displacer. The separations were subsequently transferred directly to the continuous annular chromatography system (0.5-L column) using the loading factor concept. Separations of similar quality in terms of final product purity and recovery yield were obtained using the continuous system. Typical phenomena like the influence of the displacer concentration on the separation were analogously observed in the continuous as in the

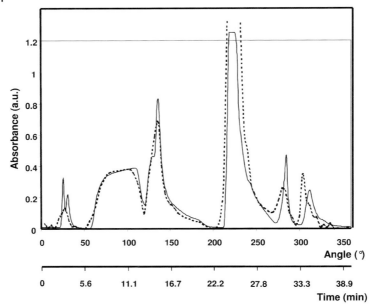

Fig. 9.5 Isolation of pDNA from bacterial (*Escherichia coli*) lysate. The separation was transferred between a 3-mL batch column and the 500-mL P-CAC using the loading factor concept. Shown is an overlay of the continuous annular chromatography separation (dashed line) and the separation on the small batch column (continuous line). (From Giovannini and Freitag [28]; reprinted with permission.)

batch separation (Fig. 9.6). Given the fact that displacement chromatography is inherently sensitive to even minor changes in the chromatographic conditions, this successful transfer can be considered a strong argument in favor of the loading factor concept for straightforward method transfer from the small (few milliliters) batch column to the 0.5-L annular column.

Further scale-up, e.g. by increasing the feed concentration, is limited by the viscosity difference between feed and eluent. As a rule of thumb, a viscosity factor of 2 is the limit before flow inhomogeneities and "viscous fingering" occur [18]. In continuous annular chromatography, scale-up is also possible via the annulus width (see above for a discussion of this factor on the bed stability), but this requires a suitable inlet design [32]. In addition, while it should be possible to approximate the minimum geometry (annulus radius and width, bed height) of an annular chromatograph for the separation of the components, the reaction constants must be known for the CIP and adsorbent regeneration steps in order to determine the limiting factor for the final optimized geometry [33].

Tab. 9.2 Plasmid DNA separation from clarified lysate; optimized conditions for a batch column (4 × 250 mm) and for the continuous annular chromatography system (15 cm outer diameter; 0.5 cm annulus width) [28]

Batch column	
feed injection	0–5.87 min
160 mM phosphate pH 6	13.58 min
10 mM phosphate pH 6	16.76 min
400 mM phosphate pH 6	22.63 min
10 mM phosphate pH 6	24.20 min
CIP (0.5 M NaOH)	27.16 min
10 mM phosphate pH 6	35.37 min
regeneration (400 mM phosphate pH 6)	36.53 min
10 mM phosphate pH 6	40.00 min
Continuous annular chromatography	
feed rate	15.0 mL min−1
first eluent (160 mM phosphate pH 6)	19.0 mL min−1
second eluent (400 mM phosphate pH 6)	15.0 mL min−1
CIP (0.5 M NaOH)	7.5 mL min−1
regeneration (400 mM phosphate pH 6)	3.0 mL min−1
main eluent (10 mM phosphate pH 6)	41.5 mL min−1
rotation rate	540° h−1
column flow	101.5 mL min−1
	4.46 cm min−1
	0.207 CV min−1
bed height	215 mm

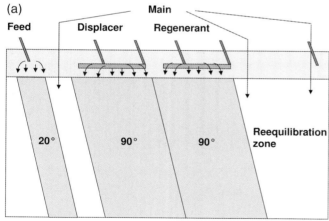

(a)

Feed   Displacer   Main   Regenerant

20°   90°   90°   Reequilibration zone

Fig. 9.6 Displacement separation of a commercial β-lactoglobulin mixture and soybean trypsin inhibitor. (a) Schematic presentation of the theoretical distribution of the various liquid zones in the P-CAC. (b) Batch displacement using a Macro-Prep Q 25 column (4 × 250 mm), feed: 5 mg mL^{-1} of each protein (volume 1 mL); carrier: 20 mM Tris–HCl pH 7.2; displacer: 100 mM GBS; flow rate: 0.3 mL min^{-1}. (c) Continuous displacement using a 500-mL annular column (5 × 126 mm) packed with Macro-Prep Q 25; feed flow rate: 2.6 mL min^{-1}; displacer flow rate: 12 mL min^{-1}; main buffer flow rate: 21 mL min^{-1}; regenerant (2 M NaCl) flow rate: 12 mL min^{-1}; column rotation rate: 360° h^{-1}; all other conditions as given for (b). (From Giovannini and Freitag [30]; reprinted with permission.)

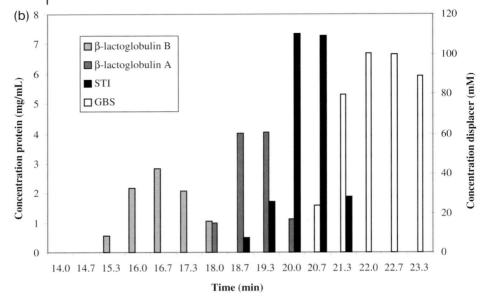

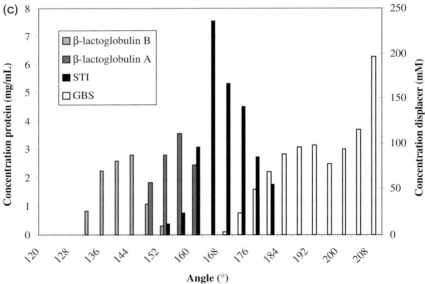

Fig. 9.6 (Continued)

9.6.3
Throughput and Productivity

It is common wisdom in chemical engineering that continuous operations have certain advantages over the corresponding batch operations, e.g. with regard to productivity, due to more consistent product quality and less "down time" between cycles. In the case of continuous annular chromatography it has been argued based on purely theoretical considerations, e.g. by Seidel-Morgenstern and coworkers [34, 35], that in terms of production rate and eluent consumption continuous annular chromatography should not differ from conventional (noncontinuous) chromatography. In this comparative evaluation a mathematical model was used in order to optimize these two parameters for a minimal purity of the collected fractions of 99% and minimum recovery of 95%.

From a more practical point of view, the group of Jungbauer in collaboration with the Austrian blood plasma fractionation company Octapharma has investigated productivity and buffer consumption for the separation of monomeric IgG from its aggregates in an immunoglobulin concentrate by batch and continuous annular size-exclusion chromatography using preparative-grade Superdex 200 to pack the annular column [36, 37]. Size-exclusion chromatography is a popular separation and, especially, polishing step in biochromatography, as it allows the efficient removal of aggregates, but also contributes to virus removal, for example. However, throughput is generally low in size-exclusion chromatography, as good separation can usually only be achieved at fairly low column loading (below 5%). If continuous annular chromatography has advantages over conventional batch column chromatography in terms of throughput, this should be especially important in the case of size-exclusion chromatography. Moreover, this chromatographic mode, which is not based on any form of interaction with the solid phase and only uses isocratic elution, should be especially suited to transfer to the continuous annular chromatography approach.

The IgG application was developed on an analytical column, and transferred to both a larger batch column and the continuous annular chromatography. Although this option is often cited as a principal advantage of continuous annular chromatography over conventional batch column chromatography, Jungbauer and coworkers were not able to include continuous CIP and regeneration steps in their continuous annular chromatography separation, as the entire available separation space (360°) was required to obtain an aggregate-free IgG monomer fraction. Cyclic regeneration was instead done in a discontinuous manner after each 5.5 h of continuous separation. The recovery rate was set at 85%. Under these conditions, the cycle time, which is an inverse measure of productivity, was similar in both cases during the first round of separation. In the subsequent cycles the productivity of the continuous annular chromatography was 2 times higher than that of the batch column. This is due to a simple fact. In batch column chromatography a new cycle typically begins only after the preceding one has been completely finished, i.e. all components have eluted from the column, and the column has been regenerated, cleaned (sanitized) and re-equilibrated. The cycle time is a constant for all cycles in this case. In annular chromatography all steps occur simultaneously and fresh feed may be applied to the top of the column before elution of all compounds from the previously applied feed has occurred. In fact, in this case only the time for a further application of feed contributes to the "cycle time", as the elution is performed simultaneously. From the second cycle onwards, annular size-exclusion chromatography therefore has an advantages in terms of productivity and consequently in terms of buffer consumption compared to conventional batch chromatography. Moreover, according to this investigation by Jungbauer and coworkers, for a given bed height the HETP was lower in the continuous annular chromatography compared to the batch column, an effect ascribed by the authors to a reduced extra column band spreading. The rotation speed of the continuous annular chromatography column, on the other hand, had no effect on the resolution between 200 and 900° h^{-1}.

The group of Jungbauer also investigated protein purification by continuous annular ion-exchange chromatography using a concentrate of clotting factor IX and vitronectin, a common impurity in commercial factor IX preparations, as a model system [38]. The weak anion-exchanger Toyopearl DEAE 650M was used as the solid phase. Batch column and continuous annular chromatography column separations were compared regarding enrichment, purity and productivity. Both the batch and the continuous annular chromatography process were optimized individually; in addition, the batch process as equivalent to the continuous annular chromatography process was taken into consideration. In spite of the fact that in this case continuous column regeneration could be established in the case of the continuous annular chromatography, the optimized batch process was found to be more productive than the continuous annular chromatography process (or the batch process equivalent to the continuous annular chromatography process). The reasons for this were the reduced recovery and the smaller concentration factor for factor IX in case of the continuous annular chromatography process.

9.6.4
Partial Recycling Schemes

Up to a point, the resolution can be increased in chromatographic separations by reducing the liquid-phase flow rate (higher efficiencies, lower plate heights as the van Deemter curve approaches the optimal flow rate). However, under such conditions the throughput is also reduced compared to separation performed at higher than optimal flow rate. As one solution to this dilemma, Kitakawa and coworkers [39] developed a partial recycling operation mode for continuous annular chromatography, which eventually allows the complete separation of a binary mixture, although the initial separation is done at suboptimal, i.e. "too high", flow rates causing some loss in resolution and therefore some zone overlap. The overall throughput of such a process is nevertheless higher than for an analogous process done at optimal resolution, i.e. the lower, but optimal flow rate. Fukumura and coworkers extended this method to the successful separation of a ternary amino acid mixture by step elution in combination with a partial recycling operation [40].

The principle of the partial recycling operation mode is illustrated in Fig. 9.7. A chromatographic system is used in which the eluent flow is not controlled by pressure, but by a multichannel peristaltic pump for variable eluent withdrawal [41]. Let us assume that a binary mixture, where component A is less retained by the adsorbent than component B, elutes partially separated with an overlap of X. Product streams containing pure A and B, respectively, are directly collected, while the eluate from the overlap zone is split into two fractions (Fig. 9.7a). In fraction A + B component A is in excess over component B and in fraction B + A component B is in excess over component A. Sample A + B is again injected into inlet nozzles placed after the initial feed inlet nozzle, while B + A is reinjected via an inlet nozzle placed before that of the initial feed nozzle. In this way, a complete separation of the binary mixture can be achieved as demonstrated for the case of a mixture of glutamic acid and valine (Fig. 9.7b).

Such a steady-state operation with partial recycling that leads to complete separation is possible in continuous annular chromatography, but has no equivalent in conventional batch column chromatography operation, which is intrinsically an

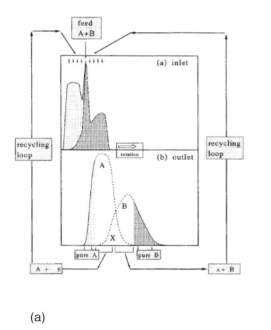

(a)

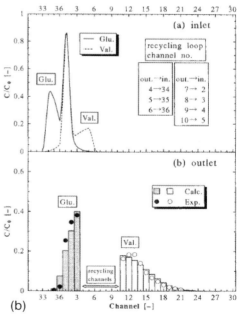

(b)

Fig. 9.7 (a) Schematic presentation of the concept of continuous annular chromatography with partial recycling. (b) Elution profiles of amino acids in continuous annular chromatography with partial recycling. (From Kitakawa and coworkers [39]; reprinted with permission.)

unsteady state process. The advantage of a partial recycling operation in continuous annular chromatography over either a batch operation performed at similar flow rate (no complete resolution) or at similar resolution (lower throughput) is obvious. However, it would be interesting to compare the performance of such a continuous annular chromatography systems with a continuous separation of a binary mixture by SMB [42], but also by the recently developed closed-loop recycling with periodic intra-profile injection (CLRPIPI [43]) method. CLRPIPI has been developed for the separation of binary mixtures. The process is repetitive and a steady-state is reached where highly purified fractions can be collected. However, contrary to SMB and continuous annular chromatography, CLRPIPI is not continuous.

9.7
Applications of Continuous Annular Chromatography Separation in Biotechnology

While the principle is viable, continuous annular chromatography separations have not entered the main stream of industrial bioseparation. In fact, while a number of potential applications have been identified, some of which are summarized in Tab. 9.1, as far as the authors know no industrial process exists at present that makes use of continuous annular chromatography for product isolation and purification. Some of the most advanced exemplary applications will be discussed below. Already mentioned was the process co-developed by the group of Jungbauer and the Austrian plasma fractionation company Octapharma for the purification of IgG monomers from an immunoglobulin concentrate by size-exclusion chromatography. Although in this case the continuous annular chromatography separation was found to be more efficient than the separation by batch column, continuous annular chromatography is at present not used routinely for steps in plasma fractionation, presumably due to problems with the long-term stability of the continuous operation. It remains to be seen whether the reliability and reproducibility of continuous annular chromatography columns and separations can be further improved in future.

From the viewpoint of bioprocess development, one of the most attractive possible applications of the continuous annular chromatography approach seems to be for the processing of large volumes of complex, low-titer product streams, e.g. in molecular or cellular biotechnology. Especially when interfaced to a continuous production process, continuous annular chromatography may considerably contribute to a reduction of the scale of the production facility. A continuous fermentation process is an efficient method for biomolecule production, the cell density of which can be significantly increased by running the process in the perfusion mode, resulting in very high space-time yields. It is therefore a logical consequence also to operate the downstream side as much as possible continuously in order to keep the residence time of the culture supernatant at a minimum and productivity at high level. Here, the integration of continuous annular chromatography into a continuous production process is certainly an option.

One industrial process, where this was investigated, is the production of a recombinant human blood coagulation Factor VIII (rhFVIII) by Bayer Biotechnology (Berkeley, CA) [12,44]. The blood factor is produced by perfusion culture (baby hamster kidney cells) and the first capture step after clarification is by ion-exchange chromatography. In the conventional process, this is done by batch-type chromatography. For this purpose, the continuously produced product stream has to be collected and stored at 4°C until a sufficient amount is available for the chromatographic step. This led to considerable product loss and the possibility of interfacing a continuous annular chromatography directly to the bioprocess was investigated. Since concentration is the main purpose of the capturing step, the standard continuous annular chromatography was modified in this case. The (low-titer) feed was introduced over a large section of the column using the main eluent port, while elution took place over a strongly focused small section of the column using the nozzle originally reserved for the feed introduction. Several stationary phases were investigated, including the Poros DE 2 material, SP Sepharose FF and Fractogel EMD DEAE 650 M. While in principle the unit performed well, the peak wobbling effect discussed above manifested itself and led to an undesirable increase in product dilution. The effect was more pronounced for the softer, more compressible solid phases such as Sepharose than for the mechanically more stable ones such as the Fractogel, where the most focused product zone was obtained [45].

Finally, the continuous annular chromatography approach has been used for separation of fructans (fructo-oligosaccharides and inulin) by Numico Research (Germany) [46]. The size-exclusion resin Toyopearl HW 40 (S) was used in this case (annulus dimensions: $10 \times 400\,mm$). The feed concentration was $100\,g\,L^{-1}$ and the system was operated at 60°C. Compared to the corresponding separations by conventional batch chromatography, the resolution of the continuous annular chromatography was lower. However, this is due to the fact that column length (bed height), which is very important in size-exclusion chromatography for the provision of a sufficient number of theoretical plates for the separation, with 0.6 m is fairly low in the commercial P-CAC system compared to typical bed heights in preparative size exclusion columns (above 2 m). In addition, viscous fingering tended to manifest itself and deteriorate the separation, especially at high feed concentration and high column rotation speeds.

9.8
Continuous Reactor/Separators

Many catalytic reactions in industry, including those catalyzed by enzymes, occur in porous media similar to those used in chromatography. At the same time, these reactions are typically reversible. Substrate and, especially, product concentrations may thus exert unfavorable inhibitory effects. The idea of combining such catalytic reactors with the direct separation of the product as it is formed thus presents itself as an attractive idea for process intensification. Another area where this principle of combining reaction with separation could putatively be of interest is that of inclusion body refolding. In such cases, the process could presumably profit further from transfer to the continuous mode, i.e. to continuous annular chromatography. A further extension of continuous annular chromatography is, hence, to use the system not only for separation, but also in combination with reaction, which may allow a significant improvement in process performance [47].

In recombinant protein technology, the overexpression of a protein in bacteria often results in protein aggregation (formation of so-called inclusion bodies). A major step in the downstream processing of such proteins is the refolding of the isolated inclusion bodies to the native state. As this is never 100% successful, the refolding step is typically followed by a separation of the refolded proteins from the residual inclusion bodies, normally by size-exclusion chromatography. A common refolding protocol includes the treatment of the inclusion bodies with chaotropic and reducing agents followed by a dilution into physiological conditions. During this treatment the protein refolds spontaneously into its native conformation. Conditions, success rates and kinetics of this reaction differ widely.

As mentioned previously, in the subsequent separation step, the native protein molecules are removed from still unfolded and aggregated protein molecules, but also from the low-molecular-weight agents added to aid refolding. Size-exclusion chromatography is obviously the method of choice for this separation (also due to the fact that mobile phases can be chosen over a wide range and exclusively in regard to the requirements of the protein). Moreover, in this case, the pronounced dilution of the feed, which is characteristic for size-exclusion chromatography, is of advantage, as refolding requires dilution in a suitable buffer. Refolding by (batch column) size-exclusion chromatography has thus become an interesting and powerful alternative to the batch dilution method used previously. Since the mechanism of refolding by size-exclusion chromatography includes more than a buffer exchange, the method has been called matrix-assisted refolding by size-exclusion chromatography.

The possibility of continuous size-exclusion refolding by continuous annular chromatography has successfully been demonstrated for a number of model proteins, including α-lactalbumin [48] and lysozyme [49]. High refolding yields and high native protein concentration at higher initial concentrations of denatured proteins than for batch dilution could be obtained. The recycling of remaining protein aggregates could be used to further increase the yield (see also Section 9.6.4). Since the components elute over a broad elution angle range, matrix-assisted refolding by continuous annular size-exclusion chromatography can effectively be performed under suboptimal flow conditions. Blocked or partially blocked outlet ports due to bed inhomogeneities average out when the fractions are pooled.

The basic concept of applying continuous reactive chromatography to enzyme-catalyzed reactions can be described as follows. It is assumed first that a reaction $A \leftrightarrow B + C$ is equilibrium limited and that the reaction takes place under diluted conditions in an inert solvent. Secondly, it is assumed that the stationary phase has a high affinity towards B, and a lower one towards A and C. Educt A is injected onto the continuously rotating annulus and the reaction takes place while both product species migrate through the annular bed at different velocities according to their affinity to the adsorbent. As a consequence of the continuous separation of the products B and C, the backward reaction is suppressed and the conversion from A to B and C is driven towards completion. Finally, pure B and pure C are eluted from the column at distinct outlets. The requirement for complete product separation is therefore an instantaneous and irreversible reaction. Furthermore, only isothermal conditions guarantee reproducibility.

Such a simultaneous biochemical reaction and separation in a rotating annular chromatograph was first described by Sarmidi and Barker [50] for sucrose inversion. Complete inversion of sucrose at feed concentrations of up to 50% (w/v) by invertase and subsequent separation of the products, i.e. glucose and fructose, could be achieved on a Dowex 50W-X4 calcium form ion-exchange resin at feed throughput of up to 15 kg sucrose m^{-3} resin h^{-1}. Figure 9.8(a) shows the elution profile of such a continuous annular reactive chromatography column in sucrose

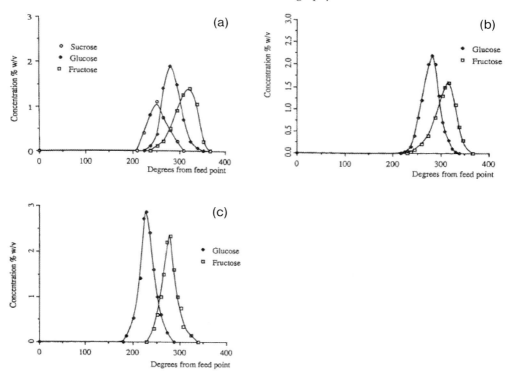

Fig. 9.8 Elution profile for continuous annular sucrose inversion and product separation. Conditions: (a) feed flow rate: 230 mL min^{-1}; feed concentration: 25% (w/v); eluent flow rate: 8 L min^{-1}; invertase activity: 60 U mL^{-1}; column rotation rate: 240° min^{-1}; (b) as (a) but invertase activity: 75 U mL^{-1}; (c) as (b) but eluent flow rate: 10 L min^{-1}. (From Sarmidi and Barker [50]; reprinted with permission.)

conversion. The feed flow rate was 230 mL min^{-1}, the feed concentration 25% (w/v), the eluent flow rate 8 L min^{-1} and the column rotation rate 240° h^{-1} in this application; 60 EU mL^{-1} invertase activity was used. Conversion is 82% under these conditions and, as Fig. 9.8(a) demonstrates, there is a significant overlapping of the elution zones of sucrose, glucose and fructose. By increasing the enzyme activity to 75 EU mL^{-1} (Fig. 9.8b), sucrose is completely converted, but the products, glucose and fructose, are still not fully separated. A much sharper elution profile can be achieved by increasing the eluent flow rate to 10 L min^{-1} (Fig. 9.8c). An additional step for separating the enzyme from the products is, however, necessary in this particular application. One can imagine avoiding this step by using, for example, a layer of beads containing the immobilized enzyme on top of the chromatographic bed – a concept that has been demonstrated by the group of Barth and coworkers [51]. Cocurrent reactive continuous annular chromatography processes with multimolecular reactions are also conceivable.

9.9
Conclusions

The principle of continuous annular chromatography has been known for several decades. Continuous annular chromatography is a continuous chromatographic mode that lends itself to the separation of multicomponent mixtures as well as of bicomponent ones. In continuous annular chromatography, the mobile and stationary phases move in a cross-current fashion, which allows transforming the typical one-dimensional batch column separation into a continuous two-dimensional one. The advances in continuous annular chromatography in recent years clearly demonstrate that this method performs for several separation *modi* at least as good as conventional batch chromatography. The tools for convenient method transfer from a batch to the continuous column or *vice versa* and for numerical profile prediction are available. The potential of continuous annular chromatography clearly lies in its continuous and steady-state character. With the exception of linear gradient elution, all chromatographic modes have at present been applied in continuous annular chromatography. Long-term studies will show whether the integration of continuous annular chromatography up- or downstream in a (continuous) process has economical advantages over conventional chromatography. The additional determining factor for the success of continuous annular chromatography will be the development of reliable continuous annular chromatography devices allowing a reproducible and truly continuous, long-term separation using chromatographic beds at industrial size.

References

1 Hilbrig, F., Freitag, R., *J Chromatogr B* **2003**, *790*, 1–15.

2 Martin, A. J. P., *Discuss Faraday Soc* **1949**, *7*, 332–336.

3 Svensson, H., Agrell, C.-E., Dehlen, S.-O., Hagdahl, L., *Sci Tools* **1955**, *2*, 17–21.

4 Fox, J. B., Calhoun, R. C., Eglinton, W. J., *J Chromatogr* **1969**, *43*, 48–54; J. B. Fox, *J Chromatogr* **1969**, *43*, 55–60.

5 Nicholas, R. A., Fox, J. B., *J Chromatogr* **1969**, *43*, 61–65.

6 Scott, C. D., Spence, R. D., Sisson, W. G., *J Chromatogr* **1976**, *126*, 381–400.

7 Canon, R. M., Sisson, W. G., *J Liquid Chromatogr* **1978**, *1*, 427–441.

8 Dunnill, P., Lilly, M. D., *Biotechnol Bioeng Symp* **1972**, *3*, 97–113.

9 Goto, M., Goto, S., *J Chem Eng Jpn* **1987**, *20*, 598–603.

10 Apostolidis, A., Lehmann, H., Schwotzer, G., Willsch, R., Prior, A., Wolfgang, J., Klimant, I., Wolfbeis, O. S., *J Chromatogr A* **2002**, *967*, 183–189.

11 Wolfgang, J., *Adv Biochem Eng/Biotechnol* **2002**, *76*, 233–255.

12 Vogel, J. H., Nguyen, H., Pritschet, M., van Wegen, R., Konstantinov, K., *Biotechnol Bioeng* **2002**, *80*, 559–568.

13 Wankat, P. C., *AIChE J* **1977**, *23*, 859–867.

14 Thiele, A., Falk, T., Tobiska, L., Seidel-Morgenstern, A., *Comput Chem Eng* **2001**, *25*, 1089–1101.

15 Howard, A. J., Carta, G., Byers, C. H., *Ind Eng Chem Res* **1988**, *27*, 1873–1882.

16 Bart, H. J., Messenböck, R. C., Byers, C. H., Prior, A., Wolfgang, J., *Chem Eng Proc* **1996**, *35*, 459–471.

17 Wolfgang, J., Prior, A., Bart, H. J., Messenböck, R. C., Byers, C. H., *Sep Sci Technol* **1997**, *32*, 71–82.

18 Byers, C. H., Sisson, W. G., DeCarli, J. P., Carta, G., *Biotechnol Prog* **1990**, *6*, 13–20.

19 Bloomingburg, G. F., Bauer, J. S., Carta, G., Byers, C. H., *Ind Eng Chem Res* **1991**, *30*, 1061–1067.

20 Carta, G., DeCarli, J. P., Byers, C. H., Sisson, W. G., *Chem Eng Comm* **1989**, *79*, 207–227.

21 Bloomingburg, G. F., Carta, G., *Chem Eng J* **1994**, *55*, B19–B27.

22 DeCarli, J. P., Carta, G., Byers, C. H., *AIChE J* **1990**, *36*, 1220–1228.

23 Sisson, W. G., Bergovich, J. M., Byers, C. H., Scott, C. D., *Prep Chromatogr* **1989**, *1*, 139–162.

24 Dalvie, S. K., Gajiwala, K. S., Baltus, R. E., *ACS Symp Ser* **1990**, *419*, 268–284.

25 Hoshimoto, K., Morishita, M., Adachi, S., Shindo, K., Shirai, Y., Tanigaki, M., *Prep Chromatogr* **1989**, *1*, 163–177.

26 Kitakawa, A., Yamanishi, Y., Yonemoto, T., Tadaki, T., *Sep Sci Technol* **1995**, *30*, 3089–3110.

27 Giddings, J. C., *Anal Chem* **1962**, *34*, 37–39.

28 Giovannini, R., Freitag, R., *Biotechnol Bioeng* **2002**, *77*, 445–454.

29 Knox, J. H., Laird, G. R., Raven, P. A., *J Chromatogr* **1976**, *122*, 129–145.

30 Giovannini, R., Freitag, R., *Biotechnol Prog* **2002**, *18*, 1324–1331.

31 Giovannini, R., Freitag, R., *Biotechnol Bioeng* **2001**, *73*, 522–529.

32 Bergovich, J. M., Sisson, W. G., *Resour Conserv* **1982**, *9*, 219–229.

33 Uretschläger, A., Jungbauer, A., *J Chromatogr A* **2000**, *890*, 53–59.

34 Seidel-Morgenstern, A., *Analusis* **1998**, *26*, M46–M55.

35 Heuer, C., Kniep, H., Falk, T., Seidel-Morgenstern, A., *Chem Ing Tech* **1997**, *69*, 1535–1546.

36 Iberer, G., Schwinn, H., Josic, D., Jungbauer, A., Buchacher, A., *J Chromatogr A* **2001**, *921*, 15–24.

37 Buchacher, A., Iberer, G., Jungbauer, A., Schwinn, H., Josic, D., *Biotechnol Prog* **2001**, *17*, 140–149.

38 Iberer, G., Schwinn, H., Josic, D., Jungbauer, A., Buchacher, A., *J Chromatogr A* **2002**, *972*, 115–129.

39 Kitakawa, A., Yamanishi, Y., Yonemoto, T., *Ind Eng Chem Res* **1997**, *36*, 3809–3814.

40 Fukumura, T., Bhandari, V. M., Kitakawa, A. Yonemoto, T. , *J Chem Eng Jpn* **2000**, *33*, 778–784.

41 Yonemoto, T., Kitakawa, A., Zheng, S. N., Tadaki, T., *Sep Sci Technol* **1993**, *28*, 2587–2605.

42 Nicoud, R. M., *LC GC Int* **1992**, *5*, 43–47.

43 Grill, C. M., *J Chromatogr A* **1998**, *796*, 101–113.

44 Vogel, J. H., Pritschet, M., Wolfgang, J., Wu, P., Konstantinov, K., In *Animal Cell Technology: From Target to Market* (E. Lindner-Olsson, N. Chatzissavidou, E. Luellau, Eds.), Kluwer, Dordrecht, **2001**, pp. 313–317.

45 Schmidt, S., Wu, P., Konstantinov, K., Kaiser, K., Kauling, J., Henzler, H.-J., Vogel, J. H., *Chem Ing Tech* **2003**, *75*, 302–305.

46 Finke, B., Stahl, B., Pritschet, M., Facius, D., Wolfgang, J., Boehm, G., *J Agric Food Chem* **2002**, *50*, 4743–4748.

47 Lode, F., Houmard, M., Migliorini, C., Mazzotti, M., Morbidelli, M., *Chem Eng Sci* **2001**, *56*, 269–291.

48 Schlegl, R., Iberer, G., Machold, C., Necina, R., Jungbauer, A., *J Chromatogr A* **2003**, *1009*, 119–132.

49 Lanckriet, H., Middelberg, A. P. J., *J Chromatogr A* **2004**, *1022*, 103–113.

50 Sarmidi, M. R., Barker, P. E., *Chem Eng Sci* **1993**, *48*, 2615–2623.

51 Herbsthofer, R., Bart, H. J., *Chem Eng Technol* **2003**, *26*, 874–879.

52 Takahashi, Y., Goto, S., *Sep Sci Technol* **1991**, *26*, 1–13.

53 Takahashi, Y., Goto, S., *J Chem Eng Jpn* **1991**, *24*, 121–123.

54 Takahashi, Y., Goto, S., *J Chem Eng Jpn* **1992**, *25*, 403–407.

55 Genest, P. W., Field, T. G., Vasudevan, P. T., Palekar, A. A., *Appl Biochem Biotechnol* **1998**, *73*, 215–230.

10
Principles of Membrane Separation Processes

Yusuf Chisti

10.1
Introduction

Semipermeable membranes that allow passage of only some of the components present in a liquid, gas or slurry held on one side of the membrane are used widely in the biotechnology industry [1–8]. The commonly used membrane separation operations include microfiltration, ultrafiltration, reverse osmosis (or nanofiltration), electrodialysis and pervaporation [8, 9]. In addition, membranes are used in various types of membrane bioreactors [4, 5, 10, 11], membrane chromatography [12, 13] and biomedical applications [14]. In bioreactor applications, the membrane confines or otherwise immobilizes a biocatalyst (e.g. enzymes, cells, tissue fragments) to enable its interaction with substrates dissolved in a fluid. This chapter discusses the principles of membrane separation processes as generally applicable to microfiltration and ultrafiltration in biotechnology applications.

10.2
Membrane Separation Processes

Microfiltration, ultrafiltration and reverse osmosis differ primarily in the dimensions of the particles or dissolved molecules that these processes are concerned with. Characteristic dimensions of some particles and solutes of relevance in biotechnology processes are shown in Tab. 10.1.

Microfiltration is used to recover particles in the typical size range of 0.1–20 μm from a solvent that may contain dissolved solutes (salts, proteins). Typical applications include recovery of animal cells, yeasts, bacteria and fungi from culture broth. Viruses are not removed by microfiltration.

Ultrafiltration is used to separate macromolecules such as proteins, starch and DNA from smaller solutes such as salts, sugars, amino acids, surfactants and water. Solutions of proteins and other dissolved polymers are commonly concentrated by ultrafiltration. Also, ultrafiltration is used in solvent- or buffer-exchange

Bioseparation and Bioprocessing. Edited by G. Subramanian
Copyright © 2007 WILEY-VCH Verlag GmbH & Co. KGaA, Weinheim
ISBN: 978-3-527-31585-7

Tab. 10.1 Characteristic dimensions of some particles and solutes

Particle	Diameter (nm)
Yeast and fungi	10^3-10^4
Human red blood cells	7000–8000
Bacteria	$300-10^3$
Viruses	20–200
Proteins and polysaccharides (10^4-10^6 Da)	2–10
Antibiotics (300–10^3 Da)	0.6–1.2
Mono- and disaccharides (200–400 Da)	0.8–1.0
Organic acids (100–500 Da)	0.4–0.8
Inorganic ions (10–100 Da)	0.2–0.4
Water (18 Da)	0.2

Based on Bailey and Ollis [15].

processes in which the solvent or buffer containing the dissolved macromolecule is replaced with a different solvent or buffer. Suitably validated ultrafiltration membranes can be used to remove viruses from a liquid or gaseous stream.

Reverse osmosis is used to separate small solute molecules such as sodium chloride from water. Reverse osmosis is commonly used in the biotechnology industry to produce high-purity water. Desalination of seawater by reverse osmosis is another well-established industrial process.

Sterile filtrations of water, nutrient media and gases are common operations in bioprocessing. Sterilizing-grade filters are generally microfilters that have been shown to meet certain particle removal criteria. Filter sterilization of liquids and gases requires filters that are nominally rated to remove particles down to 0.22 μm. Such filters are validated to produce sterile effluent when challenged with a suspension containing 10^7 cells cm^{-2} of the bacterium *Brevundimonas diminuta* ATCC 19146. *B. diminuta* is one of the smallest bacteria known. Its diameter ranges between 0.3 and 0.4 μm, and length tends to be between 0.6 and 1.0 μm.

Sterilizing-grade filters that are nominally rated to remove particles down to 0.1 μm are available. These filters are validated with the mycoplasma *Acholeplasma laidlawii* – an organism that is even smaller than *B. diminuta*.

10.3
Membrane Characteristics

Selecting a suitable membrane for a given application requires attention to a number of factors, including temperature and chemical compatibility with process materials, hydrophobicity, permeation and solute retention characteristics, fouling characteristics and cleanability, and overall costs. The first three of these factors are discussed in Sections 10.3.1–10.3.3. The remaining factors are taken up in other parts of this chapter.

10.3.1
Materials of Construction

Membrane materials must be compatible with the temperature, pH and chemicals encountered during processing. Most processes for recovering cells and proteins are conducted at 4–8 °C to prevent deterioration of thermally labile products and limit growth of contaminating microorganisms. Mostly physiological pH values of near pH 7 are encountered in bioprocessing. More severe processing conditions may occur during cleaning as discussed in Section 10.7.

Filtration membranes are most commonly made from polymers such as cellulose acetate, polyamide and polypropylene. Membranes made of ceramics are also available for microfiltration and ultrafiltration. Microfiltration membranes made of sintered stainless steels are available.

Polymer membranes do not generally withstand high temperatures. Their compatibility with chemicals depends on their chemistry. Some polymer membranes may be repeatedly sterilized at 121 °C.

Ceramic membranes are generally made of alumina, titania and zirconia. They tend to be fairly chemically inert, and resistant to solvents, acids and alkalis. These membranes can be autoclaved and used at temperatures as high as 350 °C. They withstand harsh cleaning regimens and high pressures. The high initial cost of ceramic membranes is often compensated by their long service life. The considerable weight of ceramic membranes can be a disadvantage. Flat-type ceramic membranes cannot be made very large and this limits their applicability; however, hollow fiber-type ceramic membrane modules can accommodate large filtration areas. Fluxes as high as $500–700 \, \text{L} \, \text{m}^{-2} \text{h}^{-1}$ can be attained with ceramic membranes, e.g. during microfiltration of skim milk [16].

Membranes used for filtering parentrals must comply with the US Pharmacopeia (USP) limits on acceptable levels of leached chemicals. Membrane materials that are generally able to comply with these limits include the following: polycarbonate, polyester, polyethylene, polypropylene, polysulfone, polytetrafluoroethylene (Teflon), polyvinylidene difluoride (Kynar), Viton (carbon black filled) and stainless steel (Types 304, 316 and 316 L).

10.3.2
Hydrophobic and Hydrophilic Membranes

Depending on the chemical nature of the polymer that a membrane is made of, it may be hydrophilic or hydrophobic. Polyethylene, polypropylene and polytetrafluoroethylene are examples of hydrophobic polymers. Hydrophilic polymers include nylon 6,6, polyethersulfone and cellulose acetate.

Hydrophilic membranes are easily wetted by a polar solvent such as water. Hydrophobic membranes are wetted by nonpolar solvents such as hexane. Aqueous liquids are filtered with hydrophilic membranes. Use of hydrophilic membrane ensures good wetting so that the entire surface area of the membrane is available for filtration and dry spots do not develop. Gases are generally filtered with hydrophobic polymer membranes. Ceramic and sintered metal membranes are generally easily wetted by polar solvents. Whether a membrane is hydrophilic or hydrophobic influences how readily it adsorbs foulants with large hydrophobic groups in their molecules. Hydrophobic membranes are particularly easily fouled by fatty acids, surfactants and antifoam agents.

10.3.3
Permeation and Retention

Microfiltration and ultrafiltration membranes are porous and retain particles by mechanisms similar to sieving. Porous membranes allow convective flow of solvent. Porous membranes may have a symmetric morphology in which the pore structure is uniform through the membrane. Porous structure of a symmetric polypropylene membrane is shown in Fig. 10.1.

Alternatively, a polymer membrane may have an asymmetric morphology in which the pore size and structure vary a great deal from one side of the membrane to the other side. Ultrafiltration membranes often have an asymmetric structure as shown for a hollow fiber in Fig. 10.2. The active surface of the membrane, i.e. the surface responsible for retaining the solute molecules, has extremely narrow pores. In Fig. 10.2, the active surface is on the lumen side of the hollow fiber. The pore structure widens towards the outside of the fiber. This asymmetric morphology minimizes the pressure required to force the solvent through the membrane.

Microfiltration membranes are characterized in terms of their nominal removal rating, e.g. a membrane may be rated to remove particles of 0.45 µm and larger. The nominal removal rating has no relationship to the actual pore size of the filter.

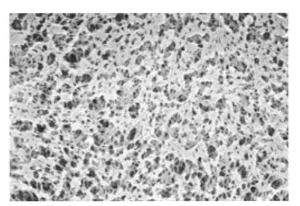

Fig. 10.1 Scanning electron micrograph of a polypropylene membrane magnified ×1000. Courtesy of Sterlitech Corp.

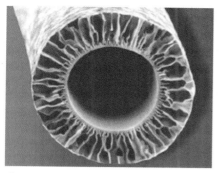

Fig. 10.2 Cross-section of a highly magnified hollow fiber. The active microporous ultrafiltration layer is on the lumen side. The outer macroporous layer provides support. Courtesy of Koch Membrane Systems, Inc.

Tab. 10.2 Molecular weights of some globular proteins

Protein	Molecular weight (Da)
Escherichia coli β-galactosidase	116 000
Bovine serum albumin	66 200
Bovine liver catalase	58 100
Chicken ovalbumin	45 000
Hen egg lysozyme	14 400

Nominal rating is established experimentally by challenging the filter with particles of defined sizes.

Ultrafiltration membranes are typically characterized in terms of their molecular weight cut-off (MWCO). A membrane that retains 90% of a globular protein with, for example, a molecular weight of 10 000 Da is said to have MWCO of 10 000. Ultrafiltration membranes range in MWCO from 1 to 500 kDa or a membrane pore size range from 0.001 to 0.1 μm. As membranes have a distribution of pore sizes, a membrane for retaining macromolecules of a given molecular weight is usually selected such that its MWCO is 50% of the molecular weight of the protein to be retained.

The hydrodynamic diameter of a globular protein molecule is roughly proportional to the molecular weight of the protein. How tightly a protein molecule is folded is, however, influenced by factors such as pH and ionic strength of the medium in which the protein is suspended. Consequently, the same protein may have somewhat different values of hydrodynamic diameter depending on the conditions in solution. Globular proteins typically range in diameter from 2 to 10 nm. MWCO based on sizes of globular proteins is not a useful guide for selecting membranes for retention of linear macromolecules such as DNA fragments. Molecular weights of some common globular proteins are shown in Tab. 10.2.

Reverse osmosis membranes may be porous or they may have a nonporous (or homogeneous) morphology. Solvent and solute transport through nonporous membranes by diffusion. Reverse osmosis membranes are characterized in terms of their ability to retain sodium chloride under specified conditions. Salt retention of membranes ranges from 95 to 99% and above.

10.3.3.1 Integrity Testing and Characterization of Membrane Pore Size

Sterilization-grade microfilters used for terminal filtration (i.e. filtration directly into final sterile containers) of parentrals must be integrity tested at the end of the filtration process to show that the filter is still intact. Filters are typically integrity tested by the bubble-point method. The bubble-point method is also the standard procedure for characterizing the pore size of microfiltration membranes.

The bubble point is the pressure required to just displace a liquid from the pores of a wetted membrane. The bubble point pressure (P_{BP}) is inversely related to the diameter of membrane pores, as follows:

$$P_{BP} = \frac{4\varphi\sigma\cos\theta}{d_p},\tag{1}$$

where φ is the shape correction factor for membrane pores, θ is the solid–liquid contact angle for the wetting fluid in contact with the membrane material, σ is the surface tension and d_p is the pore diameter. Any dilation of the membrane pores or other breakage in the membrane will reduce the value of the bubble point pressure (Eq. 1) compared to that of the intact membrane.

For measuring the bubble point pressure, the membrane must be first fully wetted so that all the pores are filled with the solvent. The wetting solvent used depends on whether the membrane is hydrophilic or hydrophobic. Water or an aqueous solution of alcohol (propanol) may be used to wet hydrophilic membranes. The wetted and drained membrane is then subjected to a gradually increasing gas pressure on the upstream side (feed side). The pressure at which the gas bubbles first emerge through the membrane is noted as the bubble point. The measured bubble point pressure must be within the range for an intact membrane. A low value of the measured bubble point indicates a defective membrane.

Different types of membranes have different bubble points. The relevant data are available from manufacturers. The bubble point pressure depends also on the wetting solvent used. Although the bubble point can be identified by visual inspection of gas flow, bubble point determination is increasingly fully automated. Test equipment such as that shown in Fig. 10.3 is used in automated determinations. The device shown applies gradually increasing gas pressure on the upstream side of the wetted membrane and measures the flow of gas through the membrane. The gas flow rate is plotted against the applied pressure (Fig. 10.4). The slope of the plot increases sharply at the instance of commencement of bubble flow. The pressure corresponding to the point of changed slope is the bubble point.

Fig. 10.3 Equipment for automated measurement of the
bubble point pressure. Courtesy of Millipore Corp.

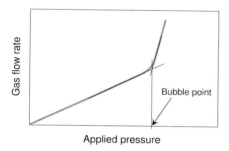

Applied pressure

Fig. 10.4 Gas flow versus applied pressure. Determination of the bubble point pressure.

10.4
Filtration Basics

The various process streams involved in membrane filtration are shown in Fig.
10.5 for filter units made of flat membranes and tubular membranes. The material
being filtered is known as the feed. The feed may be liquid or gas. Feed may
contain suspended particles, such microbial and animal cells, and dissolved solutes
that need to be separated from the fluid phase or the solvent. The part of the feed
that passes through the membrane is known as the permeate or filtrate. Permeate
generally contains only the solvent and dissolved solutes. The feed that has lost
some solvent in the permeate becomes concentrated and is known as the concentrate (or retentate). A membrane filtration module is schematically represented as
shown in Fig. 10.6, where F, P and R are the volume flow rates of feed, permeate
and retentate, respectively. In steady-state operation:

(a)

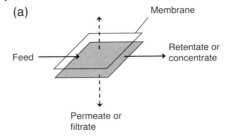

(b)

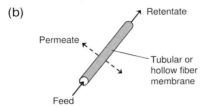

Fig. 10.5 Flat membrane (a) and tubular membrane (b) flow channels showing feed, retentate and permeate streams.

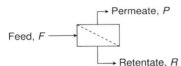

Fig. 10.6 Schematic representation of a membrane filtration process.

$$F = P + R. \tag{2}$$

Any solute that is not quantitatively retained by the membrane will appear in the permeate in addition to being present in the feed and the retentate.

Membrane filtrations can be carried out in cross-flow (or tangential-flow) and dead-end modes of operation (Fig. 10.7). In cross-flow operation the feed continuously flows parallel to the surface of the membrane and helps sweep the solute from the surface (Fig. 10.7a). This continuous sweeping of solute reduces the thickness of the solute layer on the membrane to ensure that a high rate of filtration can be maintained. The "cross-flow" terminology originates in the direction of permeate flow being perpendicular to the direction of flow of the feed (Fig. 10.7a).

In contrast to cross-flow, in dead-end filtration a solute layer of increasing thickness builds up with time on the surface of the membrane (Fig. 10.7b). The added resistance to flow offered by this layer slows the rate of filtration. Most large-volume industrial filtrations are conducted in the cross-flow mode. One exception is when filters are used to remove relatively low concentrations of microorganism and viruses from a feed in order to sterilize it.

(a)

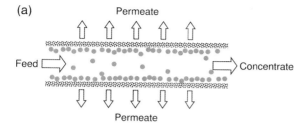

(b)

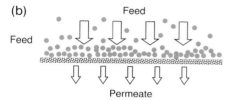

Fig. 10.7 Modes of filtration: (a) cross-flow and (b) dead-end.

Filter sterilization of gases and liquid media (see Section 10.2) is invariably carried out using dead-end filter cartridges that are commonly used only once. Dead-end sterile filtration is feasible because the fluid being filter sterilized generally has a low burden of particles. Disposable filter cartridges are simple to use, easily steam sterilized and can be reliably checked for presence of leaks.

10.4.1
Driving Forces for Flow

The permeate flow through the membrane is driven by the transmembrane pressure (ΔP_{TM}) or the average pressure difference between the feed and permeate sides of the membrane. In contrast, the flow of fluid within the feed channel is driven by the pressure difference between the inlet and outlet of the channel. These two distinct driving forces can have quite different values in a given membrane module. For example, in Fig. 10.8, the feed-side pressure drop and the transmembrane pressure are as follows:

$$\text{Feed side presure drop, } \Delta P_{Feed} = P_f - P_r = 2.0 - 1.5 = 0.5 \text{ bar} \qquad (3)$$

$$\text{Transmembrane pressure} = \frac{(P_f + P_r)}{2} - P_p = \frac{(2.0 + 1.5)}{2} - 0.1 = 1.65 \text{ bar} \qquad (4)$$

In the above equations, P_f, P_r and P_p are the pressures at the feed inlet, the retentate outlet and the permeate outlet, respectively.

Typical values of transmembrane pressures required for various types of filtrations are shown in Tab. 10.3. The required pressure depends mainly on the average

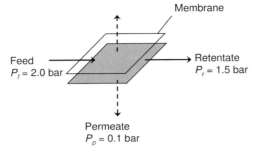

Fig. 10.8 Pressure difference driving forces for feed flow and permeate flow.

Tab. 10.3 Typically used values of transmembrane pressures

Operation	Transmembrane pressure (bar)
Microfiltration	<1
Ultrafiltration	1–10
Reverse osmosis	30–80

pore size of the filtration membrane. Microfiltration membranes with relatively large pores require the least transmembrane pressure to achieve reasonable permeate fluxes. Ultrafiltration membranes that retain large molecules such as proteins but not small solutes require significantly higher values of the transmembrane pressure to achieve permeation compared with microfiltration membranes. Reverse osmosis membranes have very fine pores or they may be of nonporous diffusive type. These membranes retain exceedingly small solutes such as sodium chloride and require high pressures to cause permeation. In addition, in reverse osmosis the transmembrane pressure must exceed the osmotic pressure difference across the membrane for permeation to occur.

10.4.2
Permeate Flux

Permeate flux J is the volume flow rate of the permeate through the membrane per unit area of membrane; thus:

$$J = P/A, \tag{5}$$

where P is the flow rate of the permeate and A is the area of the membrane. The permeate flux has the units of $m^3 m^{-2} s^{-1}$ (or $m\, s^{-1}$).

The average flux J_{av} in a batch filtration operation is calculated as follows:

$$J_{av} = \frac{V_p}{A \cdot t},$$

(6)

where V_p is the volume of the permeate collected in process time t. In batch filtration, the flux declines with time. The following empirical relationship is sometimes used to estimate the average flux from values of the initial ($J_{initial}$) or starting flux and the flux at the end (J_{final}) of the process:

$$J_{av} = J_{final} + \frac{2}{3}(J_{initial} - J_{final}).$$

(7)

Permeate flux through a membrane depends on the following main factors: porosity of the membrane (or the fraction of the total area that is occupied by pores), membrane thickness, applied transmembrane pressure, type and concentration of suspended solids, viscosity of the fluid being filtered, temperature of operation, geometry of the filtration device, and flow regime of operation.

Permeate flux declines with increasing viscosity, reduced porosity and reduced transmembrane pressure. An elevated concentration of solids, such as microbial cells, and solutes, such as proteins, will reduce permeate flux. Increasing temperature increases flux because of the effect of temperature on viscosity of liquids. Higher values of flux are attained when the flow in the membrane channel is turbulent. This requires a high rate of cross-flow.

10.4.3
Volume Concentration Factor

As solvent is removed in the permeate, the retentate becomes more concentrated. The extent of concentration is defined by the volume concentration factor VCF calculated as follows:

$$VCF = F/R,$$

(8)

where F is the volume flow rate of feed and R is the volume flow rate of retentate in steady-state filtration. In batch operation, F and R are the volumes of feed and retentate processed during filtration.

10.4.4
Solute Rejection

A semipermeable membrane allows the solvent to pass through freely but, depending on the pore size, retains some or all of the solutes. As the membrane pores have a distribution of sizes, the solutes that are larger than the average pore size of the membrane are often not retained quantitatively, that is a fraction of the solute passes through the membrane.

Ability of the membrane to prevent passage of a given solute is quantified in terms of solute rejection coefficient, SRC, defined as the fraction of the feed solute that is prevented from passing into permeate; thus:

$$SRC = \frac{C_f - C_p}{C_f} = 1 - \frac{C_p}{C_f}, \tag{9}$$

where C_f is the solute concentration in the feed and C_p is the concentration of the solute in the permeate.

If the solute is completely prevented from passing through the membrane, i.e. is totally rejected by the membrane, C_p is zero and SRC is unity. Although a value of $SRC=1$ is desired, in many cases SRC tends to be less than unity. For a given membrane, SRC is not necessarily a constant. SRC is influenced by factors such as the transmembrane pressure, solute concentration in the feed, temperature, pH, presence of other solutes and membrane characteristics (e.g. porosity, chemistry). The solute rejection coefficient is also known as solute reflection coefficient.

The ratio C_p/C_f is known as the solute permeability (PR) of the membrane. From Eq. (9), for a given solute and membrane, we can deduce the following relationship between solute rejection coefficient and solute permeability:

$$SRC + PR = 1. \tag{10}$$

Example Problem

A three-stage steady-state ultrafiltration process is used to concentrate tissue plasminogen activator (tPA) from a dilute solution. For the specified flow rates of feed (F) and permeate (P), calculate the various unknowns in the process flow diagram shown in Fig. 10.9. Assume that tPA is retained quantitatively by the membrane. What is the volume concentration factor for the entire system?

Answer: $R_1=56\,\mathrm{m^3\,h^{-1}}$; $P_1=44\,\mathrm{m^3\,h^{-1}}$; $R_2=51\,\mathrm{m^3\,h^{-1}}$; $VCF_2=1.10$; $P_3=25\,\mathrm{m^3\,h^{-1}}$; $R_3=26\,\mathrm{m^3\,h^{-1}}$; $VCF_3=1.98$; $VCF_{system}=3.85$.

If the feed contained tPA at a concentration of $0.5\,\mathrm{kg\,m^{-3}}$ and tPA was retained quantitatively in the ultrafiltration stages 1 and 2, calculate the tPA concentration in the retentate stream of the ultrafiltration stage 3, if the SRC of stage 3 is 0.98.

Answer: $1.9\,\mathrm{kg\,m^{-3}}$.

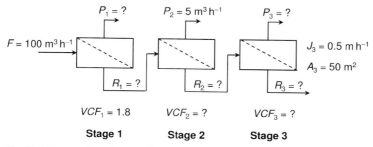

$$P_1 = ? \qquad P_2 = 5 \text{ m}^3\text{ h}^{-1} \qquad P_3 = ?$$

$$F = 100 \text{ m}^3\text{ h}^{-1}$$

$$J_3 = 0.5 \text{ m h}^{-1}$$
$$A_3 = 50 \text{ m}^2$$

$$R_1 = ? \qquad R_2 = ? \qquad R_3 = ?$$

$$VCF_1 = 1.8 \qquad VCF_2 = ? \qquad VCF_3 = ?$$

Stage 1   **Stage 2**   **Stage 3**

Fig. 10.9 Process flow diagram for the example problem. R is the retentate flow rate, J is the permeate flux, A is the membrane area and VCF is the volume concentration factor. The subscript n is the stage number.

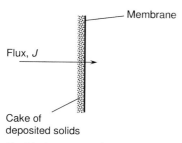

Membrane

Flux, J

Cake of deposited solids

Fig. 10.10 Permeate flow through the deposited cake and the membrane in series.

10.4.5
Resistances to Flow Through the Membrane

The permeate flux J is driven by transmembrane pressure and is directly proportional to it. During filtration the permeate must pass through a layer of the solids deposited on the surface of the membrane and then the membrane itself (Fig. 10.10). Both the solids layer and the membrane resist the flow of permeate. Similarly, viscosity of the permeate impedes flow through the porous solids cake and the membrane. The permeate flux is related to the driving force and various resistances to flow, as follows:

$$J = \frac{\Delta P_{\text{TM}}}{\mu(R_{\text{m}} + R_{\text{c}})}, \tag{11}$$

where μ is the viscosity of pure solvent. Notice that μ is not the viscosity of the feed slurry, as only the solvent permeates through the cake of deposited solids and the membrane. R_{m} and R_{c} are the absolute resistances of the membrane and the layer of deposited solids, respectively. With continuing use, a layer of fouling material accumulates on the membrane and this offers its own resistance R_f to permeate flow. Considering this additional resistance modifies Eq. (11) to the following:

$$J = \frac{\Delta P_{TM}}{\mu(R_m + R_c + R_f)}.$$ (12)

The resistance of the solids cake increases with thickness of the cake. The specific resistance (R_{sp}) of the cake, i.e. its resistance per unit thickness, is generally a constant that depends on the morphology of the solid particles. The specific resistance and the absolute resistance are related, as follows:

$$R_{sp} = R_c / \Delta x,$$ (13)

where Δx is the thickness of the cake. The specific resistance may increase with increasing pressure if the solids are compressible or pack closer together with increasing pressure.

The membrane resistance R_m can be estimated by measuring the steady-state flux of a pure solvent (no solute or suspended solids present) through the membrane at various values of the transmembrane pressure. Estimation of the cake resistance R_c requires steady-state filtration of the slurry with a membrane of known resistance R_m. The measured flux, transmembrane pressure, viscosity of the solvent and R_m are then used in Eq. (11) to calculate the cake resistance.

10.4.6
Flux versus Transmembrane Pressure

If a membrane is used to filter a pure solvent without any suspended particles or dissolved solutes present, the permeate flux increases linearly with increasing transmembrane pressure as shown in Fig. 10.11. If, however, particles or dissolved solute are present, the steady-state permeate flux increases with increasing transmembrane pressure only up to a certain point and then becomes insensitive to further increases in transmembrane pressure (Fig. 10.11). The maximum attainable value of the flux is known as the limiting flux. Existence of a limiting flux during filtration of solutes and particles is explained by a phenomenon known as concentration polarization or accumulation of solute adjacent to the membrane on the feed side (Fig. 10.12). This layer of elevated solute concentration is known as the gel layer or concentration polarization layer. While the transmembrane pressure attempts to force the solvent and dissolved solute towards the membrane, the high concentration of solute adjacent to the membrane produces a mass transfer-driven back diffusion of solute. This back diffusion counteracts the pressure driven flow towards the membrane. At some point, therefore, further increases in transmembrane pressure fail to increase permeate flow.

While limiting flux does not vary with increasing transmembrane pressure, it can be influenced by factors that reduce concentration polarization. Thus, increasing

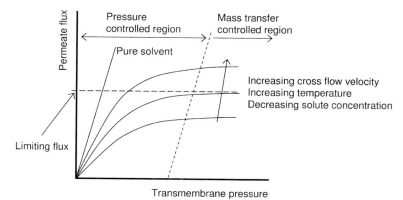

Fig. 10.11 Permeate flux versus transmembrane pressure. The cases for pure solvent and solute-containing feeds are shown.

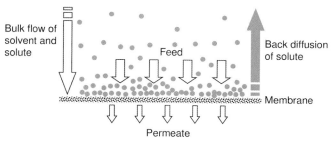

Fig. 10.12 Concentration polarization. Bulk flow of the solvent towards the membrane on the feed side carries dissolved solute (and suspended solids) to the membrane. The solute cannot go past the membrane and, therefore, its concentration in the vicinity of the membrane eventually exceeds the concentration in the bulk fluid away from the membrane. This concentration gradient causes a mass transfer-driven back diffusion of solute from the vicinity of the membrane to the bulk fluid. This back diffusion in turn counteracts the bulk transport of solute and solvent towards the membrane.

the cross-flow velocity at the surface of the membrane increases limiting flux. Similarly, a reduced concentration of solute or particles in the feed increases flux. Increasing temperature increases flux because viscosity declines with increasing temperature. In Fig. 10.11, the region in which the flux is not affected by transmembrane pressure, but is affected by factors that influence concentration polarization (i.e. affect the value of solute mass transfer coefficient at the surface of the membrane), is known as the mass transfer controlled region of filtration.

The flux-reducing effect of concentration polarization is especially pronounced in microfiltration and ultrafiltration processes. Concentration polarization has been further discussed by Sablani and coworkers [17]. Concentration polarization cannot be prevented, but its severity can be reduced in some of the following ways:

(i) Increasing the cross-flow rate on the feed side to reduce the thickness of the solute gel layer.

(ii) Installing turbulence promoters such as wire screens and other internals in the feed flow channel in order to disturb the gel layer.

(iii) Using pulsating flow of feed to disturb the gel layer.

(iv) Using elevated temperature to increase diffusivity of solute and reduce viscosity of the solvent.

(v) Using dimpled or corrugated membrane to create eddies that disturb the gel layer.

(vi) Scouring of the gel layer through the action of a small concentration of dense coarse inert solid particles suspended in the feed.

10.4.7
Relationship between Concentration Polarization and Mass Transfer of Solute in the Gel Layer

Concentration polarization is related to the mass transfer coefficient of solute at the surface of the membrane. All those factors that enhance the mass transfer coefficient reduce concentration polarization and enhance permeate flux. A quantitative relationship between concentration polarization and mass transfer coefficient can be derived from a steady-state mass balance of solute at the surface of the membrane.

The steady-state solute concentration profile in the vicinity of the membrane is shown in Fig. 10.13. The solute concentration at the surface of the membrane on the feed side, i.e. C_m, is higher than the bulk solute concentration in the feed, i.e. C_b). The solute flux towards the membrane on the feed side is simply the solvent flux J multiplied by the solute concentration C. On the permeate side, the solute flux away from the membrane is JC_p, where C_p is the solute concentration in the

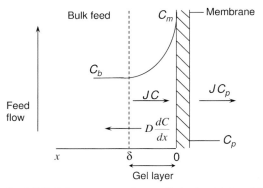

Fig. 10.13 Solute concentration profile in the vicinity of the membrane during steady-state filtration.

permeate. As the concentration C_m is greater than the bulk solute concentration C_b, mass transfer driven back diffusion of solute occurs from the surface of the membrane back into the bulk feed. The rate of this back diffusion is $D(dC/dx)$, where D is the molecular diffusivity of the solute. The concentration polarization layer, or the gel layer, of thickness δ is defined as that region adjacent to the membrane where the solute concentration is appreciably greater than the solute concentration in the bulk feed.

The steady-state solute balance at the membrane is as follows:

Solute transport rate towards membrane = rate of back diffusion + solute removal rate in permeate,

or:

$$\underbrace{JC}_{\substack{\text{Solute transport} \\ \text{to membrane}}} = \underbrace{-D\frac{dC}{dx}}_{\substack{\text{back diffusion} \\ \text{of solute}}} + \underbrace{JC_p}_{\substack{\text{solute removal} \\ \text{in permeate}}}. \tag{14}$$

The above equation can be rearranged to the following:

$$\frac{dC}{C_p - C} = \left(\frac{J}{D}\right)dx. \tag{15}$$

At steady-state, the flux J is constant and Eq. (15) can be integrated over the thickness of the gel layer, thus:

$$\int_{C_m}^{C_b} \frac{dC}{(C_p - C)} = \left(\frac{J}{D}\right)\int_0^\delta dx \tag{16}$$

or:

$$\frac{C_m - C_p}{C_b - C_p} = \exp\left(\frac{J\delta}{D}\right). \tag{17}$$

where C_m is the solute concentration at membrane, C_b is the solute concentration in the bulk feed and δ is the thickness of the gel layer.

By definition [18], the mass transfer coefficient k is:

$$k = \frac{D}{\delta}. \tag{18}$$

Therefore, Eq. (17) can be written in terms of k, as follows:

$$\frac{C_m - C_p}{C_b - C_p} = \exp\left(\frac{J}{k}\right). \tag{19}$$

Usually, there is no solute in the permeate, i.e., $C_p = 0$, and Eq. (19) becomes:

$$\frac{C_m}{C_b} = \exp\left(\frac{J}{k}\right).$$ (20)

The ratio c_m/c_b is also known as the polarization modulus.

In view of the definition of the solute rejection coefficient (Eq. 9), Eq. (20) can be written in terms of SRC, as follows:

$$\frac{C_m}{C_b} = 1 - SRC + SRC \cdot \exp\left(\frac{J}{k}\right).$$ (21)

A rearrangement of Eq. (20) produces the following equation:

$$J = k \ln\left(\frac{C_m}{C_b}\right).$$ (22)

Thus, the steady-state flux declines with increasing solute concentration in the bulk feed. Furthermore, for otherwise fixed conditions, the value of the steady-state flux can be increased by affecting the value of the mass transfer coefficient k.

10.4.7.1 Mass Transfer Coefficient

The mass transfer coefficient k depends on the Reynolds number in the feed channel, the geometry of the channel and the properties of the feed. This dependence is generally represented in terms of dimensionless Sherwood number (Sh), Reynolds number (Re) and Schmidt number (Sc), as follows:

$$\underbrace{\frac{kd_h}{D}}_{\substack{\text{Sherwood} \\ \text{number, Sh}}} = \alpha \underbrace{\left(\frac{\rho d_h U_L}{\mu_L}\right)^b}_{\substack{\text{Reynolds} \\ \text{number, Re}}} \underbrace{\left(\frac{\mu_L}{\rho D}\right)^{1/3}}_{\substack{\text{Schmidt} \\ \text{number, Sc}}},$$ (23)

where d_h is the hydraulic diameter of the flow channel, D is the diffusivity of the solute, ρ is the density of the bulk feed, μ_L is the viscosity of the bulk feed and U_L is the velocity of feed flow. The hydraulic diameter of any flow channel is calculated using the following equation:

$$d_h = \frac{4 \times \text{cross-sectional area for flow}}{\text{wetted perimeter of flow channel}}.$$ (24)

During ultrafiltration b (Eq. 23) is 0.5 in laminar flow and about 1.0 in turbulent flow. In cross-flow microfiltration of particles and cells, $b = 0.8$ in laminar flow, but increases to about 1.3 in turbulent flow [19]. The α value is 0.023 in turbulent flow. Other expressions for Sherwood number are:

$$\text{Sh} = 1.62\left(\text{Re} \cdot \text{Sc} \frac{d_h}{L}\right)^{0.33}$$ (25)

$$Sh = 1.86\left(Re \cdot Sc \frac{d_h}{L}\right)^{0.33} \tag{26}$$

$$Sh = 0.023\, Re^{0.875} Sc^{0.25}. \tag{27}$$

Equations (25) and (26) are for laminar flow in tubes and channels, respectively; Eq. (27) is for turbulent flow.

For fully developed laminar flow in ultrafiltration k may be related to shear rate γ by the Porter equation:

$$k = 0.816\gamma^{0.33} D^{0.67} L^{-0.33}, \tag{28}$$

where L is the length of the flow channel and the shear rate depends on the channel geometry:

$$\gamma = \frac{8U_L}{d} \quad \text{for tubes with diameter } d$$

$$\gamma = \frac{6U_L}{h} \quad \text{for rectangular channels of height } h.$$

The cross-flow velocity is the principal operating variable for enhancing the performance of a given filtration membrane module. The optimal cross-flow velocity depends on the product being filtered and the configuration of the filtration module. For tubular microfiltration membranes with around 5.5 mm inner diameter, the optimal cross-flow velocity is about 2.5–5 m s^{-1} [20].

All methods that have been pointed out previously (Section 10.4.6) for reducing concentration polarization increase the mass transfer coefficient in membrane processes. Another method of mass transfer enhancement is the use of dynamic filtration systems with rotating membranes or agitators placed in close proximity to the membrane [21–23].

In both ultrafiltration and microfiltration the mass transfer coefficient tends to be quite small because of the small diffusivities of the cells and macromolecules [18]. Concentration polarization reduces the permeate flux in microfiltration to only about 5% of the pure water flux. Unlike microfiltration and ultrafiltration, pervaporation processes use nonporous homogeneous membranes. Typically, the solute flux is low and the mass transfer coefficient k is relatively large in view of the higher diffusivities of small solutes such as ethanol. For further information on mass transfer in membrane processes, see Refs. [2, 24–26].

Sensitive microorganism and cells of higher eukaryotes may be affected by the intensity of hydrodynamic shear forces in membrane channels [27, 28]. Shear rates in centrifugal pumps that are sometimes used for circulating the feed, can be particularly harmful [27].

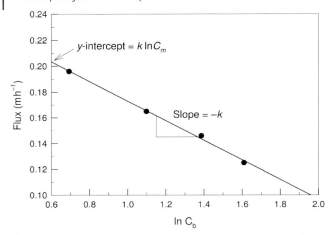

Fig. 10.14 Steady-state permeate flux versus concentration (C_b) of solute in the feed.

10.4.7.2 Experimental Estimation of Solute Concentration in the Gel Layer

The solute concentration C_m in the gel layer at the surface of the membrane can be determined by measuring the steady-state permeate flux at different concentrations of the solute in the feed during continuous flow filtration. Care must be taken that the range of solute concentration used does not affect the properties of the feed, and the same values of cross-flow rate and transmembrane pressure are used in filtering the different feed compositions. The measured flux values are then plotted against the natural log of feed solute concentrations to obtain a straight line (Fig. 10.14). The slope of this line is used to calculate the mass transfer coefficient k; thus:

$$k = -\text{slope}. \tag{29}$$

The y-intercept of the line (Fig. 10.14) and the already calculated k-value are used to determine the solute concentration in the gel layer:

$$C_m = \exp\left(\frac{y\text{-intercept}}{k}\right). \tag{30}$$

10.5
Membrane Modules

Typically, membranes are folded or otherwise packed in housings or modules of various configurations to provide a large filtration area in a compact volume. Commonly used membrane module types include the plate-and-frame, spiral-wound, tubular and hollow fiber.

10.5.1
Plate-and-Frame Configuration

Membranes are often deployed as flat sheets in a plate-and-frame housing configuration (Fig. 10.15) in which the height of the flow channel is 0.5–1.5 mm [20]. Feed flows in alternate parallel channels while the remaining channels are occupied by permeate (Fig. 10.5a). A common header supplies the feed to the various channels and permeate from all channels is collected in a separate collection pipe. Many membrane sheets are generally sealed in a cartridge (Fig. 10.15). Membrane area is easily increased by assembling multiple cartridges with common systems for distribution of feed and collection of permeate and retentate (Fig. 10.16).

10.5.2
Spiral-wound Membranes

A spiral-wound configuration (Fig. 10.17) of what are otherwise flat membranes, greatly increases the membrane area that can be packed in a given volume. Spiral-wound membranes are used in microfiltration, ultrafiltration and reverse osmosis. A single membrane module may exceed 0.45 m in diameter and 1.5 m in length. The average flux for spiral-wound ultrafiltration membranes ranges between 0.01 and 0.05 $m^3 m^{-2} h^{-1}$ in applications such as concentration of skim milk and whey protein [16]. Use of microfiltration technology in the dairy industry is further discussed by Saboya and Maubois [29].

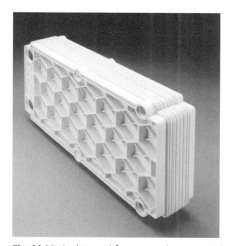

Fig. 10.15 A plate-and-frame membrane cartridge. Courtesy of Millipore Corp.

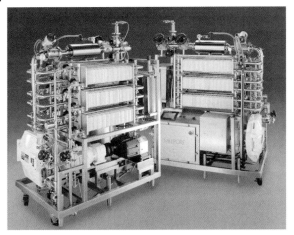

Fig. 10.16 Multiple plate-and-frame membrane cartridges assembled to produce a filtration unit. Courtesy of Millipore Corp.

Fig. 10.17 A spiral-wound membrane cartridge. Courtesy of TriSep Corp.

10.5.3
Tubular Membranes

Tubular membrane modules consist of porous tubes that are typically 5–15 mm in diameter. The filtration membrane is cast on the inside walls of these supporting tubes. Feed flows inside the tubes and permeate is collected in an outer shell that surrounds a bundle of many tubes. Velocity of flow inside tubes tends to be as high as $6\,\mathrm{m\,s^{-1}}$, to ensure highly turbulent flow and minimize concentration polarization. An average flux of $0.15\,\mathrm{m^3\,m^{-2}\,h^{-1}}$ has been reported for tubular ultrafiltration membranes during clarification of apple juice [16]. Tubular membranes can handle fluids with higher particle loadings than can spiral-wound membranes [16]. Due to a relatively large diameter of tubes, the membrane surface area that can be packed in a single module is small. Consequently, tubular modules are relatively expensive per unit membrane area.

10.5.4
Hollow Fiber Bundles

Hollow fibers used in processing liquids range in outer diameter from 200 to 1250 µm. The diameter of the flow channel is of the order of 20 µm. The membranes are asymmetric with the pore size changing dramatically from the inside of the membrane to the outside (Fig. 10.2). The active layer of finer pores may be located on the inside or outside of the fiber. The macroporous zone serves to support the active zone. Large pores in the support layer minimize the transmembrane pressure required to achieve permeate flow. Fibers are packed into modules by the millions to provide sufficient membrane area (Fig. 10.18).

As the fibers are self-supporting, they cannot withstand high transmembrane pressures. Typically used transmembrane pressure values range from 2 to 4 bar [16]. Hollow fibers are susceptible to plugging and require the feed to be prefiltered through at least a 100-µm filter. Hollow fibers are used mostly in ultrafiltration of clean streams.

A type of hollow fiber known as hollow fine fiber has been developed for use in gas separation and pervaporation. These have outer diameters in the range of 25–250 µm. The wall thickness ranges from 5 to 50 µm. Hollow fine fibers can withstand transmembrane pressures up to 70 bar [16]. Hollow fiber membrane filters are discussed further by Gabelman and Hwang [30].

Fig. 10.18 Ultrafiltration systems with hollow fibers housed within cylindrical shells. Courtesy of GE Water & Process Technologies.

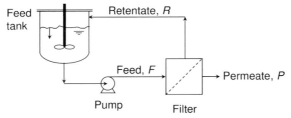

Fig. 10.19 Typical flow arrangement for batch cross-flow microfiltration and ultrafiltration.

10.6
Operation of Membrane Filtration Systems

10.6.1
Batch Filtration

In batch filtration, a given volume of feed is recycled through the filter module back to the feed tank (Fig. 10.19). As the permeate is withdrawn, the concentration of solute in the feed increases with filtration time. The permeate flux declines with time because of increasingly severe concentration polarization as a result of increasing concentration of the solute in the feed. Often, the viscosity of feed increases with time as it becomes more concentrated. Batch filtration is commonly used for solvent recovery and solute concentration when the volume to be pro-cessed is relatively small. Batch operation is not suitable for processing large volumes, as the continuously declining flux results in extended filtration times.

During filtration, the various flow rates are always related as shown in Eq. (2). The rate of decrease in feed volume (V_f) in the tank equals the rate of removal of permeate, thus:

$$-\frac{dV_f}{dt} = P. \tag{31}$$

If the initial feed volume is V_{f0} and the volume at any time t is V_f, Eq. (31) can be integrated, thus:

$$-\int_{V_{f0}}^{V_f} dV_f = \int_0^t P dt, \tag{32}$$

or:

$$V_{f0} - V_f = \int_0^t P dt. \tag{33}$$

As the permeate flow rate and the flux are related by Eq. (5), Eq. (33) can be written as follows:

$$V_f = V_{f0} - \int_0^t JA dt. \tag{34}$$

If no solute is lost in the permeate, a solute mass balance leads to the following equation:

$$\underbrace{C_f V_f}_{\substack{\text{Total solute} \\ \text{at time } t}} = \underbrace{C_{f0} V_{f0}}_{\substack{\text{total solute} \\ \text{at time } 0}}, \tag{35}$$

and, therefore, the solute concentration in the feed at any time can be calculated using:

$$C_f = \frac{C_{f0} V_{f0}}{V_f}. \tag{36}$$

10.6.2
Continuous Flow Filtration

In continuous flow filtration that is normally operated at a steady-state, the feed flows over the membranes on a once-through basis (Fig. 10.20). The cross-flow rate of the feed and its composition remain constant. The retentate and permeate are withdrawn at fixed rates. The permeate flux and the composition of the various streams remains unchanged over long periods. This type of operation is suitable for handling large volumes of feed. As the cross-flow rate must be high, the feed tank would quickly run dry if only a small volume was available for processing.

As the feed is drawn out, the volume of the feed in the tank declines with time, as follows:

$$-\frac{dV_f}{dt} = F. \tag{37}$$

If the initial feed volume is V_{f0} and the volume at any time t is V_f, Eq. (37) can be integrated:

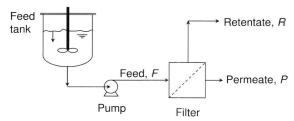

Fig. 10.20 Typical flow arrangement for steady-state cross-flow filtration.

$$-\int_{V_{f0}}^{V_f} \mathrm{d}V_f = F\int_0^t \mathrm{d}t, \tag{38}$$

or:

$$V_{f0} - V_f = Ft, \tag{39}$$

Therefore the volume remaining at any time is calculated using:

$$V_f = V_{f0} - Ft. \tag{40}$$

At steady-state the solute entering the filter module must equal the solute leaving the module. Therefore, a solute mass balance can be written as follows:

$$C_f F = C_r R + C_p P, \tag{41}$$

where C_f, C_r and C_p are the mass concentrations of the solute in the feed, retentate and permeate, respectively. When the solute is retained fully, $C_p=0$ and Eq. (41) can be rearranged to obtain the solute concentration in the retentate, thus:

$$C_r = C_f\left(\frac{F}{R}\right). \tag{42}$$

Multiple filter modules are used in relatively high-volume continuous flow steady-state processes that require large membrane areas. Multiple modules may be arranged in parallel or in series (Fig. 10.21). In the latter arrangement, the number of identical filter modules used decreases progressively downstream because the total feed flow declines as permeate is lost in each upstream module. In the in-series arrangement, the feed pressure is sometimes boosted by installing pumps between upstream and downstream modules.

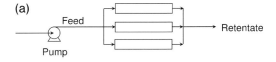

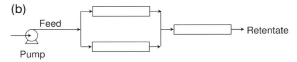

Fig. 10.21 Increasing membrane area by using multiple filter modules in parallel (a) and in series (b).

10.6.3
Continuous Feed and Bleed Recycle

In continuous flow filtration schemes such as that shown in Fig. 10.20, the cross-flow rate on the surface of the membrane cannot be varied independently of the feed flow rate. This is a significant limitation because the high feed flow rate required to minimize concentration polarization will quickly empty the feed reservoir.

A continuous feed and bleed recycle scheme (Fig. 10.22) has been developed to get around this problem. In this mode of filtration, part of the retentate is continuously recycled at a high flow rate through a loop (Fig. 10.22) to minimize concentration polarization on the surface of the membrane. Fresh feed is fed continuously to the recycle loop and the retentate is bled off from the loop. At steady-state, the flow of feed equals that of the retentate and permeate. If necessary, multiple membrane modules can be used in series to increase filtration area in a feed and bleed recycle scheme, as shown in Fig. 10.23.

10.6.4
Diafiltration

Microfiltration and ultrafiltration processes are sometimes operated in diafiltration mode. During diafiltration, the retentate is returned to the feed tank and fresh

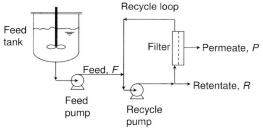

Fig. 10.22 Continuous feed and bleed recycle flow arrangement to enable the cross-flow rate to be varied independently of the feed flow rate.

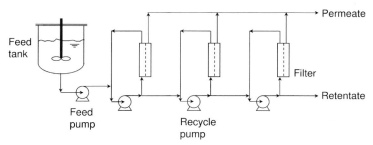

Fig. 10.23 Continuous feed and bleed filtration in series.

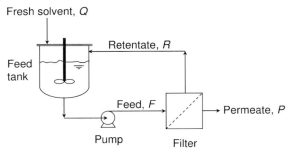

Fig. 10.24 Flow arrangement for diafiltration.

solvent is continuously added to the tank (Fig. 10.24) at the same rate as the rate of removal of permeate. Consequently, the volume of the feed in the tank remains constant, but its composition changes with time. Diafiltration is commonly used to wash microbial and other cells with buffers to remove unwanted solutes. Diafiltration is also used for buffer exchange in protein solutions and washing away of membrane-permeable small molecules from a protein solution.

If the SRC of the diafiltration membrane is unity, then of course all the initial solute will be recovered. If, however, the SRC value is less than unity then some solute will be lost in the permeate. The solute concentration remaining in the final solution is estimated using the equation:

$$C_{final} = C_i e^{-N(1-SRC)},\qquad(43)$$

where N is the number of the initial feed volumes of the wash buffer used, C_i is the initial concentration of the solute in the feed being dialyzed and C_{final} is the final concentration of solute in the feed tank. Equation (43) assumes that the material in the feed tank is always well mixed and that the volume in the tank remains constant. Often, the SRC will be nil for the solutes being washed out. Equation (43) applies to solutes being washed out provided that the solute in question is not present in the buffer used in dialysis.

10.7
Membrane Fouling and Cleaning

10.7.1
Fouling

Foulants are materials that adsorb to membranes to reduce effective pore size and porosity, consequently diminishing permeate flux. Proteins, lipids and polysaccharides act as foulants. Other foulants include cell debris that penetrates membrane pores, and is trapped there to reduce pore size and the number of available pores. Antifoaming agents that are commonly used to control foaming in micro-bial and animal cell culture broths tend to adsorb on and severely foul polymeric filtration membranes. Silicon oil-based antifoams are generally less fouling than the other types. Potentially, surfactants can be used to also improve the filtra-tion performance of membranes [31], but this is not generally the practice in bioprocessing.

Hydrophilic membranes (both polymeric and ceramic) adsorb antifoaming agents less readily and are generally less susceptible to fouling than hydrophobic membranes. Depending on the nature of the foulant and the membrane material, fouling may be reversible or irreversible. In the event of reversible fouling, flux can be recovered by cleaning the membrane. Irreversible fouling is not removed by cleaning. Some degree of irreversible fouling always occurs and as a conse-quence membranes must be replaced periodically. Membrane fouling has been discussed extensively in the literature in relation to membrane bioreactors [32, 33], ultrafiltration and reverse osmosis [34], and the dairy industry [35].

10.7.2
Cleaning of Membranes

Membrane filtration modules require periodic cleaning. Many biotechnology pro-cesses are operated in batches and the filter module is cleaned at the end of the batch. Cleaning consists of a hydraulic cleaning stage followed by a multistep chemical cleaning stage.

Hydraulic cleaning simply uses once-through cross-flow of the pure solvent to dislodge and wash away gross soil. In addition, during hydraulic cleaning of microfiltration membranes and open pore ultrafiltration membranes, the perme-ate flow may be reversed to dislodge particles from the pores. This is known as backflushing and involves flow of clean solvent from the permeate side to the feed side (Fig. 10.25). The transmembrane pressure used during backflushing is usually much smaller than that used during filtration. This is because membranes gener-ally withstand much greater pressure difference in the normal flow direction, but not in the opposite direction. Hydraulic cleaning of polymer membranes in bio-technology processing will typically involve a clean water flush at 40–50 °C. The flush volume is usually about $45\,L\,m^{-2}$ of membrane area. The cross-flow velocity during flushing is typically $1.5\,m\,s^{-1}$ or higher to ensure intense turbulence and good removal of gross soil.

Chemical cleaning involves the use of chemicals to aid removal and/or digestion of foulants. Specific chemicals are used to attain specific cleaning objectives. Cleaning agents used include acids such as phosphoric acid, alkalis such as sodium hydroxide, nonionic alkaline detergents, enzymes such as proteases and amylases, complexing agents such as ethylenediaminetetraacetic acid, and disin-fectants such as hydrogen peroxide and sodium hypochlorite. All cleaning agents must be compatible with the filter membranes at concentrations used. In addition, the cleaning agents should be selected to be effective against the soil that is char-acteristic of a given process.

(a) Normal filtration

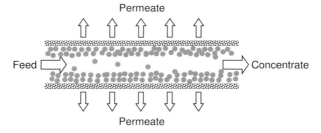

(b) Backflush

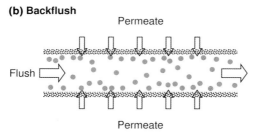

Fig. 10.25 Removal of solute accumulated on the membrane and within membrane pores during normal filtration (a) by flow reversal of permeate during backflushing (b).

A multistep cleaning regimen is almost always used after the initial hydraulic cleaning. The cleaning agent(s) are typically recirculated through the filter module at 50 °C for 30–60 min each. The specific chemical cleaning protocol used depends on the nature of the fouling problems. For example, microfilter membranes used to process animal cell culture broths are typically cleaned by recirculating sodium hydroxide (1 M, pH 14) to digest proteins, lipids and carbohydrates. This is followed by a clean water flush and recirculation of phosphoric acid (0.1 N, pH 1) to remove inorganic deposits. The alkali recirculation step may be replaced with recirculation of alkaline sodium hypochlorite (300 ppm, pH 10–11).

A filter that has been used to process a bacterial fermentation broth may require a cleaning regimen consisting of recirculating the following chemicals interspersed with clean water flushes: sodium hydroxide (0.1 M, pH 13), alkaline sodium hypochlorite (300 ppm, pH 10), an enzyme preparation such as Terg-A-zyme® (0.2%, pH 10) and a detergent wash (0.1% SDS or Tween 80, pH 5–8).

Chemical cleaning is followed by a further flush with clean water, as previously described. After cleaning, the membrane modules are generally filled with a disinfectant for storage. A further water flush, or a complete cleaning sequence, is performed just prior to processing of the next batch.

No cleaning regimen can assure complete removal of adsorbed proteins from polymer membranes. Therefore, if cross-contamination between different products is a concern, product-dedicated filter modules are generally used in bio-technology processes.

10.8
Economics of Membrane Processes

Capital costs of sanitary installations of various types of membrane systems are shown in Tab. 10.4. The data in Tab. 10.4 are for complete installations that are fully automated, and include pumps, piping, tanks and membranes for sanitary operation. The attainable permeate flux depends greatly on the kind of membrane system used and therefore the total area of the filter installation is affected by the choice of membrane configuration. This affects the total installed cost per unit volume of feed filtered.

10.8.1
Membrane Replacement Costs

Membranes have a finite life and require periodic replacement. Typical costs of replacing the membranes are shown in Tab. 10.5. In addition to being least expensive, the spiral-wound membranes require the least amount of labor for replacement [16].

Tab. 10.4 Capital costs of sanitary membrane installations [16]

Membrane system	Cost ($ m^{-2})
Spiral-wound	150–600
Tubular	1 000–1 500
Hollow fiber	1 500–2 000
Plate-and-frame	1 500–5 000
Ceramic membranes	5 000–15 000

Tab. 10.5 Cost of replacing membranes [16]

Membrane system	Cost ($ m^{-2})
Spiral-wound	30–80
Tubular	100–200
Plate-and-frame	110–160
Hollow fiber	300–700
Ceramic membranes	2 000–2 500

10.8.2
Other Operational Costs

Cost of energy is the main contributor to the direct cost of filtration in membrane processes. The energy consumption depends on the operating pressure and cross-flow rate. For any cross-flow rate Q (m³h⁻¹), the power required for pumping can be estimated using the equation:

$$Po = \frac{Q\Delta P_{pump}}{36\xi}, \tag{44}$$

where Po (kW) is the power consumed, ΔP_{pump} is the pressure increase across the pump and ξ is the efficiency of the pump and motor. Typically, ξ has a value of 0.85 for positive displacement pumps and 0.65 for centrifugal pumps.

Reverse osmosis processes for producing potable water from seawater consume 9–12 kW m⁻³ of potable water produced [16]. This is similar to energy consumption of cross-flow microfiltration per unit of permeate produced. Energy consumption in ultrafiltration is about 2–5 kW m⁻³ of permeate generated. Energy input from pumps can cause significant heating of the process fluids and consequently cooling may be required to prevent temperature rise.

Cost of cleaning makes a further significant contribution to the operating cost.

10.9
Conclusions

Membrane filtration processes provide powerful methods for recovering solvents, replacing buffers in solutions and concentrating solutes of a great range of sizes. Products are typically recovered without subjecting them to adverse temperatures, detrimental pH values and damaging solvents as encountered in some extraction and precipitation processes. Selection of an appropriate membrane separation process for a given application requires a good understanding of the membranes available, the possible operational schemes, the effects of the operational variables on productivity and the necessary cleaning regimens. In addition, pilot testing is necessary to ensure that the equipment and operational conditions that are eventually selected will be satisfactory in a production scenario.

References

1 McGregor, W. C. (Ed.), *Membrane Separations in Biotechnology*. Dekker, New York, **1986**.

2 Brindle, K., Stephenson, T., The application of membrane biological reactors for the treatment of wastewaters. *Biotechnol Bioeng* **1996**, *49*, 601–610.

3 Meltzer, T. H., Jornitz, M. W. (Eds.), *Filtration in the Biopharmaceutical Industry*. Dekker, New York, **1998**.

4 Drioli, E., Giorno, L., *Biocatalytic Membrane Reactors: Applications in Biotechnology and the Pharmaceutical Industry*. Taylor & Francis, London, **1999**.

5 Visvanathan, C., Ben Aim, R., Parameshwaran, K., Membrane separation bioreactors for wastewater treatment. *Crit Rev Environ Sci Technol* **2000**, *30*, 1–48.

6 Wang, W. K. (Ed.), *Membrane Separations in Biotechnology*, 2nd edn. Dekker, New York, **2001**.

7 van Reis, R., Zydney, A., Membrane separations in biotechnology. *Curr Opin Biotechnol* **2001**, *12*, 208–211.

8 Vane, L. M., A review of pervaporation for product recovery from biomass fermentation processes. *J Chem Technol Biotechnol* **2005**, *80*, 603–629.

9 Lipnizki, F., Hausmanns, S., Laufenberg, G., Field, R., Kunz, B., Use of pervaporation–bioreactor hybrid processes in biotechnology. *Chem Eng Technol* **2000**, *23*, 569–577.

10 Reij, M. W., Keurentjes, J. T. F., Hartmans, S., Membrane bioreactors for waste gas treatment. *J Biotechnol* **1998**, *59*, 155–167.

11 Crespo, J. G., Velizarov, S., Reis, M. A., Membrane bioreactors for the removal of anionic micropollutants from drinking water. *Curr Opin Biotechnol* **2004**, *15*, 463–468.

12 Charcosset, C., Purification of proteins by membrane chromatography. *J Chem Technol Biotechnol* **1998**, *71*, 95–110.

13 Zeng, X. F., Ruckenstein, E., Membrane chromatography: preparation and applications to protein separation. *Biotechnol Prog* **1999**, *15*, 1003–1019.

14 Leegsma-Vogt, G., Janle, E., Ash, S. R., Venema, K., Korf, J., Utilization of *in vivo* ultrafiltration in biomedical research and clinical applications. *Life Sci* **2003**, *73*, 2005–2018.

15 Bailey, J. E., Ollis, D. F., *Biochemical Engineering Fundamentals*, 2nd edn. McGraw-Hill, New York, **1986**.

16 Nielsen, W. K. (Ed.), *Membrane Filtration and Related Molecular Separation Technologies*. APV Systems, Silkenborg, **2000**.

17 Sablani, S. S., Goosen, M. F. A., Al-Belushi, R., Wilf, M., Concentration polarization in ultrafiltration and reverse osmosis: a critical review. *Desalination* **2001**, *141*, 269–289.

18 Chisti, Y., Mass transfer, in *Encyclopedia of Bioprocess Technology: Fermentation, Biocatalysis and Bioseparation*, Flickinger, M. C., Drew, S. W. (Eds.). Wiley, New York, **1999**, vol. 3, pp. 1607–1640.

19 Tutunjian, R. S., Scale-up considerations for membrane processes. *Biotechnol* **1985**, *3*, 615–626.

20 Ripperger, S., Engineering aspects and applications of crossflow microfiltration. *Chem Eng Technol* **1988**, *11*, 17–25.

21 Kroner, K. H., Nissinen, V., Dynamic filtration of microbial suspensions using an axially rotating filter. *J Membr Sci* **1988**, *36*, 85–100.

22 Chisti, Y., Moo-Young, M., Fermentation technology, bioprocessing, scaleup and manufacture, in *Biotechnology: the Science and The Business*, Moses, V., Cape, R. E. (Eds.). Harwood Academic, New York, **1991**, pp. 167–209.

23 Vogel, J. H., Kroner, K. H., Controlled shear filtration: a novel technique for animal cell separation. *Biotechnol Bioeng* **1999**, *63*, 663–674.

24 Ho, W. S. W., Sirkar, K. K. (Eds.), *Membrane Handbook*. Van Nostrand Reinhold, New York, **1992**.

25 Mulder, M. H. V., Polarization phenomena and membrane fouling, in *Membrane Separations Technology: Principles and Applications*, Noble, R. D., Stern, S. A. (Eds.). Elsevier, Amsterdam, **1995**, pp. 45–84.

26 Cheryan, M., *Ultrafiltration and Microfiltration Handbook*. Technomic, Lancaster, **1998**.

27 Chisti, Y., Shear sensitivity, in *Encyclopedia of Bioprocess Technology: Fermentation, Biocatalysis and Bioseparation*, Flickinger, M. C., Drew, S. W. (Eds.). Wiley, New York, **1999**, vol. *3*, pp. 2379–2406.

28 Chisti, Y., Hydrodynamic damage to animal cells. *Crit Rev Biotechnol* **2001**, *21*, 67–110.

29 Saboya, L. V., Maubois, J. L., Current developments of microfiltration technology in the dairy industry. *Lait* **2000**, *80*, 541–553.

30 Gabelman, A., Hwang, S. T., Hollow fiber membrane contactors. *J Membr Sci* **1999**, *159*, 61–106.

31 Xiarchos, I., Doulia, D., Gekas, V., Tragardh, G., Polymeric ultrafiltration membranes and surfactants. *Sep Purif Rev* **2003**, *32*, 215–278.

32 Chang, I. S., Le Clech, P., Jefferson, B., Judd, S., Membrane fouling in membrane bioreactors for wastewater treatment. *J Environ Eng* **2002**, *128*, 1018–1029.

33 Pollice, A., Brookes, A., Jefferson, B., Judd, S., Sub-critical flux fouling in membrane bioreactors – a review of recent literature. *Desalination* **2005**, *174*, 221–230.

34 Goosen, M. F. A., Sablani, S. S., Ai-Hinai, H., Ai-Obeidani, S., Al-Belushi, R., Jackson, D., Fouling of reverse osmosis and ultrafiltration membranes: a critical review. *Sep Sci Technol* **2004**, *39*, 2261–2297.

35 D'Souza, N. M., Mawson, A. J., Membrane cleaning in the dairy industry: a review. *Crit Rev Food Sci Nutr* **2005**, *45*, 125–134.

11
Affinity Precipitation

Frank Hilbrig and Ruth Freitag

11.1
Introduction

Precipitation is a commonly used unit operation in the downstream processing of biologicals. Moreover, precipitations are used for this purpose at various scales from the very small [microliters, e.g. plasmid DNA (pDNA) preparation] to the extremely large (e.g. technical enzymes). Perhaps the most impressive application of the principle of precipitation in the general area of bioseparation is that of plasma fractionation, i.e. the process used to produce certain therapeutic plasma proteins from human blood donations. Plasma fractionation today is carried out at a scale that surpasses any other established biopharmaceutical process. The worldwide manufacturing scale for human albumin, immunoglobulins (IgGs) and other plasma-derived therapeutic proteins is of the order of hundreds of tons annually [1], i.e. much larger than even that of recombinant human insulin.

The plasma fractionation process was originally developed by Cohn and colleagues in 1946 [2] and is still used in this form today in the USA, while the modified version introduced by Kistler and Nitschmann [3] in 1962 is preferred in Europe. The process fractionates the human plasma proteins in a series of steps deploying the unique physicochemical properties of each plasma protein, i.e. its solubility (hydrophobicity) and its isoelectric point (pI). Following a cryoprecipitation (depletion) step to capture certain sensitive blood factors, successive fractions of the plasma proteins are precipitated via adjustment of the pH, temperature, ionic strength and ethanol concentration in the environment. The precipitating proteins first form colloidal particles, which then grow and aggregate. The particle size distribution of the resulting precipitate as well as its mechanical strength and stability are controlled by a dedicated aging treatment. In the case of the most common batch process, the aging treatment consist of mixing at defined shear force and time [4]. This way the efficient separation of the precipitate by the subsequent centrifugation or body-feed filtration is ensured. The supernatants obtained in the various steps after separation of the precipitate are subjected to additional extraction/precipitation steps carried out in the same manner, but with

Bioseparation and Bioprocessing. Edited by G. Subramanian
Copyright © 2007 WILEY-VCH Verlag GmbH & Co. KGaA, Weinheim
ISBN: 978-3-527-31585-7

modified parameters. Sequential plasma fractionation according to this scheme is a relatively simple and (cost)-effective purification method. Yields of albumin and IgGs are in the range of 60–70% (at 95% purity) and 50–60%, respectively [1, 5]. Plasma fractionation thus combines a simple operation at large scale with considerable purification and good yields. The use of ethanol is attractive as this is also a powerful virus clearant. The safety record of albumin products obtained according to this protocol with respect to virus transmission over the past 50 years has been excellent. Chromatographic purification steps have also been tested in plasma fractionation resulting in higher yields and purity, e.g. for albumin [1], but required additional virus-inactivation steps [6].

Other commonly found types of precipitation-based separations in the area of biopolymer and in particular protein separation are the addition of nonionic polymers such as poly(ethylene glycol) (PEG) or poly(vinyl pyrolidone) (PVP), but also the so-called salting-out methods. The latter approach is a commonly used first-capture and concentration step for many fairly high-titer protein products. The method is based on the fact that protein aggregation and subsequent precipitation can be enforced by the addition of salt. The effect is generally interpreted as the result of a hydrophobic interaction and the salting-out potential of a given salt, thus depends on its position in the Hofmeister series [7]. Ammonium sulfate is a very strong salting-out agent and is widely used as such in preparative protein separation. Salting-out (and precipitation in general) is typically applied during the early stages of the bioseparation process, as it is quite tolerant to impurities/feed composition and convenient to use even at large scale. That the product fraction is concentrated and that some partial purification usually takes place are additional benefits of this method. In sequential combination with caprylic acid it was even possible to achieve a crude antibody purification by precipitation alone [8].

The limits of the outlined precipitation methods are generally the stability of the target molecule and the lack of specificity of the method. In the case of low-titer products, the yields are also often quite low. In such cases precipitation is typically not even considered, instead an approach combining enrichment with specific capture is chosen, most commonly affinity chromatography. Affinity chromatography has supplied the bioseparation community for more than 50 years with an efficient one-step technique for the specific isolation *cum* concentration of proteins. The method is based on immobilizing so-called affinity ligands, i.e. molecules capable of a (bio)specific interaction (molecular recognition) to the chromatographic support. These ligands then selectively retain and separate the target molecules even from a very diluted and complex feed. Elution can be achieved by a buffer, which no longer supports the noncovalent affinity interaction. Often a pH shift or a chaotropic, respectively, competing agent is used.

Affinity chromatography is very popular in research laboratories as well as in bioproduction plants; however, compared to a precipitation, the use of a chromatographic column for separation during early stages of the downstream process is also know to be beset with certain difficulties. Scale-up, column fouling and flow-rate limitations frequently cause problems. A particular problem with affinity chromatography is the sometimes inconveniently slow association rate of the target protein molecule with the immobilized biospecific ligand as a result of pronounced mass transfer limitations, but also the restriction of the available stationary phase capacity in the case of the commonly used porous beads because of steric hindrances (access to the pores). Affinity chromatography is also awkward to use with many raw process streams. Consequently, some effort is still spent on the development of alternative affinity techniques, which retain the principle of biospecific interaction up to the use of the same affinity ligands, but which might overcome some of the known disadvantages of affinity chromatography yet retain its unparalleled selectivity.

In the late 1970s, these efforts led to the development of two concepts for "affinity precipitation" – one by the group of Klaus Mosbach and the other by Michel Schneider and coworkers. The two principles have little in common, but the same nomenclature was used. Precipitation is an attractive concept in this context, since the required solid–liquid separation is extremely well understood and can be handled at various scales in the production as well as in the research environment. The two concepts developed for affinity precipitation were later distinguished as "primary-effect" and "secondary-effect" affinity precipitation (see Ref. [9] for a review). At present, however, the term "affinity precipitation" is used almost exclusively for processes employing the more general concept developed by Schneider and coworkers, i.e. the formerly called secondary-effect or indirect affinity precipitation. This type of separation process will therefore be the topic of most of the remainder of this chapter, while primary-effect affinity precipitation will only be briefly discussed in the next section.

11.2
Primary-effect Affinity Precipitation

As already discussed for plasma protein precipitation, size increase will usually entail precipitation. A protein molecule, which becomes insoluble due to changed environmental conditions (forming thus a small colloidal particle), acts as a nucleation center for further diffusion controlled growth via the attachment of others (aggregation). Larger particle can then be formed by collision-induced agglomeration. Finally, particles will reach a sufficient size to form a macroscopically observable "precipitate", which spontaneously separates from the liquid phase. In analogy, the concept of primary-effect affinity precipitation is based on the controlled growth of protein molecules to a network, which becomes insoluble once a given size limit is surpassed. Such a controlled network formation is possible if there are at least two points of specific interaction between the involved molecules. The principle is illustrated in Fig. 11.1. A prerequisite for the formation of large networks according to this concept is an approximately 1:1 ratio of the binding sites. If one species dominates, the formation of small complexes where all binding sites of the "minority" molecules are saturated by individual molecules of the other species and no network formation takes place, becomes possible.

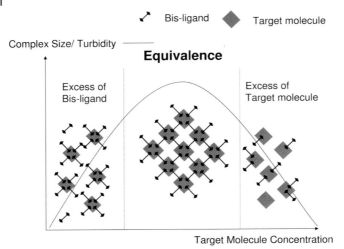

Fig. 11.1 Concept of primary-effect affinity precipitation.

Mosbach and coworkers exploited this concept in the late 1970s for a form of affinity precipitation [10]. Its proof-of-feasibility was the demonstration that the tetrameric enzyme lactate dehydrogenase (LDH) and the bifunctional molecule $N_2,N_{2'}$-adipodihydrazido-bis-(N^6-carbonylmethyl-NAD) (bis-NAD) form at low temperature and in a certain concentration ratio a cross-linked, macromolecular network, which becomes insoluble and precipitates when sufficiently large. In primary-effect affinity precipitation the formation of a precipitate is therefore the direct consequence of the affinity interaction between the multivalent target molecule (typically an enzyme) and the bifunctional ligand ("cross-linker"). The spacer length between the two NAD units in the bis-ligand has to be chosen in such a way that it allows the simultaneous interaction with the active sites of two LDH molecules, but not the competing intramolecular cross-linking reaction of the bis-NAD molecule itself. It was later found useful to add substrate analogs during complex formation in order to increase the strength and the specificity of the binding between the enzyme and the NAD affinity ligand, by exploiting the so-called "locking-on" effect of ternary complex formation [11]. After separation of the precipitate by centrifugation the aggregates could be redissolved by the addition of NADH, which competes with the bis-NAD for the binding sites of the enzyme. Once the complex had dissociated, contaminations such as the bis-NAD, NADH and ternary complex agents such as pyruvate or oxalate could be removed by a subsequent gel-filtration step.

In order to reach the maximum precipitation yield in such a primary-effect affinity precipitation, the molecular ratio adjusted between the LDH and the bis-NAD has to be optimized. A yield in the range of 90% can be achieved under nearly equinormal conditions (1.1 NAD equivalents per LDH subunit). As to be expected, small deviations from the optimal ratio tend to significantly diminish the extent of precipitation. In case of the above-cited example, the effect was more pronounced for under- than for over-saturation with bis-NAD. If the bis-NAD concentration was too low, no precipitation was observed. The enzyme concentration itself, but also the substrate analog concentration were also found critical for the precipitation yield [12, 13].

NAD is a natural coenzyme and its ability to specifically interact with the corresponding enzymes is exploited in a number of affinity-based separation. With slight modifications, primary-effect affinity precipitation exploiting bis-NAD could thus be used for the purification of a number of other enzymes besides LDH. Yeast alcohol dehydrogenase is, like LDH a tetrameric enzyme, which was found to precipitate slowly in the presence of bis-NAD, pyrazole and NaCl. Without NaCl, no precipitation occurred. Hexameric glutamate dehydrogenase, by contrast, precipitates spontaneously with high precipitation yield over a wide bis-NAD ratio. The resulting precipitate is, however, difficult to redissolve, requiring high NAD concentrations. The dimeric liver alcohol dehydrogenase, on the other hand, does not form cross-linked macromolecular networks even when pyrazole is added, apparently due to a preference for the dead-end formation of a dimeric complex of two bis-NAD and two enzyme molecules [14].

A number of triazine dyes, most prominently Cibacron Blue, imitate the interaction of the NAD and exhibit similar affinities to NAD-binding enzymes. Cibacron Blue-based bis-ligands were therefore also successfully used for primary-effect affinity precipitation. Ninety percent of LDH could be recovered under optimized conditions, i.e. in the presence of pyruvate [15]. For BSA and chymosin the yields were 50 and 20%, respectively. Rabbit muscle LDH, but not the pig heart isoenzyme, could be efficiently precipitated at 97% yields (purification factor of 6) with a methoxylated derivative of the triazine dye Procion Blue H-B. In this case the methoxytriazinyl ring and the terminal p-aminobenzenesulfonate ring are presumed to interact both with the target molecule, while the central p-phenylenediaminesulfonate ring residue presumably serves as the interlinking spacer [16]. ATP is another natural molecule that plays a role in many enzyme-catalyzed reactions and therefore can bind to specific sites in such enzyme molecules. Bis-ATP has, for example, been employed for primary-effect affinity precipitation of bovine heart phosphofructokinase from tissue extracts (precipitation in the presence of citrate) [17]. The application of bis-ATP in general is, however, limited by its instability.

In addition, a number of miscellaneous applications of the principle of primary-effect affinity precipitation have been reported. Bis-borate ligands, for example, have been tested for cell separation [18]. Borate was also used as additive to induce the precipitation of polysaccharide Protein A–IgG affinity complexes [19]. Biotinylated phospholipids act as strong binding ligands for the tetrameric avidin molecule. In aqueous solution such phospholipids form micelles by hydrophobic interaction, which can also be used for the separation of a target molecule from solution. Phospholipids have been successfully used for the isolation both of avidin [20] and of antibodies [21]. So-called polyligands have also been proposed for the same purpose [22].

The manifold prerequisites for performing primary-effect affinity precipitation efficiently, in particular the prerequisite for multivalency in the target protein, but also the limited number of useful bis-ligands, reduces the application potential of this method mainly to the dehydrogenases. In an attempt to increase the versatility of primary-effect affinity precipitation, Van Dam and coworkers synthesized bis-copper chelates, which were supposed to cross-link the target proteins via accessible surface histidine residues into a precipitating macromolecular network [23]. At a copper/histidine ratio near 1, human hemoglobin, which has 26 histidine residues, is precipitated at 100% with $Cu(II)_2EGTA$. Sperm whale myoglobin has six histidine residues and requires a copper/histidine ratio near 50 for a 100% precipitation yield. The precipitation efficiency increases with increasing spacer length of the bis-ligand, as shown by using $Cu(II)_2PEG\ 20000(IDA)_2$. In the case of precipitating hemoglobin, however, Cu^{2+} ions ($CuSO_4$) alone were just as effective in precipitating this particular protein as the bis-ligand $Cu(II)_2EGTA$. Horse cytochrome c (one histidine residue), on the other hand, could not be precipitated by either the bis-ligand or $CuSO_4$. Concanavalin A (Con A) is also known for its unusually strong affinity for copper ions and specific precipitation using metal-charged EGTA [glycol-bis(2-aminoethylether)-N,N,N',N'-tetraacetic acid] has been reported [24]. The application range of metal-affinity precipitation can be extended further by genetic engineering of proteins, which allows adding so-called histidine (His)-tags to the recombinant proteins. Lilius and coworkers [25] fused five histidine residues as an affinity tail to galactose dehydrogenase, an enzyme that forms homodimers. Addition of $(Zn)_2EGTA$ then resulted in affinity precipitation by linear polymer chain formation. A yield of 90% and a purification factor of 11 were achieved. The amount of metal chelates added was found to be critical. Above 10 mM, the native protein precipitated spontaneously.

A few preparative-scale applications of primary-effect affinity precipitation are known. LDH was recovered from a crude mixture with recovery yields above 90% and a purification factor of 40 [14]. However, several elaborate pilot precipitation tests were necessary in order to determine the optimal ratio between the LDH concentration/activity and that of the bis-NAD. Deviation from the optimum ratio caused significant product loss. At the same time it was found that a bis-NAD addition based only on enzyme activity measurements (the easiest and most common "quantification" method for an enzyme) gives only a low precipitation yield [26]. These difficulties render the standardization of primary-effect affinity precipitation at preparative-scale rather complex. Concomitantly, the kinetic of complex formation and dissociation are slow [27], and entrapment of impurities in the aggregates is a very common problem.

11.3
Affinity Precipitation by Stimuli-responsive Materials

Affinity interaction and precipitation are directly linked in primary-effect affinity precipitation. While this makes the process very straightforward, it is also the cause of many of the above-mentioned disadvantages of this bioseparation method. In secondary-effect or indirect affinity precipitation, these two aspects are no longer linked; hence, they can be performed and controlled independently. Secondary-effect affinity precipitation, from here onward simply called "affinity precipitation" as it has become the more ubiquitous form of affinity precipitation, makes instead use of stimuli-responsive materials to bring about precipitation. Stimuli-responsive or "intelligent" materials react by pronounced property changes to a small change in an environmental parameter. The effect is usually fully reversible, i.e. the material reverts to its original state once the stimulus has been removed. For application in the context of protein isolation, materials that undergo pronounced changes in their water compatibility (hydrophobicity) are very interesting. Materials responding to a wide variety of stimuli have been described, and many of them have been already used for bioseparation purposes and will be discussed below.

If an affinity ligand is coupled to such a stimuli-responsive material, the resulting bioconjugate, the affinity macroligand (AML), forms the basis for a very elegant bioseparation process. The principal concept of this method is illustrated in Fig. 11.2. In order to separate the target molecule, the AML is simply stirred into the raw solution. The affinity complex forms in solution, i.e. with very little

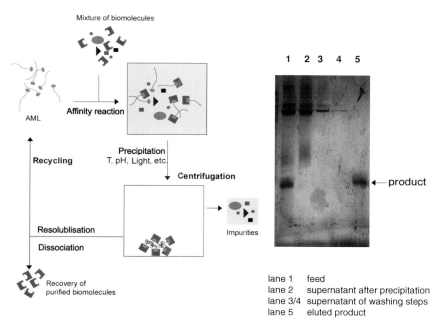

Fig. 11.2 Concept of secondary-effect or indirect affinity precipitation ("affinity precipitation"). The SDS–PAGE gel on the right-hand side demonstrates the effectiveness of the separation in case of a purification of a protein from a cell culture supernatant containing 5% FCS.

steric hindrance and a minimum of mass transfer resistance. Then the stimulus is applied, the affinity complex precipitates and can be removed, while the impurities stay in solution. Subsequently, the target molecule can be eluted either by redissolution of the complex in dissociation buffer under conditions that promote redissolution of the material followed by stimuli-induced separation of only the AML or via direct elution from the separated precipitate under conditions that promote continued precipitation of the AML. In Fig. 11.2 the efficiency of such a separation is documented for the isolation of a protein from a cell culture supernatant containing 5% fetal calf serum (FCS) by an analytical sodium dodecylsulfate–polyacrylamide gel electrophoresis (SDS–PAGE) gel of the raw culture supernatant, the supernatant after removal of the target protein by affinity precipitation, the supernatants of the two washing steps and the eluted target protein after removal of the AML.

Affinity and stimulus-responsiveness are both embodied in the AML (Fig. 11.3) – generally a polymeric substance to which one or several affinity ligands are linked and which precipitates reversibly from aqueous solution upon external stimulation, e.g. a change in the ambient pH or temperature. In his pertinent review [28], Mattiasson discussed the design criteria for an ideal AML, especially the requirements on the side of the polymer, which should:

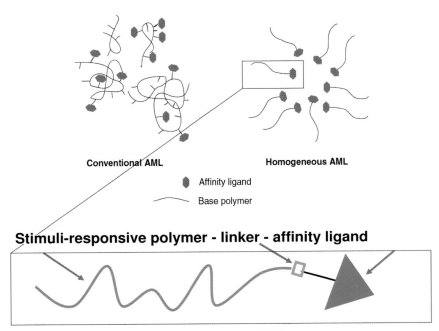

Fig. 11.3 Schematic presentation of AMLs: conventional AML (heterogeneous in size, structure and affinity ligand distribution) and oligomeric "homogeneous" AML (small, narrow size distribution, affinity ligand in terminal position).

- be available and cheap
- have a narrow molecular mass distribution (few heterogeneities in general) as well as contain reactive groups suitable for ligand coupling
- not interact strongly with the affinity ligand or impurities in the feed, to make the affinity ligand available for interaction with the target molecule, and reduce nonspecific interaction and entrapment of impurities
- give complete phase separation upon application of the stimulus with a well-characterized, sharp phase transition and facile resolubilization
- form compact precipitates to facilitate recovery
- enable validated recycling of the AML

However, to date, this ideal remains somewhat elusive and many successful affinity precipitations were performed with AMLs that failed to meet one or several of these criteria.

11.3.1
Stimulus: pH

The first description of a stimuli-induced affinity precipitation was published in 1981 by Schneider and coworkers [29]. The target molecule was trypsin, i.e. again an enzyme, but in this case an enzyme that contained only a single binding site for the ligand (monovalent enzyme) and hence could not have been purified by primary-effect affinity precipitation, where multivalency is required. The AML in this case was based on a ter-polymer composed of acrylamide, *N*-acryloyl-*p*-aminobenzoic acid and *N*-acryloyl-*m*-aminobenzamidine. The benzamidine units in this molecule served as affinity ligands, as depending on the pH they represent strong and specific inhibitors of the protease trypsin. The entire AML was water-soluble above a pH of 4. Below a pH of 4, the acidic residues on the polymer backbone became neutralized (protonation) and hydrophobic interactions between the then uncharged polymer backbones caused aggregation. As a result, the AML precipitated whenever the pH was reduced below a value of 4. The process was reversible and the AML–affinity complex did resolubilize when the pH was again raised.

The related bioseparation process for the purification of the protease trypsin from a crude pancreatic extract at pH 8 can be considered typical for pH-induced affinity precipitation and is therefore given in some detail here. In particular, the AML was added to the extract at a load of 0.5 wt%. The biospecific interaction between the trypsin and its inhibitor (AML) took place in homogeneous solution. The AML–trypsin complex was then:
  (i)   precipitated by lowering the pH to 4
  (ii) separated by centrifugation from the supernatant, in which putative impurities remained dissolved

(iii) washed once with water

(iv) resuspended at pH 2 for dissociation of the affinity complex

The latter point is an interesting variant, since at pH 2 the affinity interaction is no longer possible, while the AML still forms a precipitate. In this process variant the product is thus eluted from the existing precipitate. Alternatively, in some pH-induced affinity precipitation schemes, the complex is first redissolved and afterwards the affinity interaction is interrupted ("elution") followed by reprecipitation of only the AML. In both cases, however, the purified product stays in solution after elution, while the precipitated AML is easily removed from this solution, e.g. by centrifugation. In the case of the trypsin purification, a supernatant containing 83% of the initial trypsin activity and 7% of the initial chymotrypsin activity was collected after the removal (centrifugation) of the AML. The AML itself was recovered at 99% yield by this step. Afterwards the AML could be resolubilized at pH 8 and reused for subsequent processing of crude trypsin batches. Under these circumstance (repeated use of the AML) a loss of approximately 1% of AML per cycle was observed. The AML performance was found to deteriorute only slightly with recycling.

Affinity precipitation as designed by Schneider and coworkers became the basic concept for stimuli-induced affinity precipitation, permitting in essence the one-step purification of a wide variety of biomolecules by homogeneous affinity interaction with suitable AML. In 1994, a first generic and commercially available AML precursor was introduced for this type of affinity precipitation. Kumar and Gupta [30] reported the purification of trypsin from a crude sample using a pH-responsive conjugate of Eudragit S-100 and soybean trypsin inhibitor (STI) as AML; 83% of the enzyme activity could be recovered in this process. Eudragit S-100 is a commercially available anionic copolymer composed of methacrylic acid and methyl methacrylate units (molecular weight $135\,000$ g mol^{-1}) with a $1:2$ ratio of carboxyl and ester groups. It precipitates fully reversibly at a pH below 4.8 and is soluble at pH values higher than 5.5 (phase transition hysteresis).

Eudragit S-100 is capable of both electrostatic and hydrophobic interactions, and thus adsorbs by itself, i.e. without conjugation of any affinity ligand, a number of enzymes with varying efficiency, which then coprecipitate with the Eudragit S-100 upon lowering the pH. Examples include LDH and xylanase. The need for non-ionic detergents (Tween, Triton) for elution of the LDH indicates that hydrophobic interactions between the polymer and LDH [31] are involved in the interaction, whereas for the purification of xylanase the presence of salt prevents the enzyme binding [32], arguing for an interaction dominated by electrostatic effects. In case of xylanase, microwave treatment of Eudragit S-100 improved the purification [33]. For efficient enzyme purification the Eudragit:target molecule ratio is critical for initial complex formation prior to precipitation by hydrophobic interaction; the dissociation of the complex can be described by a first-order reaction [34, 35].

Conjugation of Eudragit S-100 with affinity ligands shifts the phase transition to higher pH values, because conjugation reduces the number of charged carboxylic acid groups in the Eudragit backbone, and thus changes the balance of charged and hydrophobic groups on the modified polymer. The same effect occurs after addition of ammonium sulfate or PEG 8000 to a Eudragit-containing solution [36]. In cases where the enforcement of a precipitation by low pH is not possible, e.g. in the case of a pH-sensitive product, Eudragit S-100-based AMLs can also be phase-separated by the addition of Ca^{2+} together with an increase of the temperature to 40 °C. This has been demonstrated by Guoqiang and coworkers for the purification of yeast alcohol dehydrogenase in connection with zinc ion-promoted binding (affinity interaction) of the enzyme with Eudragit–Cibacron Blue [37]. Other examples of Eudragit-based affinity precipitation are the purification of monoclonal antibodies by conjugated lipase [38, 39], of Con A by conjugated *p*-amino-phenyl-α-D-glucopyranoside [34, 40] and of Protein A by conjugated IgG ligands [41]. Ethanolamine-modified Eudragit S-100 has been used to precipitate protein impurities in order to increase the yield of the target enzyme [42].

Eudragit S-100 tends to precipitate gel-like, which allows separation by centrifugation, but not by sedimentation/filtration. The combination of high temperature and low calcium concentration or calcium and high concentration of organic solvent has a positive effect on the separation efficiency. In addition, Eudragit S-100-based AMLs are known to show a high tendency for nonspecific protein binding during the affinity precipitation process. These nonspecific interactions are generally enforced or even caused by the modifications of the polymer backbone structure during AML synthesis. The comparison with pure Eudragit S-100 in a control process therefore does not give the correct picture with regard to the extent to which to expect nonspecific binding in an actual application [43]. Depending on the application, the addition of high salt concentration, organic compounds or PEG may decrease the nonspecific electrostatic or hydrophobic interaction between protein impurities and the polymer (the AML). Nonspecific binding can also be kept low, if the minimum AML load is used. While nonspecific interactions generally present a nuisance in affinity purifications, the effect can be exploited, as mentioned above, in a beneficial way, e.g. for LDH or xylanase enrichment [31, 44].

In addition to Eudragit, several pH-sensitive polymers are known in nature, which have also been adapted to the affinity precipitation of proteins. These include chitosan [44–47] and derivatized cellulose [48–50]. Chitosan, which is partly deacetylated chitin, is soluble at pH below 6.5 and precipitates at higher pH values. The purification of wheatgerm agglutinin [46], chitinase [51], *N*-acetyl glycosamine [47] as well as trypsin by STI-conjugated chitosan [45] has been reported. Since chitosan is positively charged and Eudragit S-100 is negatively charged, while both polymers bind xylanase and LDH, a purification scheme of sequential affinity precipitation steps can be designed [52]. In another application it was shown that IgG immobilized on hydroxypropylmethyl cellulose acetate succinate adsorbs Protein A at pH 7.2 and elutes it at pH 2.5; precipitation occurs at pH 4.2 [49].

A general drawback of pH-sensitive polymers including Eudragit S-100 is that they precipitate at a pH that is outside the stability range of many proteins. Polyelectrolyte complex formation (PEC) was proposed as a means to shift the precipitation pH of such polymers into the physiological range [53, 54]. In an exemplary application, a polyethyleneimine–Cibacron Blue AML was used in such a way for the purification of LDH [53]. Upon the addition of the negatively charged polyacrylate, the AML precipitated at a pH between 6.5 and 8.9 depending on the AML : polyacrylate ratio and the dilution factor (ionic strength). The recovery of the PEC was, however, low, and the coprecipitation of negatively charged protein impurities and nucleic acids was esteemed very likely [55]. Concomitantly, the addition of polyelectrolytes to achieve precipitation of the affinity complex via PEC formation stands in contradiction to the general philosophy of downstream processing, which says to keep the number of intentionally added substances to the absolute minimum.

11.3.2
Stimulus: Salt Addition

Another common means to bring about the precipitation of certain water-soluble (bio)polymers is the addition of salts. Alginate, for example, is a soluble polysaccharide that precipitates reversibly from solution upon the addition of bivalent ions such as calcium (cross-linking agent). This natural polymer is composed of linear, unbranched blocks of guluronic acid and mannuronic acid. The size and sequence of these blocks determine the chemical and physical properties of the alginate. Alginate possesses an inherent biological affinity for some enzymes such as amylase, pectinase and lipase. An overview of applications of alginate for protein purification is given in Ref. [56]. Examples of alginate-based AML for trypsin and lectin purification are discussed in Refs. [57, 58]. As alginate is of natural resource, lot-to-lot variability and contamination, e.g. by endotoxins, are unavoidable. The separation of alginate and endotoxin, which are both negatively charged, is laborious and finally makes alginate for bioseparation purposes an expensive material [59].

κ-Carrageenan is another natural polysaccharide that precipitates from solution upon addition of potassium ions. The purifications of pullulanase and yeast alcohol dehydrogenase by Cibacron Blue 3GA-conjugated κ-carrageenan have been reported [60, 61]. Other natural polymers suitable for affinity precipitation by salt addition are carboxymethyl cellulose and starch. Carboxymethyl cellulose precipitates upon addition of calcium ions plus PEG. Cibacron Blue 3GA-conjugated carboxymethyl cellulose has been shown to be useful for the purification of LDH [48]. Starch precipitates upon addition of ammonium sulfate and adsorbs cyclodextrin glycosyltransferase, for example [62].

A somewhat analogous affinity precipitation (via cross-linking of the affinity complexes) was described by Senstad and Mattiasson in 1989 for the enzyme LDH [63]. In this case, Blue Dextran 2000, i.e. a dextran modified with the triazine dye, was used to interact with the LDH in solution. The precipitation and separation of the affinity complex was then initiated by the addition of Con A, a tetrameric lectin known for its ability to crosslink carbohydrate units. In the considered case the Con A was used to crosslink the dextran molecules into an insoluble network.

Compared to direct primary-effect affinity precipitation of LDH, this method allowed the separation of affinity complex formation on the one side and precipitation on the other, and was thus much less dependent on the target molecule's concentration. In particular, a titration of the LDH concentration in the raw solution prior to the separation was no longer necessary. Elution of 25% of the original LDH activity from the precipitate was achieved by adding KCl. Handling of the precipitate turned out to be difficult, however, since it had a gel-like structure and was rather voluminous.

11.3.3
Stimulus: Temperature

Another class of stimuli-responsive materials with potential for affinity precipitation are the thermo-responsive ones, i.e. polymers that are soluble in (cold) water, but precipitate once a certain critical solution temperature (CST) is surpassed. Thermo-responsiveness is often observed in amphiphilic polymer molecules, where dissolution is characterized by a negative dissolution entropy together with a negative dissolution enthalpy. Such molecules dissolve well in cold water. As the temperature increases, the unfavorable dissolution entropy starts to dominate the behavior and at a certain temperature (the CST) – the formerly soluble macromolecules become insoluble; macroscopically, this manifests itself as precipitation. The fact that the H-bridges, which typically aid the dissolution of such polymers in water, become weaker as the temperature increases also contributes to this effect. Since thermo-responsive polymers bear little to no charges, while they only show moderate hydrophobicity, their tendency for nonspecific interaction with biological molecules is low – another advantage for their application in affinity precipitation.

A number of poly-N-alkylacrylamides such as poly-N-isopropylacrylamide (polyNIPAM) and poly-N,N-diethylacrylamide possesses this property [64]. Most cosolutes lower the critical solution temperature. The pH, on the other hand, is usually only of very little influence. This fact has important consequences for the application of temperature-induced affinity precipitation as a bioseparation tool. In such processes it becomes possible to disconnect the pH as a means to modulate the affinity interaction (association/elution) and the temperature as a stimulus for reversible precipitation/solubilization of the AML (the affinity complex). Fong and coworkers, on the other hand, described a possible strategy for the sequential affinity separation of proteins using pH- and thermo-responsive AMLs [65].

A number of pertinent applications of thermo-responsive AMLs for affinity precipitation have been described in the literature. Chen and Hoffman [66] synthesized a statistical copolymer of NIPAM and N-acryloxysuccinimide having a molar mass of approximately $24\,000\,\mathrm{g\,mol^{-1}}$, an average of 5.5 activated ester groups per polymer chain and a CST of 32 °C in pure water. Conjugation of Protein A to this copolymer enabled the capture and recovery of IgG. The dissociation constant of the corresponding affinity complex was 3×10^{-6} M and the binding capacity of the AML 25% of the theoretical one. The authors concluded that sterical interference of the Protein A-binding sites with the polymer might be the reason for the low binding capacity of the copolymer–AML, whereas they could exclude the occurrence of nonspecific binding between IgG and the polymer backbone. Thermo-responsive AMLs based on polyNIPAM copolymer functionalized with glycidyl methacrylate groups were also used in the Ca^{2+}-depending affinity complex formation between *p*-aminophenylphosphorylcholine and rabbit C-reactive protein, which formed the basis for the purification of this protein at high yield [67]. Further applications of AMLs based on poly-functionalized, thermo-responsive copolymers were the purification of trypsin [68], Protein A [69], alkaline protease [70] and egg white lysozyme [71].

The early applicants of temperature-induced affinity precipitation generally chose to introduce the interactive groups (functionalization) into the thermo-responsive polymer by statistical radical copolymerization, e.g. of NIPAM (in the case of polyNIPAM-based thermo-responsive structure) with suitable hydrophilic or hydrophobic acryl monomers. The result was a heterogeneous AML preparation with wide variation in size and structure, and a statistical distribution of the monomeric units and affinity ligands. This approach soon reached some limits, mainly because high precipitation efficiency and predictable precipitation behavior were difficult to achieve in such heterogeneous AML preparations. An example is the attempt to use a Cu(II)–AML for the purification of α-amylase [72] and His-tagged single-chain Fv antibody fragments [73]. High salt concentrations were necessary in this case to achieve precipitation at all. Due to the different microenvironments, the binding strength of the individual affinity ligands also tended to vary considerably in such heterogeneous AML preparations, while the experimentally observed "average affinity" to the target molecule was often orders of magnitude lower than that of the free affinity ligand [30]. This presented a particular problem in affinity precipitation, which as a one-stage process theoretically requires binding constants for efficient protein capture that should be at least one order of magnitude higher than those used in affinity chromatography [74]. In addition, the high-molar-mass polymers of the first generation of thermo-responsive AMLs were known to form highly coiled and entangled (viscous) structures in solutions that hinder the access of incoming macromolecules, such as the target molecules, and hence contribute to the loss in affinity. Vaidya and coworkers [75] found that spatial separation of the affinity ligands from the polymer backbone by introduction of spacers increased the binding strength to the target proteins. This effect increases with increasing spacer chain length in synergism with a more favorable microenvironment for the affinity ligand. Another option to increase the binding strength between target and AML is the introduction of multipoint interactions (concept of avidity [72]).

A ligand accessibility comparable to that of a free molecule can only be achieved by coupling the ligand, preferably in terminal position, to a low-molar-mass second-generation AML precursor (oligomer). Then the crowding effect is almost negligible [74–76]. The use of low-molar-mass AML precursors also reduces the probability of steric interference of the mobile polymer chains with the ligand-binding sites [77]. The second generation of thermo-responsive polymers for affinity precipitation became available as a result of the developments in polymer synthesis. In particular, the development of a number of methods to create by a form of radical polymerization linear, homogeneous, low-molar-mass polymers containing one reactive end-group opened new possibilities for the creation of AMLs. Such oligomers can, for example, be synthesized by group-transfer [74] or chain-transfer polymerization [78]. In the latter case the polymer chain length is essentially controlled by the molar ratio between the monomer and the chain transfer agent [79]. Reactive end-groups are also easily integrated into such oligomers via the chain transfer agent. When 3-mercaptopropionic acid is used as chain-transfer agent, for example, the synthesized oligomers carry a carboxylic acid end-group, which may easily be used for the conjugation to the affinity ligand via carbodiimide coupling. Interestingly and contrary to the above-mentioned statistical copolymer-based AML, such end-group-linked affinity ligands have little to no influence on the precipitation behavior of the AML/affinity complex. This is even the case when the ligand is large and/or charged [80, 81]. Presumably such bioconjugates resemble block copolymers, where the behavior of each part is determined by its own physicochemical character and independent of the others. Statistical copolymers, on the other hand, show a "new" behavior that is determined by the sum of the molecule.

Oligomeric polyNIPAM usually has a CST at 32–34 °C in pure water similar to that of polymeric polyNIPAM. In addition. oligomeric polyNIPAM has been synthesized in a more ordered isotactic structure with a CST of 41–42 °C in pure water [82, 83]. Garret-Flaudy and Freitag [76, 84] were among the first to report the application of oligomeric AML for protein purification. The AML precursor was a polyNIPAM synthesized by chain-transfer polymerization and characterized by low molar mass (M_n < 5000 g mol^{-1}) and a low polydispersity (M_w/M_n = 1.15). Conjugation of 2-iminobiotin to this AML precursor resulted in an AML that could be used very successfully for the purification of avidin from a cell culture supernatant containing 5% FCS. The avidin recovery was 90% and the enrichment factor 14. The remaining protein contaminations of the product were below the detection limit. The affinity between the AML and the avidin could be modulated by the pH. Binding required a high pH (above 9), while elution becomes possible at low pH (below 4). The pH did not affect the CST in the range from 1 to 12. The phase transition of the AML (precursor) as well as that of the affinity complex was very sharp, despite the low molar mass of the oligomer involved. Around 99% of the AML molecules could be recovered by thermo-precipitation and centrifugation. The AML could be reused several times and nonspecific adsorption was of no concern. A high off-rate streptavidin mutant AML has been proposed for the purification of biotinylated biomolecules by affinity thermo-precipitation [85].

In another early application, Takei and coworkers [77] coupled polyNIPAM (M_w 6100 g mol^{-1}; M_w/M_n = 1.22) to the outer surface of anti-HSA anti-human serum albumin) goat IgG. Upon temperature increase a rapid-response precipitation of the conjugates occurred at approximately 34 °C. They also found that if the polyNIPAM conjugation is not too excessive, the IgG antigen-binding specificity and capacity remained unchanged compared to the free affinity ligand. However, especially when relatively long oligomers were used at high density, these were found to sterically interfere with the IgG binding sites, presumably due to a restriction of the mobility or the accessibility by the free oligomer end. Hoshino and coworkers proposed poly-N-acryloylpiperidine as a thermo-responsive AML precursor [86]. Due to the low CST of this precursor (4–8 °C), poly-N-acryloylpiperidine-based AMLs are especially useful for the purification of thermo-labile proteins. In an exemplary application, 68% of the α-glycosidase activity and 80% of the Con A were recovered from their crude solutions by using maltose as an affinity tag on such AMLs. The affinity constant of association of this tag to the targets was in both cases approximately 10^3 M^{-1}.

The concept of using such linear, low-mass and homogeneous oligomers in affinity precipitation was recently extended to the purification of pDNA [83]. Oligomers with a molar mass of 2300 g mol^{-1} and a polydispersity below 1.1 were used to create AML for triple-helix affinity precipitation of the target DNA. The oligomers were first activated by avidin and then conjugated to biotinylated 21mer oligonucleotides. The ligand efficiency of such AML was an order of magnitude higher than the one normally measured in triple-helix affinity chromatography, confirming Eggert's thesis [74] that an affinity ligand placed in a terminal position in a low molar mass AML retains a maximum of activity.

Low-molar-polymer-mass AMLs are therefore superior to the previously described polymeric molecules in terms of homogeneity, ligand efficiency and flexibility, making affinity precipitation a very valuable purification method. The fact that the addition of such oligomeric molecules to the raw solution will not lead to a pronounced increase in viscosity as is the case for the corresponding polymeric agents constitutes another strong argument for the use of small AMLs, especially at large scale, where mixing may otherwise become a problem. A exemplary comparison between an oligomeric polyNIPAM-WGL AML and a commercial affinity column (HiTrap WGL; Amersham Biosciences, USA), both activated by wheatgerm lectin (WGL) as affinity ligand, showed a 10 times higher ligand concentration and a 3 times higher binding capacity per unit volume for ovomucoid in the case of the AML compared to the affinity column [87].

The utilization of group-specific capturing steps is increasingly applied in downstream processing of recombinant proteins at an industrial scale. This trend can also be observed in affinity precipitation. In a recent publication on the use of such group-specific ligands in affinity precipitation the potential of a number ligands was discussed [88], i.e. Ni-NTA for the capture of His-tagged proteins, the use of an antibody that recognizes a proprietary peptide sequence for capture of the corresponding fusion proteins and Protein A for antibody capture. In another contribution the reuse of the AML and reversibility of the separation through three cycles of precipitation and dissolution has been demonstrated for a Fv antibody fragment–polymer conjugate [89].

Affinity precipitation will in future continue to require the development of new "intelligent" materials. Highly branched thermo-responsive NIPAM-based copolymers, synthesized by reversible addition–fragmentation chain transfer (RAFT) polymerization and containing imidazole functionality in the polymer chain-ends, were, successfully tested for metal-affinity precipitation of a His-tagged protein [90]. Another fairly new type of responsive material, which recently found applications in affinity precipitation, is elastin-like polypeptide (ELP) [91]. ELPs are biopolymers consisting of repeats of the pentapeptide VPGVG that can undergo a reversible phase transition from the water-soluble forms into aggregates upon increasing temperature. The transition temperature can be controlled by the chain length and peptide sequence, and is also responsive to pH, ionic strength, pressure and covalent modifications of amino acid residues. ELPs can be overexpressed in *Escherichia coli* to high yields, and they are easily purified by thermo-precipitation. ELPs have been successfully fused to other peptides or proteins while retaining the temperature responsive property as well as the functionality of the fusion partner. Genetically engineered ELP–Protein A fusion [92] and chemical ELP–carbohydrate conjugation [93] are reported applications.

11.3.4
Stimulus: Photo

Affinity precipitation is an extremely versatile purification method, because a broad range of stimuli-responsive materials can be used, where each material has its specific chemical and physical properties. In addition to pH- and temperature-responsive materials and polymers, which precipitate upon an increase of ionic strength, a fourth option in affinity precipitation has been reported by Desponds and Freitag [94], particularly the use of photo-responsive polymers as AML precursors. Strictly speaking, the proposed photo-responsive molecules are variants of the thermo-responsive ones [95]. Chain-transfer copolymerization of NIPAM and *N*-acryloxysuccinimide using a biotinylated chain-transfer agent and conjugation of (3-aminopropyloxy)azobenzene chromophores to the activated polymer backbone produced a prototype of a photo-responsive biotin–AML. This AML shows a critical solution temperature in pure water of 16 °C when the azo groups in the side-chains are predominately in the (stable) *trans*-state. Irradiation with ultraviolet light (330 nm) switches the azo group into the more hydrophilic *cis*-state and the critical solution temperature rises to 18 °C. Irradiation with visible light (above 440 nm) switches the group back to the *trans*-state. Adjusting the temperature to an intermediate level, the photo-AML was used to demonstrate the concept of photo-affinity precipitation, i.e. the specific capture and recovery by light-induced precipitation of a target molecule (avidin) from a serum-containing cell culture supernatant. The avidin was obtained in highly purified form without nonspecific co-enrichment of protein impurities.

11.4
Application of Affinity Precipitation in Bioseparation

The attractiveness of affinity precipitation stems from the fact that the well-understood and manageable unit operation "precipitation" is combined with the specificity and efficiency of the "affinity" approach in homogeneous solution. In their perennial review, Labrou and Clonis [96] placed (secondary-effect) affinity precipitation highest among the major affinity purification technologies in terms of purification power combined with large-scale potential. Affinity precipitation also shows broad flexibility towards variation in the process parameters. Despite these advantages, however, affinity precipitation has to date hardly made a major impact on biotechnical downstream processing. The fact that until recently AML and application protocols had to be developed individually by the applicant due to a lack of commercially available solutions may in the past have contributed. Today, however, such support is available from the Swiss company polyTag Technology (www.polytag.ch), which offers a number of generic and pre-activated AML together with instrumentation and protocols for their (automated) application at varied scale. In order to introduce affinity precipitation, various aspects of the practical application and method development are discussed in the following sections.

11.4.1
Application Protocol

Compared to other bioseparation methods, affinity precipitation is fairly unknown. The knowledge of how to set up and optimize such procedures is therefore not ubiquitously available, even though process development is not necessarily more difficult than in chromatography, for example. In order to give an idea how to proceed, a protocol for systematic development of temperature-induced affinity precipitation is given below. The protocol was taken in modified form from Ref. [97]. Since at the time of writing only thermo-responsive AMLs were commercially available, we refrain from detailing protocols for other stimuli. However, the general philosophy of method development may easily be transferred to the other types of stimuli-responsive AMLs.

The most important material for setting up an affinity precipitation is the AML. A number of thermo-responsive polyNIPAM-based AMLs are commercially available from polyTag Technology (see Tab. 11.1). The company also provides an avidin-activated AML precursor to which any biotinylated affinity ligand may be linked via the protocol indicated in Protocol 1. Due to the high affinity between avidin and biotin, the resulting AML should be as stable under process conditions as any covalently conjugated one. polyTag Technology also offers a dedicated AML synthesis service. In addition, a number of synthesis procedures based on a variety of functional end-groups in the polyNIPAM and the affinity ligand can be found in the Reference section at the end of this chapter. As in affinity chromatography, the binding and dissociation (elution) buffers for the affinity precipitation are defined by the type of interaction exploited to bring about specific separation. Often no particular binding buffer is necessary in affinity precipitation; instead, the AML may be directly stirred into the raw solution. The protocol for setting up the affinity precipitation itself is indicated in Protocol 2.

Tab. 11.1 Commercially available (polyTag Technology, Männedorf, Switzerland) thermo-responsive AMLs and AML precursors

AML	Purification of
Biotin–AML	avidin
Protein A–AML	antibodies
Single-stranded DNA–AML [e.g. (CTT)$_7$– polyNIPAM], triple helix motif	pDNA (via triple-helix interaction)
Poly(T)–AML	mRNA
	fusion proteins carrying a specific peptide tag:
Anti-FLAG-tag antibody–AML	FLAG-tag
Anti-APP-tag antibody–AML	APP-tag
Ni-NTA–AML	His$_6$-tag
Avidin–AML[a]	biotinylated molecules
Other applications	
Specifically activated screening AML	phage display

a Generic AML, the avidin–biotin interaction can be used to link biotinylated affinity ligands to the thermo-responsive AML.

Protocol 1: Protocol for the conjugation of biotinylated affinity ligands to avidin-activated polyNIPAM

1. Dissolve the biotinylated ligand (1 mg mL^{-1}) in 0.5 M borate buffer (pH 9).
2. Add the solid polyNIPAM–avidin directly to this solution (molar ratio 1:1) and stir gently for 2 h at 18 °C for AML formation.
3. Add 10% (v/v) of a saturated ammonium sulfate solution (this will lower the critical solution temperature of the AML).
4. Purify the AML by repeated thermo-precipitation (30 °C) and centrifugation at 30 °C and 10 000 g (10 min) followed by redissolution in a fresh batch of buffer solution at 4 °C.
5. Go at least 3 times through steps 3 and 4.
6. Depending on the nature of the molecule used as affinity ligand, the final AML may be redissolved in pure water and lyophilized for storage.

Protocol 2: Setting up a specific bioseparation via affinity precipitation

- Chose suitable binding and dissociation (elution) buffers for the intended affinity interaction. Indications can usually be taken from the corresponding affinity chromatography protocols. Often the conditions in the raw feed solution will be suitable to affinity complex formation. In such cases the solid AML can be added directly to the raw solution in the desired concentration. Otherwise, buffer exchange must be performed.
- Stir the AML at "binding temperature"[a] into the target molecule solution using a molar excess. A 10-fold molar excess of binding places to target molecules usually works well. However, for fine-tuning of the process, e.g. in a large-scale bioseparation, it is usually worth the time to record an "adsorption isotherm". This will give a very good idea of how much of the product will be recovered for a given amount of AML. In addition the affinity constant can be determined from such a plot. Keep in mind that affinity precipitation is a one-stage process, there will always be an equilibrium between bound and unbound target molecules, although for a high-affinity AML this equilibrium may be well on the side of the affinity complex.
- Stir at binding temperature[a] for 30 min. The optimal duration of this process step is dictated by the kinetics of the affinity complex formation and may be much shorter. For fine-tuning of this parameter, the complex formation kinetics should be recorded.
- Raise the temperature at least 5 °C above the critical solution temperature of the AML in the binding environment ("precipitation temperature" [a]) and remove the precipitated affinity complex by centrifugation (10 000 g, precipitation temperature, 10 min).
- Remove entrapped impurities ("washing" steps) by repeated thermo-precipitation from fresh binding buffer (redissolution of the precipitate in 4 °C cold binding buffer, precipitation at precipitation temperature, followed by redissolution of the pellet in fresh 4 °C cold binding buffer). If some fine cellulose fibers are added in a concentration of 50% (w/w$_{polymer}$) to the solution, the precipitate will redissolve faster. The cellulose does not interfere with the affinity interaction. For precipitates recovered in the presence of cellulose, a dissolution temperature of only a few degrees below the CST may be more suitable than 4 °C. Centrifugation works well especially for small samples. At larger scale, filtration may be more useful than centrifugation for precipitate recovery. The interested reader is referred to the polyTag Technology web page (see text) for protocols and instrumentation to be used in such cases.
- For release of the target molecule, redissolve the pellet in 4 °C cold dissociation buffer.
- Remove the AML from the purified product by thermo-precipitation followed by centrifugation. It is also possible to recover the affinity complex

via filtration in the presence of cellulose. For release of the target molecule, cold dissolution buffer is passed through the filter cake thereby simultaneously redissolving the polyNIPAM and releasing the target molecule. The AML is removed from the filtrate be thermo-precipitation. Alternatively it is also possible to pass a warm (temperature above the CST) dissociation buffer through the cake. In this case the target molecule is released, while the stimuli-responsive AML stays in the cake. Depending on the intended use, but also on the biochemistry of the interaction and the release conditions, the AML may be recycled. Validation of this possibility is recommended (see related protocols for affinity chromatography).

a   "Binding temperature" and "precipitation temperature" are defined by the CST of the AML in the solution of interest. A binding temperature of 5 °C below the CST and a precipitation temperature of 5 °C above usually work well for polyNIPAM, which shows a very sharp phase transition. However, in some solutions phase transition may occur over a broader temperature interval. Keep in mind that the CST depends on the composition of the solution and that some polyNIPAM bioconjugates show a phase-transition hysteresis, i.e. dissolution may occur at a lower temperature than precipitation.

11.4.2
Scale-up and Technical Realization

The theoretical scale-up potential of affinity precipitation is indisputable. At present, however, the technical realization of such large-scale processes is difficult due to the lack of a suitable large-scale separation technology for the precipitated polymers and the affinity complex. High-speed centrifugation is an efficient method except for oligomeric materials [78]. The recovered precipitate, however, tends to present a compact gel, the resolubilization of which can be very time consuming. Entrapment of impurities in the wet gel is almost unavoidable, making three precipitation/resolubilization washing/cleaning steps necessary as a rule of thumb [76].

Senstad and Mattiasson [46] were the first to address the necessity of developing alternative modes of precipitate recovery at large scale in order to avoid the centrifugation steps. They proposed flotation as a relatively mild operation for this purpose. In particular, chitosan-based AMLs were precipitated by raising the pH of the solution above 8. Afterwards the mixture (liquid containing the precipitate) was put under pressure (3 bar, 15 min) and then transferred into a flotation chamber at atmospheric pressure. Air bubbles, which formed as result of the pressure expansion, adsorbed to the surface of the precipitated material causing the flotation of the particles and thereby their accumulation at the surface, where they could easily be recovered. Unfortunately, flotation is not applicable for large-scale separations of polymers like polyNIPAM, since polyNIPAM has a strong tendency to foam. Moreover, polyNIPAM particles, precipitated and aggregated in this manner, are quite unstable [98].

Filtration is an alternative solid–liquid separation operation, but for oligomers with a molar mass below 5000 g mol^{-1} at a concentration of ≤ 1 wt% a high loss upon precipitation has been reported [78, 86]. The recovery of precipitated oligomeric AMLs by filtration can be improved by using a higher oligomer concentration, e.g. 10 wt%. However, this will also entail a higher viscosity of the solution and mass transfer limitations similar to those observed for high-molar-mass polymers. A procedure has therefore been proposed where filtration – by the addition of filter aids and salts – becomes a very efficient separation operation for thermoresponsive oligomers [98, 99]. In particular, it has been found that the thermoprecipitation of oligomeric polyNIPAM (M_n = 2300 g mol^{-1}) from a 0.5 wt% aqueous solution in the presence of short cellulose fibers and salts gives a suspension of compact aggregates that filters (and subsequently washes) very well. The precipitated mixture redisssolves readily and the oligomer recovery is higher than 97%. Nonspecific adsorption (entrapment of impurities) is not detectable. Obviously, the self-association of the oligomeric polyNIPAM, which is already observable before macroscopic precipitation occurs [82], can be sufficiently promoted by filter aids/salts to allow efficient filtration. The type of salt used plays an important role, since besides the predictable salting-out effect [81, 82] it also determines the aggregation behavior and hence the consistency of the precipitate. The determination of the most suitable salt and its optimum concentration is, however, still done empirically.

Even under optimized conditions, only between 97 and 99% of the oligomeric polyNIPAM is recovered and separated by the filtration step. Some contamination of the product by the remainder is thus possible. However, this should be acceptable in an initial capturing and first purification step for a target protein from a cell culture supernatant. The pronounced difference in size between the oligomeric polyNIPAM and the target protein should allow easy separation during an additional chromatographic polishing step. PolyNIPAM itself is considered a safe material [100, 101] if it is sufficiently purified from the neurotoxic monomer. One point that needs to be further investigated is the mode of target molecule elution. One option is the elution directly from the filter cake. It might also be possible to remove the filter cake from the filter, redissolve the precipitate in cold dissociation buffer and subsequently retrieve the AML by thermo-precipitation in the presence of the filter aid. The final step would then be a separation by filtration of the solids (AML, filter aid) from the solution containing the highly purified target protein.

11.4.3
Protein Purification

Most previously reported affinity precipitations were designed for protein recovery. Although affinity precipitation has not yet been established for industrial protein purification, a number of processes for the isolation of (recombinant) proteins from real matrices have been described among these applications. Table 11.2 attempts to give a not necessarily comprehensive overview.

Tab. 11.2 Application of affinity precipitation for the recovery of
biomacromolecules from complex raw solutions

Target molecule/feed	AML/stimulus	Comments	Reference
Enzymes			
Trypsin/crude pancreatic extract	ter-polymer (acrylamide/ N-acryloyl-p-aminobenzoic acid/N-acryloyl-m-aminobenzamidine *stimulus*: pH	elution from the precipitate	27
Trypsin/crude extract	Eudragit S-100–STI *stimulus*: pH		29
Alcohol dehydrogenase/ crude yeast extract	κ-carrageenan–Cibacron Blue 3GA *stimulus*: K⁺ ions		60
Alcohol dehydrogenase/ yeast extract	Eudragit S100–Cibacron Blue 3GA *stimulus*: $CaCl_2$/temperature		41
(Recombinant) proteins other than enzymes			
Avidin/cell culture supernatant containing 5% FCS	polyNIPAM–iminobiotin *stimulus*: temperature		81
His-tagged proteins/ cell lysates	polyNIPAM–Ni-NTA *stimulus*: temperature	generic protein separation process	86
Concanavalin A/jack bean extract	Eudragit S 100–p-aminophenyl-α-D-glucopyranoside *stimulus*: pH		32
Protein A/ *Staphylococcus aureus* lysate	hydroxypropyl methylcellulose–IgG *stimulus*: pH		48
His-tagged single-chain Fv-antibody fragments/cell-free *E coli* culture supernatants	Cu^{2+}/Ni^{2+}-loaded copolymer of NIPAM and vinylimidazole *stimulus*: salt/temperature		71
α-Amylase inhibitor/ wheat meal	Cu^{2+}-loaded copolymer of NIPAM and vinylimidazole *stimulus*: salt/temperature	reuse of AML possible	70
Antibody/hybridoma culture supernatant	Eudragit–antigen *stimulus*: pH		30, 31
Rabbit C-reactive protein/rabbit acute-phase serum	polyNIPAM–p-aminophenylphosphorylcholine *stimulus*: temperature		65
Oligonucleotides			
mRNA/cell lysates	PolyNIPAM–T_8 *stimulus*: temperature	comparison to magnetic beads	97
Plasmid DNA/ bacterial lysates	PolyNIPAM–(CTT)₇ *stimulus*: temperature	THAP process	77

Tab. 11.2 (*Continued*)

Target molecule/feed	AML/stimulus	Comments	Reference
Miscellaneous			
Taxol/homogeneous immunoassay	Elastin-like polypeptide–Protein A *stimulus*: temperature	proof-of-feasibility of a novel immunoassay format	101
Peanut lectin/peanuts	Guar gum-linked alginate *stimulus*: addition of CaCl$_2$	cheap and efficient single-step procedure	57

11.4.4
Nucleic Acid Purification

Compared to the state-of-the-art in protein purification, the isolation and purification of nucleic acids, particularly pDNA at larger scale, is still a bottleneck in downstream processing. Given the typical sizes of pDNA molecules, mass transfer limitations and steric hindrances of chromatographic materials are even more likely for this biomacromolecule class than for the proteins. While efficient affinity (hybridization) methods exists for single-stranded oligonucleotides [see also the purification of messenger RNA (mRNA) below], the concept of biospecific affinity interactions is less often used in the case of double-stranded polynucleotides such as pDNA. However, the reversible formation of a specific triple-helical structure between an oligomeric single-stranded polypyrimidine sequence and a complementary motif in the double-stranded DNA helix has been reported. The specificity of the triple helix is mediated by Hoogsteen hydrogen bonding between base triplets of the T–AT and C$^+$–GC type. The latter interaction is pH dependent and can be exploited for reversible interaction as a basis for separation.

On the basis of this type of specific interaction, the triple-helix affinity precipitation (THAP) process [83] has been designed as a putative solution for (plasmid) DNA purification at large scale. The principle of the THAP concept is the reversible formation of a specific triple-helical structure between the oligomeric polyNIPAM-AML bearing a single-stranded polypyrimidine sequence, e.g. (CTT)$_7$, and the complementary motif in the double-stranded DNA helix, which allows the specific coprecipitation of the target molecule from the raw solution. Figure 11.4 shows the analytical agarose gel of this separation (lysate, supernatant after affinity precipitation, eluted target molecule).

Using the THAP approach, highly pure (plasmid) DNA was routinely prepared from bacterial lysate at yields between 70 and 90%. Nonspecific coprecipitation

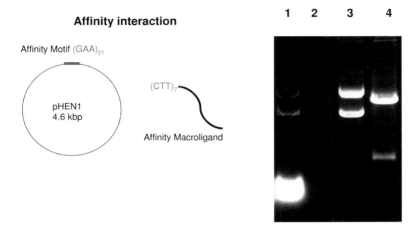

Affinity interaction

Affinity Motif (GAA)$_{21}$

pHEN1
4.6 kbp

(CTT)$_7$

Affinity Macroligand

1   2   3   4

**1: bacterial lysate, 2: supernatant after
affinity precipitation, 3: recovered plasmid,
4 digested plasmid**

Fig. 11.4 Affinity interaction and agarose gel summarizing the isolation of pDNA from a bacterial lysate via the THAP process.

of contaminants by the AML was below 7% and presumably due to physical entrapment of these molecules in the wet precipitate. Ligand efficiencies calculated for the THAP process were at least one order of magnitude higher than in triple-helix affinity chromatography. The AML/plasmid complex could be directly transfected into mammalian cells (transient transfection, Lipofectamine protocol) with success rates similar or superior to those obtained with pure pDNA obtained by conventional preparation kits [102]. The viability of the cells was not affected, which incidentally also demonstrates the low cytotoxicity of the oligomeric polyNIPAM bioconjugates. Another strategy for DNA isolation by affinity precipitation is the use of affinity ligands, which intercalate into the DNA double strand [103–105].

While some effort for the integration of a specifically interactive sequence into the molecule is necessary in the case of DNA purification by affinity interaction, single-stranded oligonucleotides can be specifically captured via simple hybridization. One example for such a purification, which is routinely used in many molecular biology laboratories, is the preparation of mRNA from eukaryotic cells and tissues. This process is an important step in many genetic engineering protocols, e.g. for the creation of recombinant production organisms, but also in the analysis of gene structure and regulation. In the case of eukaryotic mRNA this isolation typically relies on the poly(A) tail, which is present on most mature eukaryotic mRNAs, whereas the other RNA (transfer RNA, ribosomal RNA) species normally do not carry such a tag. In a typical separation process, biotinylated oligo(dT) probes are first mixed into the crude mRNA preparations. Once the probes and the poly(A) mRNA have annealed, the complexes can be specifically captured by any (strept)avidin-based separation technique. Streptavidin-coated magnetic particles are especially popular in this context. After a series of high-stringency washing steps, water is used to release the poly(A) mRNA into solution.

Recently, the efficient capture of mature eukaryotic poly(A) mRNA by affinity precipitation using an oligomeric avidin–AML has been described, and the advantages in terms of cost, handling and scalability for affinity precipitation compared to more conventional approaches were discussed [106]. Putative problems with RNases could be prevented by a suitable treatment of the AML preparation. The RNA yield and quality were at least equal to that of a standard approach using streptavidin-coated paramagnetic beads. The produced poly(A) mRNA proved to be an excellent target for reverse transcription-polymerase chain reaction amplification. While magnetic beads require a two-step (hybridization followed by capture) protocol, affinity precipitation could be set up as a one-step protocol using the oligo(dT)-activated AML directly in solution followed by *in situ* precipitation.

In earlier studies for RNA purification, a copolymer of NIPAM and vinyl-derivatized (dT)$_8$ was used for the separation of dA oligonucleotides from a mixture of its one-point-mismatched oligonucleotides at high NaCl concentration [107]. Longer oligo(dA)s (more than five) tended to be precipitated more efficiently than shorter ones, and the AML could be used repeatedly (5 times) without loss in precipitation efficiency. A study of separation of RNA from pDNA by metal chelate affinity precipitation showed that a copper-loaded NIPAM–vinyl imidazole copolymer interacts with exposed purine residues. RNA could be eluted in that case by addition of imidazole. The presence of imidazole after RNA elution, however, hindered copolymer reprecipitation, requiring the addition of high salt concentration for that particular step [108].

11.5
Perspectives

Affinity precipitation is a relatively simple, convenient and reproducible technique that results in high target molecule recovery at high specificity. Using AMLs based on homogeneous oligomers, the ligand efficiency is very high and the (expensive) affinity ligands are effectively used during the purification procedure. The method is robust and shows flexibility towards variation in the process parameters. Since only mixing of AMLs into the target solution is involved, the scale-up potential is high. At preparative scale the most appropriate separation operation for the precipitate is filtration. Future work should be directed to further optimization of the AML (precursor), the establishment of more applications and towards the development of a robust, integrated purification process.

References

1 Matejtschuk, P., Dash, C. H., Gascoigne, E. W., *Br J Anaesth* **2000**, *85*, 887–895.

2 Cohn, E. J., Strong, L. E., Hughes Jr., W. L., , Mulford, D. J., Ashworth, J. N., Melin, M., Taylor, H. L., *J Am Chem Soc* **1946**, *68*, 459–475.

3 Kistler, P., Nitschmann, H. S., *Vox Sang* **1962**, *7*, 414–424.

4 Harrison, R. G., Todd, P., Rudge, S. R., Petrides, D. P., *Bioseparations Science and Engineering*. Oxford University Press, Oxford, **2003**.

5 Buchacher, A., Iberer, G., *Biotechnol J* **2006**, *1*, 148–163.

6 Nydegger, U., *Pipette* **2006**, *2*, 13–15.

7 Melander, W., Horvath, C., *Arch Biochem Biophys* **1977**, *183*, 200–215.

8 Temponi, M., Kekish, U., Ferrone, S., *J Immunol Methods* **1988**, *115*, 151–152.

9 Freitag, R., *Curr Trends Polymer Sci* **1998**, *3*, 63–79.

10 Larsson, P.-O., Mosbach, K., *FEBS Lett* **1979**, *98*, 333–338.

11 Irwin, J. A., Tipton, K. F., *Essays Biochem* **1995**, *29*, 137–156.

12 Graham, L. D., Griffin, T. O., Beatty, R. E., McCarthy, A. D., Tipton, K. F., *Biochim Biophys Acta* **1985**, *828*, 266–269.

13 Irwin, J. A., Tipton, K. F., *Biochem Soc Trans* **1995**, *23*, 365S.

14 Flygare, S., Griffin, T., Larsson, P.-O., Mosbach, K., *Anal Biochem* **1983**, *133*, 409–416.

15 Hayet, M., Vijayalakshmi, M. A., *J Chromatogr* **1986**, *376*, 157–161.

16 Pearson, J. C., Burton, S. J., Lowe, C. R., *Anal Biochem* **1986**, *158*, 382–389.

17 Beattie, R. E., Tipton, K. F., Buchanan, M. G., *Biochem Soc Trans* **1987**, *15*, 1043–1046.

18 Burnett, T. J., Peebles, H. C., Hageman, J. H., *Biochem Biophys Res Commun* **1980**, *96*, 157–162.

19 Bradshaw, A. P., Sturgeon, R. J., *Biotechnol Tech* **1990**, *4*, 67–71.

20 Powers, D. D., Willard, B. L., Carbonell, R. G., Kilpatrick, P. K., *Biotechnol Prog* **1992**, *8*, 436–453.

21 Powers, D. D., Carbonell, R. G., Kilpatrick, P. K., *Biotechnol Bioeng* **1994**, *44*, 509–522.

22 Morris, J. E., Hoffman, A. S., Fisher, R. R., *Biotechnol Bioeng* **1993**, *41*, 991–997.

23 Van Dam, M. E., Wuenschell, G. E., Arnold, F. H., *Biotechnol Appl Biochem* **1989**, *11*, 492–502.

24 Naeem, A., Khan, R. H., Saleemuddin, M., *Biochemistry (Mosc)* **2006**, *71*, 56–59.

25 Lilius, G., Persson, M., Bülow, L., Mosbach, K., *Eur J Biochem* **1991**, *198*, 499–504.

26 Larsson, P.-O., Flygare, S., Mosbach, K., *Methods Enzymol* **1984**, *104*, 364–369.

27 Linné Larsson, E., Galaev, I. Y., Mattiasson, B., *J Mol Recognit* **1998**, *11*, 236–239.

28 Mattiasson, B., Kumar, A., Galaev, I. Yu., *J Mol Recognit* **1998**, *11*, 211–216.

29 Schneider, M., Guillot, C., Lamy, B., *Ann NY Acad Sci* **1981**, *369*, 257–263.

30 Kumar, A., Gupta, M. N., *J Biotechnol* **1994**, *37*, 185–189.

31 Guoqiang, D., Kaul, R., Mattiasson, B., *Bioseparation* **1993**, *3*, 333–341.

32 Gupta, M. N., Guoqiang, D., Kaul, R., Mattiasson, B., *Biotechnol Tech* **1994**, *8*, 117–122.

33 Roy, I., Mondal, K., Sharma, A., Gupta, M. N., *Biochim Biophys Acta* **2005**, *1747*, 179–187.

34 Linné Larsson, E., Galaev, I. Yu., Lindahl, L., Mattiasson, B., *Bioseparation* **1996**, *6*, 273–282.

35 Linné Larsson, E., Galaev, I. Yu., Mattiasson, B., *Bioseparation* **1996**, *6*, 283–291.

36 Gupta, M. N., Kaul, R., Guoqiang, D., Dissing, U., Mattiasson, B., *J Mol Recognit* **1996**, *9*, 356–359.

37 Guoqiang, D., Benhura, M. A. N., Kaul, R., Mattiasson, B., *Biotechnol Prog* **1995**, *11*, 187–193.

38 Taipa, M. A., Kaul, R., Mattiasson, B., Cabral, J. M. S., *J Mol Recognit* **1998**, *11*, 240–242.

39 Taipa, M. A., Kaul, R.-H., Mattiasson, B., Cabral, J. M. S., *Bioseparation* **2000**, *9*, 291–298.

40 Linné Larsson, E., Mattiasson, B., *Biotechnol Tech* **1994**, *8*, 51–56.

41 Kamihara, M., Kaul, R., Mattiasson, B., *Biotechnol Bioeng* **1992**, *40*, 1381–1387.

42 Shu, H.-C., Guoqiang, D., Kaul, R., Mattiasson, B., *J Biotechnol* **1994**, *34*, 1–11.

43 Kumar, A., Gupta, M. N., *Mol Biotechnol* **1996**, *6*, 1–6.

44 Gupta, M. N., Guoqiang, D., Mattiasson, B., *Biotechnol Appl Biochem* **1993**, *18*, 321–327.

45 Senstad, C., Mattiasson, B., *Biotechnol Bioeng* **1989**, *33*, 216–220.

46 Senstad, C., Mattiasson, B., *Biotechnol Bioeng* **1989**, *34*, 387–393.

47 Tyagi, R., Kumar, A., Sardar, M., Kumar, S., Gupta, M. N., *Isol Purif* **1996**, *2*, 217–226.

48 Lali, A., Balan, S., John, R., D'Souza, F., *Bioseparation* **1999**, *7*, 195–205.

49 Tanigushi, M., Kobayashi, M., Natsui, K., Fujii, M., *J Ferment Bioeng* **1989**, *68*, 32–36.

50 Tanigushi, M., Tanahashi, S., Fujii, M., *J Ferment Bioeng* **1990**, *69*, 362–364.

51 Teotia, S., Lata, R., Gupta, M. N., *J Chromatogr A* **2004**, *1052*, 85–91.

52 Agarwal, R., Gupta, M. N., *Protein Expr Purif* **1996**, *7*, 294–298.

53 Dissing, U., Mattiasson, B., *J Biotechnol* **1996**, *52*, 1–10.

54 Dainiak, M. B., Izumrudov, V. A., Muronetz, V. I., Galaev, I. Yu., Mattiasson, B., *Bioseparation* **1999**, *7*, 231–240.

55 Dissing, U., Mattiasson, B., *Bioseparation* **1999**, *7*, 221–229.

56 Jain, S., Mondal, K., Gupta, M. N., *Artif Cells Blood Substit Immobil Biotechnol* **2006**, *34*, 127–144.

57 Linné, E., Garg, N., Kaul, R., Mattiasson, B., *Biotechnol Appl Biochem* **1992**, *16*, 48–56.

58 Tyagi, R., Agarwal, R., Gupta, M. N., *J Biotechnol* **1996**, *46*, 79–83.

59 Wandrey, C., Vidal, D. S., *Ann NY Acad Sci* **2001**, *944*, 187–198.

60 Roy, I., Gupta, M. N., *J Chromatogr A* **2003**, *998*, 103–108.

61 Mondal, K., Roy, I., Gupta, M. N., *Protein Expr Purif* **2003**, *32*, 151–160.

62 Rosso, A., Ferrarotti, S., Miranda, M. V., Krymkiewicz, N., Nudel, B. C., Cascone, O., *Biotechnol Lett* **2005**, *27*, 1171–1175.

63 Senstad, C., Mattiasson, B., *Biotechnol Appl Biochem* **1989**, *11*, 41–48.

64 Ito, S., Mizogushi, K., Suda, Y., *Bull Res Inst Polym Text* **1984**, *144*, 7–12.

65 Fong, R. B., Ding, Z., Long, C. J., Hoffman, A. S., Stayton, P. S., *Bioconjug Chem* **1999**, *10*, 720–725.

66 Chen, J. P., Hoffman, A. S., *Biomaterials* **1990**, *11*, 631–634.

67 Mori, S., Nakata, Y., Endo, H., *Protein Expr Purif* **1994**, *5*, 153–156.

68 Luong, J. H. T., Male, K. B., Nguyen, A. L., *Biotechnol Bioeng* **1988**, *31*, 439–446.

69 Nguyen, A. L., Luong, J. H. T., *Biotechnol Bioeng* **1989**, *34*, 1186–1190.

70 Pécs, M., Eggert, M., Schügerl, K., *J Biotechnol* **1991**, *21*, 137–142.

71 Vaidya, A. A., Lele, B. S., Kulkarni, M. G., Mashelkar, R. A., *J Biotechnol* **2001**, *87*, 95–107.

72 Kumar, A., Galaev, I. Yu., Mattiasson, B., *Biotechnol Bioeng* **1998**, *59*, 695–704.

73 Kumar, A., Wahlund, P.-O., Kepka, C., Galaev, I. Y., Mattiasson, B., *Biotechnol Bioeng* **2003**, *84*, 494–503.

74 Eggert, M., Baltes, T., Garret-Flaudy, F., Freitag, R., *J Chromatogr A* **1998**, *827*, 269–280.

75 Vaidya, A. A., Lele, B. S., Kulkarni, M. G., Mashelkar, R. A., *Biotechnol Bioeng* **1999**, *64*, 418–425.

76 Garret-Flaudy, F., Freitag, R., *Biotechnol Bioeng* **2001**, *71*, 223–234.

77 Takei, Y. G., Matsukata, M., Aoki, T., Sanui, K., Ogata, N., Kikushi, A., Sakurai, Y., Okano, T., *Bioconjug Chem* **1994**, *5*, 577–582.

78 Takei, Y. G., Aoki, T., Sanui, K., Ogata, N., Okano, T., Sakurai, Y., *Bioconjug Chem.* **1993**, *4*, 42–46.

79 Costioli, M. D., Berdat, D., Freitag, R., André, X., Müller, A. H. E., *Macromolecules* **2005**, *38*, 3630–3637.

80 Chen, J. P., Yang, H. J., Hoffman, A. S., *Biomaterials* **1990**, *11*, 625–630.

81 Freitag, R., Garret-Flaudy, F., *Langmuir* **2001**, *17*, 4711–4716.

82 Baltes, T., Garret-Flaudy, F., Freitag, R., *J Polym Sci A* **1999**, *37*, 2977–2989.

83 Costioli, M. D., Fisch, I., Garret-Flaudy, F., Hilbrig, F., Freitag, R., *Biotechnol Bioeng* **2003**, *81*, 535–545.

84 Freitag, R., Garret-Flaudy, F., *US Patent Application* **2001**, *6,258,275*.

85 Malmstadt, N., Hyre, D. E., Ding, Z., Hoffman, A. S., Stayton, P. S., *Bioconjug Chem* **2003**, *14*, 575–580.

86 Hoshino, K., Taniguchi, M., Kitao, T., Morohashi, S., Sasakura, T., *Biotechnol Bioeng* **1998**, *60*, 568–579.

87 Pan, L.-C., Chien, C.-C., *J Biochem Biophys Methods* **2003**, *55*, 87–94.

88 Hilbrig, F., Stocker, G., Schläppi, J.-M., Kocher, H., Freitag, R., *Trans IChemE C* **2006**, *84*, 28–36.

89 Fong, R., Ding, Z., Hoffman, A. S., Stayton, P. S., *Biotechnol Bioeng* **2002**, *79*, 271–276.

90 Carter, S., Rimmer, S., Sturdy, A., Webb, M., *Macromol Biosci* **2005**, *5*, 373–378.

91 Kostal, J., Mulchandani, A., Gropp, K. E., Chen, W., *Environ Sci Technol* **2003**, *37*, 4457–4462.

92 Kim, J.-Y., O'Malley, S., Mulchandani, A., Chen, W., *Anal Chem* **2005**, *77*, 2318–2322.

93 Sun, X.-L., Haller, C. A., Wu, X., Conticello, V. P., Chaikof, E. L., *J Proteome Res* **2005**, *4*, 2355–2359.

94 Desponds, A., Freitag, R., *Biotechnol Bioeng* **2005**, *91*, 584–591.

95 Desponds, A., Freitag, R., *Langmuir* **2003**, *19*, 6261–6270.

96 Labrou, N., Clonis, Y. D., *J Biotechnol* **1994**, *36*, 95–119.

97 Freitag, R., Hilbrig, F., *Methods Mol Biol* **2007**, in press.

98 Hilbrig, F., Freitag, R., *Swiss Patent Application 0875/01*, **2001**.

99 Hilbrig, F., Freitag, R., *J Chromatogr B* **2003**, *790*, 79–90.

100 Kopecek, J., Sprincl, L., Bazilova, H., Vacik, J., *J Biomed Mater Res* **1973**, *7*, 111–121.

101 Pavia, A. A., Pucci, B., Riess, J. G., Zarif, L., **1997**, *US Patent Application 5,703,126*.

102 Costioli, M. D., Hilbrig, F., Freitag, R., *React Funct Polym* **2005**, *65*, 351–365.

103 Maeda, M., Nishimura, C., Inenaga, A., Takagi, M., *React Polym* **1993**, *21*, 27–35.

104 Soh, N., Umeno, D., Tang, Z., Murata, M., Maeda, M., *Anal Sci* **2002**, *18*, 1295–1299.

105 Umeno, D., Kawasaki, M., Maeda, M., *Bioconjug Chem* **1998**, *9*, 719–724.

106 Stocker, G., Vandevyver, C., Hilbrig, F., Freitag, R., *Biotechnol Progr* **2006**, *22*, 1621–1629.

107 Mori, T., Umeno, D., Maeda, M., *Biotechnol Bioeng* **2001**, *72*, 261–268.

108 Balan, S., Murphy, J., Galaev, I., Kumar, A., Fox, G. E., Mattiasson, B., Willson, R. C., *Biotechnol Lett* **2003**, *25*, 1111–1116.